£9.95

J. O. Bird
B.Sc. (Hons), F. Coll...
F.I.M.A., T.Eng M.I.Elec. I.E.

A. J. C. May
B.A., C.Eng., M.I.Mech.E., F.I.T.E., M.B.I.M.

Mathematics for electrical technicians
Levels 4 and 5

Longman
Scientific &
Technical

Longman Scientific & Technical
Longman Group UK Limited,
Longman House, Burnt Mill, Harlow,
Essex CM20 2JE, England
and Associated Companies throughout the world.

First published 1981
Fifth impression published by Longman Scientific & Technical 1987

British Library Cataloguing in Publication Data

Bird, J O
    Mathematics for electrical technicians
    Levels 4 and 5. – (Longman technician series:
    mathematics and sciences).
    1. Electric engineering – Mathematics
    I. Title II. May, A J C
    510'.2'46213   TK153   79-41351

ISBN 0-582-41760-0

Produced by Longman Singapore Publishers (Pte) Ltd.
Printed in Singapore

General editor   Mathematics and Sciences

D.R. Browning, B.Sc., F.R.I.C., A.R.I.C.S.

Books already published in this sector of the series:

Technician mathematics Level 1 Second edition   *J O Bird and A J C May*
Technician mathematics Level 2 Second edition   *J O Bird and A J C May*
Technician mathematics Level 3   *J O Bird and A J C May*
Technician mathematics Levels 4 and 5   *J O Bird and A J C May*
Mathematics for electrical technicians Levels 4 and 5   *J O Bird and A J C May*
Calculus for technicians   *J O Bird and A J C May*
Statistics for technicians   *J O Bird and A J C May*
Algebra for technicians   *J O Bird and A J C May*
Algebra and calculus for technicians   *J O Bird and A J C May*
Engineering science in SI units *E Hughes and C Hughes*
Organic chemistry for higher education   *J Brockington and P J Stamper*
Inorganic chemistry for higher education   *J Brockington and P J Stamper*
Mathematics for scientific and technical students   *H G Davies and G A Hicks*
Mathematical formulae   *J O Bird and A J C May*
Technician physics Level 2   *E Deeson*

Contents

Preface xi
Acknowledgements xiii

Chapter 1   Solution of equations by iterative methods
1

1   Introduction 1
2   An algebraic method of successive approximation 2
3   The Newton–Raphson method 5
4   Further problems 10

Chapter 2   Partial fractions
13

1   Introduction 13
2   Type 1. Denominator containing linear factors 14
3   Type 2. Denominator containing repeated linear factors 17
4   Type 3. Denominator containing a quadratic factor 19
5   Summary 20
6   Further problems 21

Chapter 3   Matrix arithmetic and the determinant of a matrix
23

1   Introduction 23
2   Addition, subtraction and multiplication of third order matrices 24
3   The determinant of a 3 by 3 matrix 28
4   The inverse or reciprocal of a matrix 31
5   Further problems 35

Chapter 4   The general properties of 3 by 3 determinants and the solution of simultaneous equations
38

1   The properties of third order determinants 38
2   The solution of simultaneous equations having three unknowns 42
3   Further problems 47

Chapter 5   Maclaurin's and Taylor's series   51

1   Maclaurin's series 51
2   Taylor's series 58
3   Approximate values of definite integrals by using series
     expansions 62
4   Further problems 64

Chapter 6   De Moivre's theorem   67

1   The exponential form of a complex number 67
2   De Moivre's theorem 68
3   Expressing $\cos n\theta$ and $\sin n\theta$ in terms of powers of $\cos \theta$ and
     $\sin \theta$ 72
4   Expressing $\cos^n \theta$ and $\sin^n \theta$ in terms of sines and cosines of
     multiples of θ 74
5   Further problems 76

Chapter 7   Hyperbolic functions   79

1   Definitions of hyperbolic functions 79
2   Some properties of hyperbolic functions 80
3   Evaluation of hyperbolic functions 81
4   Graphs of hyperbolic functions 84
5   Hyperbolic identities − Osborne's rule 89
6   Differentiation of hyperbolic functions 92
7   Solution of equations of the form $a \operatorname{ch} x + b \operatorname{sh} x = \mathrm{c}$ 94
8   Series expansions for $\cosh x$ and $\sinh x$ 96
9   Further problems 98

Chapter 8   The relationship between trigonometric and
         hyperbolic functions and hyperbolic identities   102

1   The relationship between trigonometric and hyperbolic
     functions 102
2   Hyperbolic identities 104
3   Summary of trigonometric and hyperbolic identities 108
4   Further problems 109

Chapter 9   Differentiation of implicit functions   112

1   Implicit functions 112
2   Differentiating implicit functions 112
3   Further problems 117

Chapter 10   Differentiation of functions defined parametrically   120

1   Parametric representation of points   120
2   Differentiation in parameters   122
3   Further problems   128

Chapter 11   Logarithmic differentiation   131

1   The laws of logarithms applied to functions   131
2   Logarithmic differentiation   135
3   Further problems   138

Chapter 12   Differentiation of inverse trigonometric and inverse hyperbolic functions   141

1   Inverse functions   141
2   Differentiation of inverse trigonometric functions   142
3   Logarithmic forms of the inverse hyperbolic functions   149
4   Differentiation of inverse hyperbolic functions   151
5   Further problems   160

Chapter 13   Partial differentiation   164

1   Differentiating a function having two variables   164
2   First order partial derivatives   165
3   Second order partial derivatives   168
4   Further problems   171

Chapter 14   Total differential, rates of change and small changes   175

1   Total differential   175
2   Rates of change   177
3   Small changes   179
4   Further problems   182

Chapter 15   Integration using substitution and partial fractions   186

1   Introduction   186
2   Integration using algebraic substitutions   187
3   Integration using trigonometrical and hyperbolic identities and substitutions   191
4   Change of limits of integration by a substitution   202
5   Integration using partial fractions   204
6   Further problems   209

Chapter 16   Integration by parts   **219**

1   Introduction 219
2   Application of the integration by parts formula 220
3   Further problems 225

Chapter 17   First order differential equations by separation of the variables   **228**

1   Introduction
2   Solution of differential equations of the form $\frac{dy}{dx} = f(x)$ 229

3   Solution of differential equations of the form $\frac{dy}{dx} = f(y)$ 231

4   Solution of 'variables separable' type of differential equations 233
5   Further problems 236

Chapter 18   Homogeneous first order differential equations   **241**

1   Solution of differential equations of the form $P\frac{dy}{dx} = Q$ 241

2   Further problems 245

Chapter 19   Linear first order differential equations   **246**

1   Solution of differential equations of the form $\frac{dy}{dx} + Py = Q$ 246

2   Further problems 250

Chapter 20   The solution of linear second order differential equations of the form $a\frac{d^2y}{dx^2} + b\frac{dy}{dx} + cy = 0$ **253**

1   Introduction 253
2   Types of solution of second order differential equations with constant coefficients 254
3   Summary of the procedure used to solve differential equations of

the form $a\frac{d^2y}{dx^2} + b\frac{dy}{dx} + cy = 0$ 259
4   Further problems 264

Chapter 21 The solution of linear second order differential equations of the form $a \dfrac{d^2y}{dx^2} + b \dfrac{dy}{dx} + cy = f(x)$ **268**

1   Complementary function and particular integral 268
2   Methods of finding the particular integral 269
3   Summary of procedure to solve differential equations of the type $a \dfrac{d^2y}{dx^2} + b \dfrac{dy}{dx} + cy = f(x)$ 279
4   Further problems 280

Chapter 22 The Fourier series for periodic functions of period 2π **285**

1   Introduction 285
2   Periodic functions 286
3   Evaluation of definite integrals involving trigonometric functions 290
4   Fourier series 291
5   Further problems 299

Chapter 23 The Fourier series for a non-periodic function over range 2π **301**

1   Expansion of non-periodic functions 301
2   Further problems 305

Chapter 24 The Fourier series for even and odd functions and half range series   **308**

1   Even and odd functions 308
2   Fourier cosine and Fourier sine series 311
3   Half range Fourier series 316
4   Further problems 320

Chapter 25 Fourier series over any range   **324**

1   Expansion of a periodic function of period l 324
2   Half range Fourier series for functions defined over range l 328
3   Further problems 332

Chapter 26 A numerical method of harmonic analysis   **334**

1   Introduction 334
2   Harmonic analysis on data given in tabular or graphical form 334
3   Complex waveform considerations 337
4   Further problems 342

Chapter 27 Introduction to Laplace transforms   **346**

1   Introduction 346
2   Definition of the Laplace transform 347
3   Linearity property of the Laplace transform 347
4   Laplace transforms of elementary functions 348
5   Further problems 354

Chapter 28 Properties of Laplace transforms   **356**

1   The Laplace transform of $e^{at} f(t)$ 356
2   Laplace transforms of functions of the type $e^{at} f(t)$ 356
3   The Laplace transforms of derivatives 360
4   The initial and final value theorems 362
5   Further problems 365

Chapter 29 Inverse Laplace transforms and the use of Laplace transforms to solve differential equations   **368**

1   Definition of the inverse Laplace transform 368
2   Inverse Laplace transforms of simple functions 368
3   Inverse Laplace transforms using partial fractions 372
4   Use of Laplace transforms to solve second order differential
      equations with constant coefficients 374
5   Further problems 380

Index 383

Preface

This textbook is one of a series which deal simply and carefully with the fundamental mathematics essential in the development of technicians.

The material for *Mathematics for Electrical Technicians Levels 4 and 5* has been selected from the Business and Technician Education Council Mathematics bank of objectives. Twenty-nine general objectives have been selected and each general objective is covered in a separate chapter.

Since it is recommended by the Business and Technician Education Council that a unit of design length 60 hours should consist of about 16 general objectives and a half unit consist 8 general objectives it is anticipated that sufficient material is contained within the text to meet the demands of most college-devised units.

Each topic considered in the text is presented in a way that assumes in the reader only the knowledge attained in the various BTEC level 3 mathematics units.

This practical book contains over 50 illustrations, over 220 detailed worked problems, followed by some 650 further problems with answers. The text need not be confined to students studying on Higher BTEC courses for it contains material suitable for study at H.N.C., H.N.D. or for degree courses. There is a companion textbook in this series, *Technician Mathematics Levels 4 and 5*.

The authors would like to thank Mr David Browning for his continued valuable assistance in his capacity as General Editor of the Mathematics and Science Sector of the Longman Technician Series. They would also like to express their appreciation for the friendly co-operation and helpful advice given to them by the publishers.

Thanks are also due to Mrs Elaine Mayo for the excellent typing of the manuscript.

Finally, the authors would like to add a word of thanks to their wives, Elizabeth and Juliet, for their marvellous patience, help and encouragement during the preparation of this series of books.

J. O. Bird A. J. C. May
Highbury College of Technology
Portsmouth

Acknowledgements

We are grateful to Macmillan, London and Basingstoke for permission to reproduce material from *Four Figure Mathematical Tables* by Frank Castle. Reproduced by permission.

Cover photograph by Paul Brierley

Chapter 1

Solution of equations by iterative methods

1 Introduction

The solution of equations of the form $f(x) = 0$, where the function is linear, can be achieved by applying the basic rules of algebra or trigonometry. Other equations can be solved by applying a formula, as in the case of solving quadratic equations, where the equation $ax^2 + bx + c = 0$ can be solved by applying the formula $x = \dfrac{-b \pm \sqrt{(b^2 - 4ac)}}{2a}$. More difficult functions may be solved graphically, although this can be a long, tedious process if an accurate result is required. A more analytical approach is to use methods of successive approximation, in which a sequence of calculations is repeated as many times as necessary in order to give a result to the required degree of accuracy. For example, the equation $3x = 7$ can be solved as follows:

Try various integer values until the equation is approximately true; in this case, if $x = 2$, $3x = 6$ and if $x = 3$, $3x = 9$. This shows that the root lies between 2 and 3, and since $3x = 6$ is nearer to 7 than $3x = 9$, the root is nearer $x = 2$ than $x = 3$. Now try $x = 2.3$, giving $3x = 6.9$. When $x = 2.4$, $3x$ is equal to 7.2, thus the root is nearer 2.3 than 2.4, and so on. This process can be repeated as many times as it is necessary to give the required degree of accuracy and is called a **method of successive approximation.** Clearly this technique would not be used in the example given but only where straight-forward algebraic methods fail. It is particularly suitable for using in cases where the solution is to be determined by a computer.

Several methods of successive approximation may be used to determine the value of the roots of an equation to a specified degree of accuracy, and two of these are introduced in this chapter.

2 An algebraic method of successive approximation

Methods of successive approximation may be used to solve equations of the form $a + bx + cx^2 + dx^3 + \ldots = 0$. Thus, they are suitable for solving not only polynomials of the form shown above, but also for the solution of other equations which can be expressed in polynomial form by means of, say, Maclaurin's series, introduced in Chapter 5. The method relies initially on an approximate value of a root being estimated, called the first approximation, and then a more accurate result being determined by the repetition of some procedure. The more accurately the first approximation is obtained, the less the number of repetitive cycles, called **iterations**, needed to obtain a given degree of accuracy. The first approximation can either be estimated graphically or by a method involving functional notation as shown below.

The approximate value of x at the point where the curve of $f(x) = 0$ crosses the x-axis is used as the first approximation. This occurs when the value of $f(x)$ changes from positive to negative or from negative to positive.

Consider the equation $x^3 - 2x - 5 = 0$.

When $x = 0, f(0) = (0)^3 - 2(0) - 5 = -5$,
when $x = 1, f(1) = (1)^3 - 2(1) - 5 = -6$,
when $x = 2, f(2) = (2)^3 - 2(2) - 5 = -1$, and
when $x = 3, f(3) = (3)^3 - 2(3) - 5 = 16$.

Since the sign of $f(x)$ changes from a negative value at $f(2)$ to a positive value at $f(3)$, then the first approximation is between $x = 2$ and $x = 3$. If a straight line is drawn between co-ordinates $(2, -1)$ and $(3, 16)$, it will cut the x-axis very near to 2. So a **first approximation** of x is taken as, say, 2.1.

The repetitive procedure used to determine the value of the root more accurately is shown below.

Second approximation

Let the true value of the root to be $x_1 = 2.1 + \delta_1$.

Substituting $2.1 + \delta_1$ for x in the original equation, $x^3 - 2x - 5 = 0$, gives:

$$(2.1 + \delta_1)^3 - 2(2.1 + \delta_1) - 5 = 0$$

If the first approximation is reasonably accurate, then δ_1 is small and δ_1^2 and δ_1^3 are very small. Using the binomial expansion to obtain the terms of $(2.1 + \delta_1)^3$ and neglecting terms containing δ_1^2 and δ_1^3 gives:

$$(2.1^3 + 3 \times 2.1^2 \times \delta_1 + \ldots) - 2(2.1 + \delta_1) - 5 \simeq 0$$

i.e. $9.261 + 13.23\,\delta_1 - 4.2 - 2\,\delta_1 - 5 \simeq 0$

$$\delta_1 \simeq -\frac{0.061}{11.23} \simeq -0.005$$

Thus a second (i.e. better) approximation to the root is $2.1 - 0.005$, that is, 2.095.

Third approximation

Let the true value of the root be $x_2 = 2.095 + \delta_2$.

Substituting $2.095 + \delta_2$ for x in the original equation, gives

$$(2.095 + \delta_2)^3 - 2(2.095 + \delta_2) - 5 = 0$$

Using the binomial series and neglecting terms containing δ_2^2 and δ_2^3 gives:

$(2.095^3 + 3 \times 2.095^2\, \delta_2 + \ldots) - 2\,(2.095 + \delta_2) - 5 \simeq 0$

i.e. $9.195 + 13.167\delta_2 - 4.19 - 2\delta_2 - 5 \simeq 0$

$$\delta_2 \simeq \frac{-0.005}{11.167} \simeq -0.000\,4$$

Thus, a third approximation to the root is $2.095 - 0.000\,4$, that is, $2.094\,6$.

This procedure can be repeated until the value of the required root on two consecutive iterations does not change when expressed to the stipulated degree of accuracy. The three approximations obtained of the root of the equation $x^3 - 2x - 5 = 0$ are: first approximation 2.1, second approximation 2.095 and third approximation $2.094\,6$. Since the value of the root, when expressed correct to four significant figures, does not change for the second and third approximations, then the value of the root is 2.095 correct to four significant figures.

A cubic equation can have one, two or three roots.

In this equation when x is larger than 3, the x^3 term predominates and $f(x)$ becomes large and positive. Hence there are no other roots when x is positive. When x is negative

$f(-1) = (-1)^3 - 2\,(-1) - 5 = -4$

$f(-2) = (-2)^3 - 2\,(-2) - 5 = -9$

As x becomes large and negative, the x^3 term again predominates and $f(x)$ becomes large and negative. Hence $x = 2.095$ correct to four significant figures is the only root of the equation $x^3 - 2x - 5 = 0$.

When solving a polynomial equation by an algebraic successive approximation method, the procedure can be summarised as follows:

(i) Determine x, the approximate value of the root required, either graphically or by using a functional notation method.
(ii) Let the true value of the root, x_1, be $(x + \delta_1)$.
(iii) Estimate the approximate value of δ_1 by using $f(x + \delta_1) = 0$ and neglecting terms containing δ_1^2 and higher powers of δ_1.
(iv) A better approximation of the root is $(x_1 + \delta_1)$. Repeat for $x_2, x_3, \ldots$ until the value of the root does not change on two consecutive iterations when expressed to the stipulated degree of accuracy.

Worked problems on an algebraic method of successive approximation

---

Problem 1. The equation $x^3 - 3x^2 + 4x - 7 = 0$ has only one positive root. Determine the value of this root, correct to four significant figures.

$$f(x) = x^3 - 3x^2 + 4x - 7.$$
When $x = 0$, $f(0) = (0)^3 - 3(0)^2 + 4(0) - 7 = -7$,
when $x = 1$, $f(1) = (1)^3 - 3(1)^2 + 4(1) - 7 = -5$,
when $x = 2$, $f(2) = (2)^3 - 3(2)^2 + 4(2) - 7 = -3$,
when $x = 3$, $f(3) = (3)^3 - 3(3)^2 + 4(3) - 7 = 5$

Thus, the value of the positive root is between $x = 2$ and $x = 3$ and dividing this interval in the ratio of 3 to 5 gives a value of the first approximation as $x = 2.4$.

Let the true value of the root x_1 be $(2.4 + \delta_1)$.

To estimate the approximate value of δ_1,

$$(2.4 + \delta_1)^3 - 3(2.4 + \delta_1)^2 + 4(2.4 + \delta_1) - 7 = 0$$

$$2.4^3 + 3 \times 2.4^2 \delta_1 - 3 \times 2.4^2 - 3 \times 2 \times 2.4 \times \delta_1 + 4 \times 2.4 + 4 \times \delta_1 - 7 \simeq 0$$

$$13.824 + 17.28\delta_1 - 17.28 - 14.4\delta_1 + 9.6 + 4\delta_1 - 7 \simeq 0$$

$$\delta_1 \simeq \frac{0.856}{6.88} \simeq 0.124$$

A second approximation to the root is $2.4 + 0.124 = 2.524$.

Let the true value of the root x_2 be $(2.524 + \delta_2)$

$$(2.524 + \delta_2)^3 - 3(2.524 + \delta_2)^2 + 4(2.524 + \delta_2) - 7 = 0$$

$$2.524^3 + 3 \times 2.524^2 \times \delta_2 - 3 \times 2.524^2 - 3 \times 2 \times 2.524\,\delta_2 + 4 \times 2.524$$
$$+ 4 \times \delta_2 - 7 \simeq 0$$

$$16.079 + 19.112\delta_2 - 19.112 - 15.114\delta_2 + 10.096 + 4\delta_2 - 7 \simeq 0$$

$$\delta_2 \simeq \frac{-0.063}{7.998} \simeq -0.008$$

A third approximation to the root is $2.524 - 0.008$, that is, 2.516, correct to four significant figures.

Let the true value of the root x_3 be $(2.516 + \delta_3)$

$$2.516^3 + 3 \times 2.516^2 \delta_3 - 3 \times 2.516^2 - 6 \times 2.516\delta_3 + 4 \times 2.516 + 4\delta_3$$
$$- 7 \simeq 0$$

$$15.926\,9 + 18.990\,8\,\delta_3 - 18.990\,8 - 15.096\delta_3 + 10.064 + 4\delta_3 - 7 \simeq 0$$

$$\delta_3 \simeq \frac{0.000\,1}{7.894\,8} \simeq -0.000\,01$$

A fourth approximation to the root is $2.516 - 0.000\,01$, that is, 2.516 correct to four significant figures.

Since the third approximation is equal to the fourth approximation, correct to the stipulated degree of accuracy, then the required root is **2.516, correct to four significant figures.**

Degree of accuracy of calculations

When determining the solution of equations iteratively, the accuracy of the calculations differs. In general, the accuracy to which the various calculations are done is arrived at intuitively. For example, there is little point in calculating δ_1 and δ_2 in the above problem to an accuracy of more than two or three decimal places. However, as the true value of the root is approached more closely, greater accuracy is needed and it can be seen that δ_3 is calculated correct to five decimal places. If small errors are introduced, the worst result is that it may take, say, one extra iteration to achieve the stipulated degree of accuracy. Also, since terms containing δ^2 and higher powers of δ are neglected, the values obtained in calculations are only approximate values in any case.

Problem 2. Determine the value of the root of the equation

$$x^4 - 3x^2 - 2x + 3 = 0$$

in the interval $x = 0$ to $x = 1$, correct to three decimal places.

$$f(0) = (0)^4 - 3(0)^2 - 2(0) + 3 = 3$$
$$f(1) = (1)^4 - 3(1)^2 - 2(1) + 3 = -1$$

Let the first approximation be, say $x = 0.8$
Let the true value of the root be $x_1 = (0.8 + \delta_1)$
Substituting $x_1 = (0.8 + \delta_1)$ in the original equation gives

$$(0.8 + \delta_1)^4 - 3(0.8 + \delta_1)^2 - 2(0.8 + \delta_1) + 3 = 0$$

Using the binomial series and neglecting terms in δ_1^2, δ_1^3 and δ_1^4, gives

$$0.8^4 + 4 \times 0.8^3 \delta_1 - 3 \times 0.8^2 - 6 \times 0.8 \delta_1 - 2 \times 0.8 - 2\delta_1 + 3 \simeq 0$$
$$0.409\ 6 + 2.048\delta_1 - 1.92 - 4.8\delta_1 - 1.6 - 2\delta_1 + 3 \simeq 0$$

i.e. $\delta_1 \simeq - \dfrac{0.110\ 4}{4.752} \simeq -0.023$

Hence, the second approximation is $0.8 - 0.023$, i.e. 0.777.
Let the true value of the root be $x_2 = (0.777 + \delta_2)$.
Substituting $x_2 = (0.777 + \delta_2)$ in the original equation, gives

$$(0.777 + \delta_2)^4 - 3(0.777 + \delta_2)^2 - 2(0.777 + \delta_2) + 3 = 0$$

Using the binomial series and neglecting terms in δ_2^2, δ_2^3 and δ_2^4, gives

$$0.777^4 + 4 \times 0.777^3 \delta_2 - 3 \times 0.777^2 - 6 \times 0.777\delta_2 - 2 \times 0.777 - 2\delta_2 + 3 \simeq 0$$
$$0.364\ 5 + 1.876\ 4\delta_2 - 1.811\ 2 - 4.662\delta_2 - 1.554 - 2\delta_2 + 3 \simeq 0$$

$\delta_2 \simeq - \dfrac{0.000\ 7}{4.785\ 6} \simeq -0.000\ 1$

Hence the third approximation is $0.777 - 0.000\ 1$, i.e. 0.777 correct to three decimal places. Since the second and third approximations are the same to the degree of accuracy required, the root is **0.777, correct to three decimal places.**

Further problems on an algebraic method of successive approximation may be found in Section 4 (Problems 1 to 7), page 10.

3 The Newton-Raphson method

Another iterative method of determining the values of the roots of an equation is the Newton-Raphson method, often referred to as 'Newton's method'. This states:

if r_1 is the approximate value for a real root of the equation $f(x) = 0$, then a closer approximation to the root, r_2, is given by:

$$r_2 = r_1 - \frac{f(r_1)}{f'(r_1)}$$

An explanation of this statement is given below.

Figure 1 shows part of a curve $y = f(x)$ in the region of the root, r, of the equation, shown as point A. Let the first approximation to the root be $(r_1, 0)$

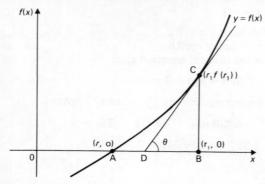

Figure 1

at point B. A vertical line BC is drawn through $(r_1, 0)$ to meet the curve at C, hence the co-ordinates of C are $(r_1, f(r_1))$. A tangent is drawn to the curve at C, cutting the x-axis at D. Let angle CDB be θ.

The slope of $CD = \dfrac{BC}{BD} = \tan\theta$,

thus $\qquad BD = \dfrac{BC}{\tan\theta}$

But $BC = f(r_1)$ and the slope of the tangent drawn to the curve at C is the differential coefficient of $f(x)$ at point r_1, that is $f'(r_1)$.

Therefore, $\quad BD = \dfrac{f(r_1)}{f'(r_1)}$

Now $\qquad OD = OB - BD$

$$= r_1 - \frac{f(r_1)}{f'(r_1)}$$

It can be seen from Fig. 1 that D is nearer to the root at A than B and hence is a better approximation.

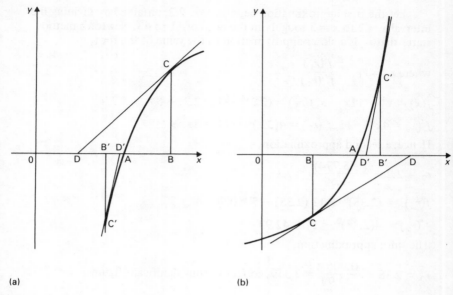

(a)   (b)

Figure 2

It will be found in practice that Newton's method sometimes does not give a closer approximation to the root. The reason for this is illustrated in Fig. 2. Occasions may occur where distance AB is less than AD, depending on the curvature, and the closeness of the first approximation to the true value of the root. In cases where successive approximations obtained by Newton's method are diverging from the root, a new value of r_1 should be chosen so that $f(r_1)$ has the same sign as $f''(r_1)$. With reference to Fig. 2(a), as x increases the slope of the curve, i.e. $f'(x)$ is decreasing, hence $f''(x)$ (the change of $f'(x)$ with respect to x) is negative. If $f(r_1)$ is negative then a first approximation is to the left of point A, say at B'; the tangent is then C'D' and the reasoning becomes similar to that applied to Fig. 1. In Fig. 2(b), as x increases, $f'(x)$ is increasing and $f''(x)$ is positive. If $f(r_1)$ is positive, the first approximation is to the right of point A at say B'; the tangent is then C'D' and again the reasoning becomes similar to that applied to Fig. 1.

Worked problems on the Newton–Raphson method

Problem 1. Use Newton's method to determine the root of the equation

$$x^4 - 11x - 5 = 0$$

in the range $x = 2$ to $x = 3$, correct to four significant figures.

$$f(x) = x^4 - 11x - 5$$
$$f(2) = (2)^4 - 11(2) - 5 = -11$$
$$f(3) = (3)^4 - 11(3) - 5 = 43$$

Let the first approximation be, say, $r_1 = 2.2$, obtained by dividing the interval $x = 2$ to $x = 3$ roughly in the ratio of 11 to 43. Newton's method states that r_2 is a closer approximation to the value of the root,

where $r_2 = r_1 - \dfrac{f(r_1)}{f'(r_1)}$

$f(x) = x^4 - 11x - 5, f(r_1) = (2.2)^4 - 11\,(2.2) - 5 \simeq -5.774$

$f'(x) = 4x^3 - 11, f'(r_1) = 4(2.2)^3 - 11 \simeq 31.59$

Hence, a second approximation,

$r_2 = 2.2 - \dfrac{(-5.774)}{31.59} \simeq 2.38$

$f(r_2) = (2.38)^4 - 11\,(2.38) - 5 \simeq 0.905\,4$

$f'(r_2) = 4(2.38)^3 - 11 \simeq 42.93$

The third approximation,

$r_3 = 2.38 - \dfrac{0.905\,4}{42.93} = 2.359$, correct to four significant figures.

$f(r_3) = (2.359)^4 - 11\,(2.359) - 5 \simeq 0.018\,9$

$f'(r_3) = 4(2.359)^3 - 11 \simeq 41.51$

The fourth approximation,

$r_4 = 2.359 - \dfrac{0.018\,9}{41.51}$

$\simeq 2.358\,5,$

i.e. 2.359 correct to four significant figures.

Since the third and fourth approximations are equal when expressed to the stipulated degree of accuracy, then the required root is **2.359, correct to four significant figures.**

Problem 2. Use Newton's method to determine the root of the equation

$3 \sin x + 4x - 5 = 0,$

in the range $x = 0$ to $x = 1$, correct to four decimal places.

$f(x) = 3 \sin x + 4x - 5$
$f(0) = 3 \sin (0) + 4(0) - 5 = -5$
$f(1) = 3 \sin 1 + 4(1) - 5 \simeq 1.52$

(Note, sin 1 means the sin of 1 radian)

Dividing the interval 0 to 1 roughly in the ratio of 5 to 1.5 gives a first approximation of $r_1 \simeq 0.75$.

$f(x) = 3 \sin x + 4x - 5, f(r_1) = 3 \sin 0.75 + 4(0.75) - 5 \simeq 0.044\,9$

$f'(x) = 3 \cos x + 4, f'(r_1) = 3 \cos 0.75 + 4 \simeq 6.195$

Hence, a second approximation,

$$r_2 = 0.75 - \frac{0.044\ 9}{6.195} = 0.742\ 8,$$

correct to four decimal places.

$$f(r_2) = 3 \sin 0.742\ 8 + 4\ (0.742\ 8) - 5 \simeq 0.000\ 259$$

$$f'(r_2) = 3 \cos 0.742\ 8 + 4 \simeq 6.210$$

Hence, a third approximation,

$$r_3 = 0.742\ 8 - \frac{0.000\ 259}{6.210}$$

$$= 0.742\ 8,\ \text{correct to four decimal places.}$$

Since the second and third approximations are equal when expressed to the stipulated degree of accuracy, then the required root is **0.742 8, correct to four decimal places.**

The solution can be checked by using either a calculator or tables. When $x = 0.742\ 8$, $\sin x = \sin (0.742\ 8\ \text{radians}) = 0.676\ 4$.

Hence $3 \sin x + 4x - 5 = 2.029\ 1 + 2.971\ 2 - 5$

$$= 0.000\ 3$$

i.e. it is very near to zero, hence the answer stated is probably correct to the required accuracy.

Problem 3. The approximate value of the root of the equation

$$0.95\ e^x - 35 \ln x - 3 = 0$$

is $x = 4.0$. Determine the value of the root, correct to four significant figures.

Using functional notation to determine how close the root is to $x = 4.0$:

$$f(4.0) = 0.95e^{4.0} - 35 \ln 4.0 - 3 \simeq 0.348$$

$$f(4.1) = 0.95e^{4.1} - 35 \ln 4.1 - 3 \simeq 4.94$$

The results indicate that the root is slightly less than $x = 4.0$ and lies very near to it. Let the first approximation be, say, $r_1 = 3.99$

$$f(x) = 0.95e^x - 35 \ln x - 3,\ f(3.99) = 0.95e^{3.99} - 35 \ln 3.99 - 3 \simeq -0.08$$

$$f'(x) = 0.95e^x - \frac{35}{x},\ f'(3.99) = 0.95e^{3.99} - \frac{35}{3.99}$$

$$= 42.58,\ \text{correct to four significant figures.}$$

Hence, a second approximation is $r_2 = 3.99 - \left(\frac{-0.08}{42.58}\right)$

$$= 3.992$$

correct to four significant figures.

10

$$f(3.992) = 0.95e^{3.992} - 35 \ln 3.992 - 3 \simeq 0.004\,72$$

$$f'(3.992) = 0.95e^{3.992} - \frac{35}{3.992} \simeq 42.69$$

Hence, a third approximation, $r_3 = 3.992 - \dfrac{0.004\,72}{42.69}$

$\qquad\qquad\qquad\qquad = 3.992$, correct to four significant figures.

Since the second and third approximations are the same when expressed to the stipulated degree of accuracy, the required root is **3.992, correct to four significant figures.**

Further problems on the Newton–Raphson method may be found in the following Section (4) (Problems 8 to 17).

4 Further problems

Algebraic method of successive approximation

Use an algebraic method of successive approximation to determine the value of the roots for the equations given in problems 1 to 7.

1. $x^3 + x = 1$ in the range $x = 0$ to $x = 1$, correct to four decimal places. [0.682 3]
2. $x^3 + 5x = 11$ in the range $x = 1$ to $x = 2$, correct to four significant figures. [1.511]
3. The two positive roots of the equation

   $$x^3 - 9x + 1 = 0,$$

   correct to four significant figures. [0.111 3, 2.943]
4. The negative root of the equation

   $$x^3 - 6x^2 + 12 = 0,$$

   correct to three decimal places. [−1.284]
5. The root of the equation

   $$x^4 - 3x^2 - 3x + 1 = 0$$

   in the interval $x = 2$ to $x = 3$, correct to two decimal places. [2.05]
6. The two roots of the equation

   $$x^4 - 12x = 5,$$

   correct to three significant figures. [−0.414, 2.41]
7. The four roots of the equation

   $$x^4 - 3x^3 - 4x^2 + 2x + 1 = 0,$$

   correct to three significant figures. [−1.19, −0.337, 0.644, 3.88]

The Newton–Raphson method

8. Determine the value of the root of the equation

   $2 \ln x + x = 2,$

   correct to two decimal places.   [1.37]
9. Determine the two roots of the equation

   $x^2 = 4 \cos x$

   correct to three significant figures.   [± 1.20]
10. Determine the roots of the equation

    $5 - \theta = 5e^{-\theta},$

    correct to three significant figures.   [0, 4.97]
11. The value of the root of the equation

    $2 \tan x - 3 \ln 2x + 4e^{x/3} = 6.5$

    is roughly 1. Determine the value of this root, correct to three decimal places.   [0.978]
12. The motion of a particle in an electrostatic field is described by the equation

    $y = x^3 + 3x^2 + 5x - 28$

    When $x = 2$, y is approximately equal to 0. Determine the value of x when $y = 0$, correct to two decimal places, using Newton's method.   [1.93]
13. The velocity v mm s^{-1} of a point on an eccentric cam at a certain instant is given by $v = \frac{1}{3} x - \ln x$, where x is the displacement in mm. The velocity becomes zero when the displacement is between 4 mm and 5 mm. Determine this value of displacement for the velocity to be zero, correct to four significant figures.   [4.536 mm]
14. The amplitude of the ripple, y, on part of a waveform displayed on a C.R.O. at time t μs is of the form

    $y = (t - \pi/3) + \sin t$

    Determine the time when the amplitude of the ripple is zero, correct to four decimal places.   [0.536 3 μs]
15. The solution to a differential equation associated with the path taken by a projectile for which the resistance to motion is proportional to the velocity, is

    $y = 2.5 (e^x - e^{-x}) + x - 25$

    Determine the value of x, correct to two decimal places, for which the value of y is zero.   [2.22]
16. An equation found in torsional oscillation problems is

    $\cot C = \dfrac{C}{2} - \dfrac{1}{2C}$

Use Newton's method to determine the first two positive roots of the equation correct to four significant figures. [1.307, 3.673]

17. The critical speeds of oscillation (λ) of a loaded beam are given by the equation

$$\lambda^3 - 3.250\,\lambda^2 + \lambda - 0.063 = 0$$

Determine the value of λ which is approximately equal to 3.0, by Newton's method, correct to four decimal places. [2.914 3]

Chapter 2

Partial fractions

1 Introduction

Consider the following addition of algebraic fractions:

$$\frac{1}{x+1} + \frac{1}{x+2} = \frac{(x+2)+(x+1)}{(x+1)(x+2)} = \frac{2x+3}{x^2+3x+2}$$

If we start with the expression $\frac{2x+3}{x^2+3x+2}$, and find the fractions whose

sum gives this result, the two fractions obtained (i.e. $\frac{1}{x+1}$ and $\frac{1}{x+2}$) are

called the **partial fractions** of $\frac{2x+3}{x^2+3x+2}$.

This process of expressing a fraction in terms of simpler fractions, called resolving into partial fractions, is used as a preliminary to integrating certain functions (see Chapter 15, Section 5) and the techniques used are explained by example later in this chapter. However, before attempting to resolve an algebraic expression into partial fractions, the following points must be considered and appreciated:

(a) **The denominator of the algebraic expression** must **factorise**. In the above example the denominator $x^2 + 3x + 2$ factorises as $(x + 1)(x + 2)$.

(b) In the above example, the numerator, $2x + 3$, is said to be of degree one since the highest powered x term is x^1. The denominator, $x^2 + 3x + 2$, is said to be of degree two since the highest powered x term is x^2.
In order to resolve an algebraic expression into partial fractions, the

numerator **must be at least one degree less than the denominator.** When the degree of the numerator is equal to or higher than the degree of the denominator, the denominator must be divided into the numerator until the remainder is of lower degree than the denominator. For example, $\dfrac{x^2 + x - 5}{x^2 - 2x - 3}$ cannot be resolved into partial fractions as it stands, since the numerator and denominator are of the same degree. Dividing $x^2 + x - 5$ by $x^2 - 2x - 3$ gives:

$$
\begin{array}{r}
1 \\
x^2 - 2x - 3 \ \overline{\smash{)}\ x^2 + x - 5} \\
\underline{x^2 - 2x - 3} \\
3x - 2
\end{array}
$$

Thus: $\dfrac{x^2 + x - 5}{x^2 - 2x - 3} = 1 + \dfrac{3x - 2}{x^2 - 2x - 3}$

Since $x^2 - 2x - 3$ factorises as $(x + 1)(x - 3)$ then $\dfrac{3x - 2}{x^2 - 2x - 3}$ may be resolved into partial fractions (see Type 1, Section 2).

(c) Given an identity such as:

$$5x^2 - 3x + 2 \equiv Ax^2 + Bx + C$$

(note: $\equiv$ means 'identically equal to'), then $A = 5, B = -3$ and $C = 2$, since the identity is true for all values of x and the coefficients of x^n (where $n = 0, 1, 2 \ldots$) on the left-hand side of the identity are equal to the coefficients of x^n on the right-hand side of the identity. Similarly, if $ax^3 + bx^2 - cx + d \equiv 2x^3 + 5x - 7$ then $a = 2, b = 0, c = -5$ and $d = -7$.

2 Type 1. Denominator containing linear factors

The corresponding partial fractions of an algebraic expression $\dfrac{f(x)}{(x - a)(x - b)}$ are of the form $\dfrac{A}{(x - a)} + \dfrac{B}{(x - b)}$, where $f(x)$ is a polynomial of degree less than 2. Similarly, the corresponding partial fractions of $\dfrac{f(x)}{(x + a)(x - b)(x - c)}$ are of the form $\dfrac{A}{(x + a)} + \dfrac{B}{(x - b)} + \dfrac{C}{(x - c)}$, where $f(x)$ is a polynomial of degree less than 3, and so on.

Problem 1. Resolve $\dfrac{x - 8}{x^2 - x - 2}$ into partial fractions.

The denominator factorises as $(x + 1)(x - 2)$ and the numerator is of less degree than the denominator. Thus $\dfrac{x-8}{x^2-x-2}$ may be resolved into partial fractions.

Let $\dfrac{x-8}{x^2-x-2} = \dfrac{x-8}{(x+1)(x-2)} \equiv \dfrac{A}{x+1} + \dfrac{B}{x-2}$ where A and B are constants to be determined.

Adding the two fractions on the right-hand side gives:

$$\dfrac{x-8}{(x+1)(x-2)} \equiv \dfrac{A(x-2)+B(x+1)}{(x+1)(x-2)}$$

Since the denominators are the same on each side of the identity then the numerators must be equal to each other.

Hence $x-8 \equiv A(x-2)+B(x+1)$

There are two methods whereby A and B may be determined using the properties of identities introduced in Section 1.

Method 1.

Since an identity is true for all real values of x, substitute into the identity a value of x to reduce one of the unknown constants to zero.

Let $x=2$.　Then　$2-8 = A(0)+B(3)$
　　　　　i.e.　　$-6 = 3B$
　　　　　　　　$B = -2$

Let $x=-1$.　Then $-1-8 = A(-3)+B(0)$
　　　　　　　$-9 = -3A$
　　　　　　　　$A = 3$

Method 2.

Since the coefficients of x^n ($n = 0, 1, 2 \ldots$) on the left-hand side of an identity equal the coefficients of x^n on the right-hand side, equate the respective coefficients on each side of the identity.

Since　$x-8 \equiv A(x-2)+B(x+1)$
then　$x-8 \equiv Ax-2A+Bx+B = (A+B)x+(-2A+B)$
Thus　　$1 = A+B$ (by equating the coefficients of x)　　　(1)
and　　$-8 = -2A+B$ (by equating the constants)　　　(2)

Solving the two simultaneous equations gives $A = 3$ and $B = -2$ as before.

Thus $\dfrac{x-8}{x^2-x-2} \equiv \dfrac{3}{x+1} + \dfrac{-2}{x-2} \equiv \dfrac{3}{x+1} - \dfrac{2}{x-2}$

　　It is usually quicker and easier to adopt the first method as far as possible although with other types of partial fractions a combination of the two methods is necessary.

Problem 2. Express $\dfrac{6x^2+7x-25}{(x-1)(x+2)(x-3)}$ in partial fractions.

Let $\dfrac{6x^2 + 7x - 25}{(x - 1)(x + 2)(x - 3)} \equiv \dfrac{A}{(x - 1)} + \dfrac{B}{(x + 2)} + \dfrac{C}{(x - 3)}$

$$\equiv \dfrac{A(x + 2)(x - 3) + B(x - 1)(x - 3) + C(x - 1)(x + 2)}{(x - 1)(x + 2)(x - 3)}$$

Equating the numerators gives:

$$6x^2 + 7x - 25 \equiv A(x + 2)(x - 3) + B(x - 1)(x - 3) + C(x - 1)(x + 2)$$

Let $x = 1$. Then $6 + 7 - 25 = A(3)(-2) + B(0)(-2) + C(0)(3)$

   i.e. $-12 = -6A$

          $A = 2$

Let $x = -2$. Then $6(-2)^2 + 7(-2) - 25 = A(0)(-5) + B(-3)(-5) + C(-3)(0)$

   i.e. $-15 = 15B$

             $B = -1$

Let $x = 3$. Then $6(3)^2 + 7(3) - 25 \quad= A(5)(0) + B(2)(0) + C(2)(5)$

   i.e $50 \quad= 10C$

           $C \quad= 5$

Thus: $\dfrac{6x^2 + 7x - 25}{(x - 1)(x + 2)(x - 3)} \equiv \dfrac{2}{(x - 1)} - \dfrac{1}{(x + 2)} + \dfrac{5}{(x - 3)}$

Problem 3. Convert $\dfrac{x^3 - x^2 - 5x}{x^2 - 3x + 2}$ into partial fractions.

The numerator is of higher degree than the denominator, thus dividing gives:

$$
\begin{array}{r}
x + 2 \\
x^2 - 3x + 2 \overline{\smash{\big)}\ x^3 - x^2 - 5x } \\
\underline{x^3 - 3x^2 + 2x} \\
2x^2 - 7x \\
\underline{2x^2 - 6x + 4} \\
-x - 4
\end{array}
$$

Thus $\dfrac{x^3 - x^2 - 5x}{x^2 - 3x + 2} = x + 2 + \dfrac{-x - 4}{x^2 - 3x + 2} = x + 2 - \dfrac{x + 4}{(x - 1)(x - 2)}$

Let $\dfrac{x + 4}{(x - 1)(x - 2)} \equiv \dfrac{A}{x - 1} + \dfrac{B}{x - 2} \equiv \dfrac{A(x - 2) + B(x - 1)}{(x - 1)(x - 2)}$

Equating the numerators gives:

$$x + 4 \equiv A(x - 2) + B(x - 1)$$

Let $x = 1$.   Then $5 = -A$

             $A = -5$

Let $x = 2$.   Then $6 = B$

Thus $\dfrac{x+4}{(x-1)(x-2)} \equiv \dfrac{-5}{(x-1)} + \dfrac{6}{(x-2)}$

Thus $\dfrac{x^3 - x^2 - 5x}{x^2 - 3x + 2} \equiv x + 2 - \dfrac{-5}{(x-1)} + \dfrac{6}{(x-2)}$

$\equiv x + 2 + \dfrac{5}{(x-1)} - \dfrac{6}{(x-2)}$

3 Type 2. Denominator containing repeated linear factors

When the denominator of an algebraic expression has a factor $(x-a)^n$ then the corresponding partial fractions are $\dfrac{A}{(x-a)} + \dfrac{B}{(x-a)^2} + \ldots + \dfrac{C}{(x-a)^n}$, since $(x-a)^n$ is assumed to hide the factors $(x-a)^{n-1}, (x-a)^{n-2} \ldots (x-a)$.

Problem 1. Express $\dfrac{x+5}{(x+3)^2}$ in partial fractions.

Let $\dfrac{x+5}{(x+3)^2} \equiv \dfrac{A}{(x+3)} + \dfrac{B}{(x+3)^2} \equiv \dfrac{A(x+3)+B}{(x+3)^2}$

Equating the numerators gives:

$$x + 5 \equiv A(x+3) + B$$
Let $x = -3$. Then $-3 + 5 = A(0) + B$
i.e. $\qquad B = 2$

Equating the coefficient of x gives $A = 1$.
(Check by equating constant terms gives: $5 = 3A + B$ which is true when $A = 1$ and $B = 2$.)

Thus $\dfrac{x+5}{(x+3)^2} \equiv \dfrac{1}{(x+3)} + \dfrac{2}{(x+3)^2}$

Problem 2. Resolve $\dfrac{5x^2 - 19x + 3}{(x-2)^2(x+1)}$ into partial fractions.

The given denominator is a combination of the linear factor type and repeated factor type.

Let $\dfrac{5x^2 - 19x + 3}{(x-2)^2(x+1)} \equiv \dfrac{A}{(x-2)} + \dfrac{B}{(x-2)^2} + \dfrac{C}{(x+1)}$

$\equiv \dfrac{A(x-2)(x+1) + B(x+1) + C(x-2)^2}{(x-2)^2(x+1)}$

Equating the numerators gives:

$$5x^2 - 19x + 3 \equiv A(x-2)(x+1) + B(x+1) + C(x-2)^2$$

Let $x = 2$. Then $5(2)^2 - 19(2) + 3 \quad = A(0)(3) + B(3) + C(0)^2$

i.e. $\qquad\qquad\qquad -15 \quad = 3B$

$\qquad\qquad\qquad\qquad B \quad = -5$

Let $x = -1$. Then $5(-1)^2 - 19(-1) + 3 = A(-3)(0) + B(0) + C(-3)^2$

i.e. $\qquad\qquad\qquad\qquad 27 = 9C$

$\qquad\qquad\qquad\qquad\quad C = 3$

$$\begin{aligned}5x^2 - 19x + 3 &\equiv A(x-2)(x+1) + B(x+1) + C(x-2)^2 \\ &\equiv A(x^2 - x - 2) + B(x+1) + C(x^2 - 4x + 4) \\ &\equiv (A+C)x^2 + (-A+B-4C)x + (-2A+B+4C)\end{aligned}$$

Equating the coefficients of x^2 gives:

$$5 = A + C$$

Since $C = 3$ then $A = 2$

[Check: Equating the coefficients of x gives: $\qquad -19 \quad = -A + B - 4C$.

If $A = 2, B = -5$ and $C = 3$ then $-A + B - 4C \quad = -19 = $ L.H.S.

Equating the constant terms gives: $\qquad\qquad 3 \quad = -2A + B + 4C$.

If $A = 2, B = -5$ and $C = 3$ then $-2A + B + 4C \quad = 3 = $ L.H.S.]

Thus $\dfrac{5x^2 - 19x + 3}{(x-2)^2(x+1)} \equiv \dfrac{2}{(x-2)} - \dfrac{5}{(x-2)^2} + \dfrac{3}{(x+1)}$

Problem 3. Convert $\dfrac{2x^2 - 13x + 13}{(x-4)^3}$ into partial fractions.

Let $\dfrac{2x^2 - 13x + 13}{(x-4)^3} \equiv \dfrac{A}{(x-4)} + \dfrac{B}{(x-4)^2} + \dfrac{C}{(x-4)^3}$

$\qquad\qquad\qquad\quad \equiv \dfrac{A(x-4)^2 + B(x-4) + C}{(x-4)^3}$

Equating the numerators gives:

$$2x^2 - 13x + 13 \equiv A(x-4)^2 + B(x-4) + C$$

Let $x = 4$. Then $2(4)^2 - 13(4) + 13 = A(0)^2 + B(0) + C$

i.e. $\qquad\qquad\qquad -7 = C$

Also $2x^2 - 13x + 13 \equiv A(x^2 - 8x + 16) + B(x-4) + C$

$\qquad\qquad\qquad\quad \equiv Ax^2 + (-8A + B)x + (16A - 4B + C)$

Equating the coefficients of x^2 gives: $\quad 2 = A$

Equating the coefficients of x gives: $-13 = -8A + B$, from which $B = 3$.

(Check: Equating the constant terms gives: $13 = 16A - 4B + C$.

If $A = 2, B = 3$ and $C = -7$

then R.H.S. $= 16(2) - 4(3) - 7 = 13 = $ L.H.S.)

Thus: $\dfrac{2x^2 - 13x + 13}{(x-4)^3} \equiv \dfrac{2}{(x-4)} + \dfrac{3}{(x-4)^2} - \dfrac{7}{(x-4)^3}$

4 Type 3. Denominator containing a quadratic factor

When the denominator contains a quadratic factor of the form $px^2 + qx + r$ (where p, q and r are constants), which does not factorise without containing surds or imaginary values, then the corresponding partial fraction is of the form $\dfrac{Ax + B}{px^2 + qx + r}$, i.e. the numerator is assumed to be a polynomial of degree one less than the denominator. Hence the corresponding partial fractions of an algebraic expression $\dfrac{f(x)}{(px^2 + qx + r)(x - a)}$ are of the form $\dfrac{Ax + B}{px^2 + qx + r} + \dfrac{C}{x - a}$.

Problem 1. Resolve $\dfrac{8x^2 - 3x + 19}{(x^2 + 3)(x - 1)}$ into partial fractions.

Let $\dfrac{8x^2 - 3x + 19}{(x^2 + 3)(x - 1)} \equiv \dfrac{Ax + B}{(x^2 + 3)} + \dfrac{C}{(x - 1)}$

$\equiv \dfrac{(Ax + B)(x - 1) + C(x^2 + 3)}{(x^2 + 3)(x - 1)}$

Equating the numerators gives:

$8x^2 - 3x + 19 \equiv (Ax + B)(x - 1) + C(x^2 + 3)$

Let $x = 1$. Then $8(1)^2 - 3(1) + 19 = (A + B)(0) + C(4)$

i.e. $\qquad\qquad\qquad\qquad 24 = 4C$

$\qquad\qquad\qquad\qquad\quad C = 6$

$8x^2 - 3x + 19 \equiv (Ax + B)(x - 1) + C(x^2 + 3)$
$\equiv Ax^2 - Ax + Bx - B + Cx^2 + 3C$
$\equiv (A + C)x^2 + (-A + B)x + (-B + 3C)$

Equating the coefficients of the x^2 terms gives:

$\qquad\qquad 8 = A + C$

Since $C = 6$, $\qquad A = 2$

Equating the coefficients of the x terms gives:

$\qquad\qquad -3 = -A + B$

Since $A = 2$, $\qquad B = -1$

(Check: Equating the constant terms gives: $19 = -B + 3C$
R.H.S. $= -B + 3C = -(-1) + 3(6) = 19 =$ L.H.S.)

Hence: $\dfrac{8x^2 - 3x + 19}{(x^2 + 3)(x - 1)} \equiv \dfrac{2x - 1}{(x^2 + 3)} + \dfrac{6}{(x - 1)}$

Problem 2. Resolve $\dfrac{2 + x + 6x^2 - 2x^3}{x^2(x^2 + 1)}$ into partial fractions.

Terms such as x^2 may be treated as $(x + 0)^2$, i.e. it is a repeated linear factor type.

Let $\dfrac{2 + x + 6x^2 - 2x^3}{x^2(x^2 + 1)} \equiv \dfrac{A}{x} + \dfrac{B}{x^2} + \dfrac{Cx + D}{(x^2 + 1)}$

$\equiv \dfrac{Ax(x^2 + 1) + B(x^2 + 1) + (Cx + D)x^2}{x^2(x^2 + 1)}$

Equating the numerators gives:

$2 + x + 6x^2 - 2x^3 \equiv Ax(x^2 + 1) + B(x^2 + 1) + (Cx + D)x^2$

$\equiv (A + C)x^3 + (B + D)x^2 + Ax + B$

Let $x = 0$. Then $2 = B$

Equating the coefficients of x^3 gives:

$-2 = A + C$ (1)

Equating the coefficients of x^2 gives:

$6 = B + D$

Since $B = 2$, $D = 4$

Equating the coefficients of x gives:

$1 = A$

From equation (1), $C = -3$

(Check: Equating the constant terms: $2 = B$ as before.)

Hence $\dfrac{2 + x + 6x^2 - 2x^3}{x^2(x^2 + 1)} \equiv \dfrac{1}{x} + \dfrac{2}{x^2} + \dfrac{4 - 3x}{x^2 + 1}$

5 Summary

Type. Denominator containing	Expression	Form of partial fractions
1. Linear factors	$\dfrac{f(x)}{(x + a)(x + b)(x + c)}$	$\dfrac{A}{(x + a)} + \dfrac{B}{(x + b)} + \dfrac{C}{(x + c)}$
2. Repeated linear factors	$\dfrac{f(x)}{(x - a)^3}$	$\dfrac{A}{(x - a)} + \dfrac{B}{(x - a)^2} + \dfrac{C}{(x - a)^3}$

Type. Denominator containing	Expression	Form of partial fractions
3. Quadratic factors	$\dfrac{f(x)}{(ax^2 + bx + c)(x - d)}$	$\dfrac{Ax + B}{(ax^2 + bx + c)} + \dfrac{C}{(x - d)}$
General Example:	$\dfrac{f(x)}{(x^2 + a)(x + b)^2(x + c)}$	$\dfrac{Ax + B}{(x^2 + a)} + \dfrac{C}{(x + b)} + \dfrac{D}{(x + b)^2} + \dfrac{E}{(x + c)}$

In each of the above cases $f(x)$ must be of less degree than the relevant denominator. If it is not, then the denominator must be divided into the numerator. For every possible factor of the denominator there is a corresponding partial fraction.

6 Further Problems

Resolve the following into partial fractions:

1. $\dfrac{8}{x^2 - 4}$ $\left[\dfrac{2}{(x - 2)} - \dfrac{2}{(x + 2)} \right]$

2. $\dfrac{3x + 5}{x^2 + 2x - 3}$ $\left[\dfrac{2}{(x - 1)} + \dfrac{1}{(x + 3)} \right]$

3. $\dfrac{y - 13}{y^2 - y - 6}$ $\left[\dfrac{3}{(y + 2)} - \dfrac{2}{(y - 3)} \right]$

4. $\dfrac{17x^2 - 21x - 6}{x(x + 1)(x - 3)}$ $\left[\dfrac{2}{x} + \dfrac{8}{(x + 1)} + \dfrac{7}{(x - 3)} \right]$

5. $\dfrac{6x^2 + 7x - 49}{(x - 4)(x + 1)(2x - 3)}$ $\left[\dfrac{3}{(x - 4)} - \dfrac{2}{(x + 1)} + \dfrac{4}{(2x - 3)} \right]$

6. $\dfrac{x^2 + 2}{(x + 4)(x - 2)}$ $\left[1 - \dfrac{3}{(x + 4)} + \dfrac{1}{(x - 2)} \right]$

7. $\dfrac{2x^2 + 4x + 19}{2(x - 3)(x + 4)}$ $\left[1 + \dfrac{7}{2(x - 3)} - \dfrac{5}{2(x + 4)} \right]$

8. $\dfrac{2x^3 + 7x^2 - 2x - 27}{(x - 1)(x + 4)}$ $\left[2x + 1 - \dfrac{4}{(x - 1)} + \dfrac{7}{(x + 4)} \right]$

9. $\dfrac{2t - 1}{(t + 1)^2}$ $\left[\dfrac{2}{(t + 1)} - \dfrac{3}{(t + 1)^2} \right]$

10. $\dfrac{8x^2 + 12x - 3}{(x + 2)^3}$ $\left[\dfrac{8}{(x + 2)} - \dfrac{20}{(x + 2)^2} + \dfrac{5}{(x + 2)^3} \right]$

11. $\dfrac{6x+1}{(2x+1)^2}$ $\left[\dfrac{3}{(2x+1)} - \dfrac{2}{(2x+1)^2}\right]$

12. $\dfrac{1}{x^2(x+2)}$ $\left[\dfrac{1}{2x^2} - \dfrac{1}{4x} + \dfrac{1}{4(x+2)}\right]$

13. $\dfrac{9x^2 - 73x + 150}{(x-7)(x-3)^2}$ $\left[\dfrac{5}{(x-7)} + \dfrac{4}{(x-3)} - \dfrac{3}{(x-3)^2}\right]$

14. $\dfrac{-(9x^2 + 4x + 4)}{x^2(x^2 - 4)}$ $\left[\dfrac{1}{x} + \dfrac{1}{x^2} + \dfrac{2}{(x+2)} - \dfrac{3}{(x-2)}\right]$

15. $\dfrac{-(a^2 + 5a + 13)}{(a^2 + 5)(a-2)}$ $\left[\dfrac{2a-1}{(a^2+5)} - \dfrac{3}{(a-2)}\right]$

16. $\dfrac{3-x}{(x^2+3)(x+3)}$ $\left[\dfrac{1-x}{2(x^2+3)} + \dfrac{1}{2(x+3)}\right]$

17. $\dfrac{12 - 2x - 5x^2}{(x^2 + x + 1)(3-x)}$ $\left[\dfrac{2x+5}{(x^2+x+1)} - \dfrac{3}{(3-x)}\right]$

18. $\dfrac{x^3 + 7x^2 + 8x + 10}{x(x^2 + 2x + 5)}$ $\left[1 + \dfrac{2}{x} + \dfrac{3x-1}{(x^2+2x+5)}\right]$

19. $\dfrac{5x^3 - 3x^2 + 41x - 64}{(x^2 + 6)(x-1)^2}$ $\left[\dfrac{2-3x}{(x^2+6)} + \dfrac{8}{(x-1)} - \dfrac{3}{(x-1)^2}\right]$

20. $\dfrac{6x^3 + 5x^2 + 4x + 3}{(x^2 + x + 1)(x^2 - 1)}$ $\left[\dfrac{2x-1}{(x^2+x+1)} + \dfrac{3}{(x-1)} + \dfrac{1}{(x+1)}\right]$

Chapter 3

Matrix arithmetic and the determinant of a matrix

1 Introduction

Second-order matrices are introduced in *Mathematics for Electrical Technicians, level 3* by J. O. Bird and A. J. C. May. In this chapter, the arithmetic of matrices is extended to third order matrices, i.e. 3 by 3 matrices, and the arithmetic operations of addition, subtraction, multiplication and inversion are introduced.

The terms used in connection with third-order matrices are the same as those used for second-order matrices. For the linear simultaneous equations, say,

$$2x + 3y + 4z = 7 \tag{1}$$
$$6x - 4y + 5z = 8 \tag{2}$$
$$7x + 5y - 6z = 9 \tag{3}$$

the coefficients of x, y and z are written in matrix notation as

$$\begin{pmatrix} 2 & 3 & 4 \\ 6 & -4 & 5 \\ 7 & 5 & -6 \end{pmatrix},$$

that is, occupying the same relative positions as in equations (1), (2) and (3). The grouping of the coefficients of x, y and z in this way is called an **array** and the coefficients forming the array are called the **elements** of the matrix.

If there are m rows across an array and n columns down an array, the matrix is said to be of order 'm by n'. A matrix having a single row is called a **row matrix** and one having a single column is called a **column matrix**. For example, in equation (1) above the coefficients of x, y and z form a row matrix of (2 3 4) and the coefficients of x in equations (1), (2) and (3) form a column matrix of $\begin{pmatrix} 2 \\ 6 \\ 7 \end{pmatrix}$

A matrix having the same number of rows as columns is called a **square matrix** and a square matrix having three rows and three columns is called a third-order matrix. Matrices are generally denoted by capital letters and if the matrix representing the coefficients of x, y and z in equations (1), (2) and (3) above is A, then

$$A = \begin{pmatrix} 2 & 3 & 4 \\ 6 & -4 & 5 \\ 7 & 5 & -6 \end{pmatrix}$$

To refer to a particular element in a matrix, the notation a_{ij} is used, where i refers to the row number and j the column number. Thus in matrix A above, the element a_{22} is -4 and element a_{31} is 7, and so on.

2. Addition, subtraction and multiplication of third order matrices

A matrix does not have a single numerical value and cannot be simplified to a particular numerical answer. However, by applying the laws of matrices given in this section, they can be simplified and often the number of matrices can be reduced, for example, by adding two matrices, a single matrix is produced. Also, by comparing one matrix with another similar matrix, values of unknown elements can be determined. Matrices are of very little value in themselves but an understanding of the laws of matrices can lead to a better understanding of determinants and also enables determinants to be simplified before evaluation, saving the number of arithmetic operations and reducing their size.

Addition
As with second-order matrices, only matrices having the same number of rows and columns may be added. The sum of two matrices is the matrix obtained by adding the elements occupying corresponding positions in the two matrices. When two matrices are added, the result is a single matrix of the same order as those being added. Thus,

$$\begin{pmatrix} p & q & r \\ s & t & u \\ v & w & x \end{pmatrix} + \begin{pmatrix} a & b & c \\ d & e & f \\ g & h & i \end{pmatrix} = \begin{pmatrix} p+a & q+b & r+c \\ s+d & t+e & u+f \\ v+g & w+h & x+i \end{pmatrix}$$

This principle may be applied to more than two m by n matrices being added, the resulting matrix being a single m by n matrix.

Subtraction
As with second-order matrices, only matrices having the same number of rows and columns may be subtracted. The difference between two matrices, say $A - B$, is the matrix obtained by subtracting the elements of matrix B from those occupying the corresponding position in matrix A. Thus,

$$\begin{pmatrix} p & q & r \\ s & t & u \\ v & w & x \end{pmatrix} - \begin{pmatrix} a & b & c \\ d & e & f \\ g & h & i \end{pmatrix} = \begin{pmatrix} p-a & q-b & r-c \\ s-d & t-e & u-f \\ v-g & w-h & x-i \end{pmatrix}$$

This principle can be applied to the difference of two m by n matrices, the resulting matrix being a single m by n matrix.

Multiplication

(a) Scalar multiplication

When a matrix is multiplied by a number, the resulting matrix is one of the same order, having each element multiplied by the number. Thus if,

$$A = \begin{pmatrix} p & q & r \\ s & t & u \\ v & w & x \end{pmatrix}, \text{ then } kA = \begin{pmatrix} kp & kq & kr \\ ks & kt & ku \\ kv & kw & kx \end{pmatrix}$$

where k is a constant.

(b) Multiplication of matrices

When a 3 by 3 matrix, say A, is multiplied by a 3 by 3 matrix, say B, the resulting matrix is a 3 by 3 matrix, say C. The a_{11} element of C is the sum of the products of the first row of A and the first column of B, taken element by element. Thus, if $A \times B = C$

$$\begin{pmatrix} p & q & r \\ - & - & - \\ - & - & - \end{pmatrix} \times \begin{pmatrix} a & - & - \\ d & - & - \\ g & - & - \end{pmatrix} = \begin{pmatrix} pa + qd + rg & - & - \\ - & & - & - \\ - & & - & - \end{pmatrix}$$

The a_{12} element of C is the sum of the products obtained by taking the first row of A with the second column of B, element by element. Thus

$$\begin{pmatrix} p & q & r \\ - & - & - \\ - & - & - \end{pmatrix} \times \begin{pmatrix} - & b & - \\ - & e & - \\ - & h & - \end{pmatrix} = \begin{pmatrix} - & pb + qe + rh & - \\ - & & - \\ - & & - \end{pmatrix}$$

The remaining elements of C are calculated in a similar way, giving

$$\begin{pmatrix} p & q & r \\ s & t & u \\ v & w & x \end{pmatrix} \times \begin{pmatrix} a & b & c \\ d & e & f \\ g & h & i \end{pmatrix} = \begin{pmatrix} (pa+qd+rg) & (pb+qe+rh) & (pc+qf+ri) \\ (sa+td+ug) & (sb+te+uh) & (sc+tf+ui) \\ (va+wd+xg) & (vb+we+xh) & (vc+wf+xi) \end{pmatrix}$$

Although the laws of matrices are so formulated that they follow most of the laws which govern the algebra of numbers, frequently in the multiplication of two matrices $A \times B$ is **not** equal to $B \times A$. This is shown in worked problem 4 following.

Worked problems on addition, subtraction and multiplication of third-order matrices

In Problems 1 to 4 following, matrices A, B and C refer to:

$$A = \begin{pmatrix} 2 & 3 & -1 \\ -2 & 0 & 4 \\ 3 & 1 & 5 \end{pmatrix} \quad B = \begin{pmatrix} 4 & 3 & 0 \\ 5 & -1 & 2 \\ 5 & 3 & 4 \end{pmatrix} \text{ and } C = \begin{pmatrix} 1 & -3 & 1 \\ -5 & 7 & 4 \\ 7 & 0 & 6 \end{pmatrix}$$

Problem 1. Determine the matrix D, where $D = A + B$.

$$A + B = \begin{pmatrix} 2 & 3 & -1 \\ -2 & 0 & 4 \\ 3 & 1 & 5 \end{pmatrix} + \begin{pmatrix} 4 & 3 & 0 \\ 5 & -1 & 2 \\ 5 & 3 & 4 \end{pmatrix} = \begin{pmatrix} 2+4 & 3+3 & -1+0 \\ -2+5 & 0+(-1) & 4+2 \\ 3+5 & 1+3 & 5+4 \end{pmatrix}$$

$$D = \begin{pmatrix} 6 & 6 & -1 \\ 3 & -1 & 6 \\ 8 & 4 & 9 \end{pmatrix}$$

Problem 2. Determine the matrix E, where $E = B - A$.

$$B - A = \begin{pmatrix} 4 & 3 & 0 \\ 5 & -1 & 2 \\ 5 & 3 & 4 \end{pmatrix} - \begin{pmatrix} 2 & 3 & -1 \\ -2 & 0 & 4 \\ 3 & 1 & 5 \end{pmatrix} = \begin{pmatrix} [4-2] & [3-3] & [0-(-1)] \\ [5-(-2)] & [-1-0] & [2-4] \\ [5-3] & [3-1] & [4-5] \end{pmatrix}$$

$$E = \begin{pmatrix} 2 & 0 & 1 \\ 7 & -1 & -2 \\ 2 & 2 & -1 \end{pmatrix}$$

Problem 3. Determine matrix F, where $F = A - B + C$.

This problem may be done in two parts, in which case the single matrix representing $A - B$ can be determined as shown in Problem 2 above, and then this single matrix can be added to matrix C, as shown in Problem 1 above. Alternatively, it may be done in one step, the elements of F being obtained from the corresponding elements of A, B and C as shown below.

$$A - B + C = \begin{pmatrix} 2 & 3 & -1 \\ -2 & 0 & 4 \\ 3 & 1 & 5 \end{pmatrix} - \begin{pmatrix} 4 & 3 & 0 \\ 5 & -1 & 2 \\ 5 & 3 & 4 \end{pmatrix} + \begin{pmatrix} 1 & -3 & 1 \\ -5 & 7 & 4 \\ 7 & 0 & 6 \end{pmatrix}$$

$$= \begin{pmatrix} [2-4+1] & [3-3+(-3)] & [-1-0+1] \\ [-2-5+(-5)] & [0-(-1)+7] & [4-2+4] \\ [3-5+7] & [1-3+0] & [5-4+6] \end{pmatrix}$$

$$F = \begin{pmatrix} -1 & -3 & 0 \\ -12 & 8 & 6 \\ 5 & -2 & 7 \end{pmatrix}$$

Problem 4. (i) Determine matrix G, where $G = A \times B$.
(ii) Show that $A \times B$ is not equal to $B \times A$, in this case.

(i) The element g_{11} is obtained from the sum of the products of the first row of A and the first column of B, taken element by element. Thus
$$g_{11} = a_{11} \times b_{11} + a_{12} \times b_{21} + a_{13} \times b_{31}.$$

Since $A = \begin{pmatrix} 2 & 3 & -1 \\ -2 & 0 & 4 \\ 3 & 1 & 5 \end{pmatrix}$ and $B = \begin{pmatrix} 4 & 3 & 0 \\ 5 & -1 & 2 \\ 5 & 3 & 4 \end{pmatrix}$

then $g_{11} = 2 \times 4 + 3 \times 5 + (-1) \times 5 = 8 + 15 - 5 = 18$

Similarly, $g_{12} = a_{11} \times b_{12} + a_{12} \times b_{22} + a_{13} \times b_{32}$
$= 2 \times 3 + 3 \times (-1) + (-1) \times 3 = 6 - 3 - 3 = 0$

The remaining seven elements are found in a similar way, giving:

$$A \times B = \begin{pmatrix} [18] & [0] & [2 \times 0 + 3 \times 2 + (-1) \times 4] \\ [-2 \times 4 + 0 \times 5 + 4 \times 5] & [-2 \times 3 + 0 \times (-1) + 4 \times 3] & [-2 \times 0 + 0 \times 2 + 4 \times 4] \\ [3 \times 4 + 1 \times 5 + 5 \times 5] & [3 \times 3 + 1 \times (-1) + 5 \times 3] & [3 \times 0 + 1 \times 2 + 5 \times 4] \end{pmatrix}$$

$$G = A \times B = \begin{pmatrix} 18 & 0 & 2 \\ 12 & 6 & 16 \\ 42 & 23 & 22 \end{pmatrix}$$

(ii) $B \times A = \begin{pmatrix} 4 & 3 & 0 \\ 5 & -1 & 2 \\ 5 & 3 & 4 \end{pmatrix} \times \begin{pmatrix} 2 & 3 & -1 \\ -2 & 0 & 4 \\ 3 & 1 & 5 \end{pmatrix}$

$$= \begin{pmatrix} [4 \times 2 + 3 \times (-2) + 0 \times 3] & [4 \times 3 + 3 \times 0 + 0 \times 1] & [4 \times (-1) + 3 \times 4 + 0 \times 5] \\ [5 \times 2 + (-1) \times (-2) + 2 \times 3] & [5 \times 3 + (-1) \times 0 + 2 \times 1] & [5 \times (-1) + (-1) \times 4 + 2 \times 5] \\ [5 \times 2 + 3 \times (-2) + 4 \times 3] & [5 \times 3 + 3 \times 0 + 4 \times 1] & [5 \times (-1) + 3 \times 4 + 4 \times 5] \end{pmatrix}$$

i.e. $B \times A = \begin{pmatrix} 2 & 12 & 8 \\ 18 & 17 & 1 \\ 16 & 19 & 27 \end{pmatrix}$

Since the matrix obtained for $A \times B$ is different to that obtained for $B \times A$, this shows that in this case $A \times B \neq B \times A$.

Problem 5.

If $P = \begin{pmatrix} 2 & 3 & -1 & 4 \\ 4 & 2 & 7 & 8 \\ 3 & -2 & 10 & 0 \end{pmatrix}$ and $Q = \begin{pmatrix} 7 \\ 4 \\ 3 \\ 1 \end{pmatrix}$,

find matrix R, where $R = P \times Q$.

The previous problem dealt with multiplication of a square, 3×3 matrix. However, the principles introduced can be extended to any m by n matrix, provided the number of elements in the rows of matrix P are equal to the number of elements in the columns of matrix Q. Using the same principle as that used in Problem 4, gives:

$$R = P \times Q = \begin{pmatrix} 2 & 3 & -1 & 4 \\ 4 & 2 & 7 & 8 \\ 3 & -2 & 10 & 0 \end{pmatrix} \times \begin{pmatrix} 7 \\ 4 \\ 3 \\ 1 \end{pmatrix}$$

$$= \begin{pmatrix} 2 \times 7 + 3 \times 4 + (-1) \times 3 + 4 \times 1 \\ 4 \times 7 + 2 \times 4 + 7 \times 3 + 8 \times 1 \\ 3 \times 7 + (-2) \times 4 + 10 \times 3 + 0 \times 1 \end{pmatrix}$$

i.e. $R = \begin{pmatrix} 14 + 12 - 3 + 4 \\ 28 + 8 + 21 + 8 \\ 21 - 8 + 30 + 0 \end{pmatrix} = \begin{pmatrix} 27 \\ 65 \\ 43 \end{pmatrix}$

In general, when multiplying a matrix of dimension (m by n) by a matrix of dimension (n by q), the resulting matrix has a dimension of (m by q), i.e. (**rows** of first $\times$ **columns** of second). In this problem, matrix P has 3 **rows** and matrix Q has 1 **column**, hence the matrix resulting from the multiplication of these two must be a matrix having 3 rows and 1 column.

Further problems on addition, subtraction and multiplication of third-order matrices may be found in Section 5 (Problems 1 to 8), page 35.

3 The determinant of a 3 by 3 matrix

For a 2 by 2 matrix, say $\begin{pmatrix} a & b \\ c & d \end{pmatrix}$, the quantity $(ad - bc)$ is called the **determinant** of a matrix.

The determinant of a matrix differs from the matrix itself in that it can be evaluated to a single numerical result. To distinguish it from the matrix, it is written as $\begin{vmatrix} a & b \\ c & d \end{vmatrix}$, vertical lines being used instead of brackets.

For a 3 by 3 matrix, say $\begin{pmatrix} a_1 & b_1 & c_1 \\ a_2 & b_2 & c_2 \\ a_3 & b_3 & c_3 \end{pmatrix}$, the **minor** of element a_1 is the

determinant obtained if the row and column containing element a_1 are

considered to be covered up, i.e. row a_1, b_1, c_1 and column $\begin{pmatrix} a_1 \\ a_2 \\ a_3 \end{pmatrix}$ are covered

up, i.e. the minor of element a_1 is found from

matrix $\begin{pmatrix} a_1 & b_1 & c_1 \\ a_2 & b_2 & c_2 \\ a_3 & b_3 & c_3 \end{pmatrix}$ and is $\begin{vmatrix} b_2 & c_2 \\ b_3 & c_3 \end{vmatrix}$

Similarly, to find the minor of element, say c_2, row a_2, b_2, c_2 and
column $\begin{matrix} c_1 \\ c_2 \\ c_3 \end{matrix}$ are considered to be covered, leaving exposed the determinant
$\begin{vmatrix} a_1 & b_1 \\ a_3 & b_3 \end{vmatrix}$. It follows that each of the nine elements of a 3 by 3 matrix has
its own minor.

The sign of the minor of an element depends on the position of the element within the matrix. Each minor has a $+$ or a $-$ sign, the signs being such that they are alternatively $+$ then $-$ in both rows and columns, element a_{11} always being $+$. Thus for a 3 by 3 matrix, the sign pattern is:

$$\begin{pmatrix} + & - & + \\ - & + & - \\ + & - & + \end{pmatrix}$$

It follows that the sign of the minor of element, say c_1 is $+$, but the sign of the element, say b_3, is $-$.

A signed $-$ minor is called the **cofactor** of an element. Thus

the cofactor of element b_2, $B_2 = +\begin{vmatrix} a_1 & c_1 \\ a_3 & c_3 \end{vmatrix} = a_1 c_3 - a_3 c_1$, and

the cofactor of element b_3, $B_3 = -\begin{vmatrix} a_1 & c_1 \\ a_2 & c_2 \end{vmatrix} = -(a_1 c_2 - a_2 c_1)$

$$= a_2 c_1 - a_1 c_2.$$

The value of a 3 by 3 determinant for matrix A is written as $|A|$ and is given by the sum of the products of elements and their cofactors of **any** row or **any** column. Thus, for the matrix

$$A = \begin{pmatrix} a_1 & b_1 & c_1 \\ a_2 & b_2 & c_2 \\ a_3 & b_3 & c_3 \end{pmatrix}$$

the value of the determinant $|A|$ is given by any of the six expressions:

$|A| = a_1 A_1 + b_1 B_1 + c_1 C_1$ or $a_2 A_2 + b_2 B_2 + c_2 C_2$
or $a_3 A_3 + b_3 B_3 + c_3 C_3$ or $a_1 A_1 + a_2 A_2 + a_3 A_3$
or $b_1 B_1 + b_2 B_2 + b_3 B_3$ or $c_1 C_1 + c_2 C_2 + c_3 C_3$.

Using, say, the first row elements,

$$|A| = a_1 \begin{vmatrix} b_2 & c_2 \\ b_3 & c_3 \end{vmatrix} - b_1 \begin{vmatrix} a_2 & c_2 \\ a_3 & c_3 \end{vmatrix} + c_1 \begin{vmatrix} a_2 & b_2 \\ a_3 & b_3 \end{vmatrix}$$

$$= a_1 (b_2 c_3 - b_3 c_2) - b_1 (a_2 c_3 - a_3 c_2) + c_1 (a_2 b_3 - a_3 b_2)$$

$$= a_1 b_2 c_3 - a_1 b_3 c_2 - a_2 b_1 c_3 + a_3 b_1 c_2 + a_2 b_3 c_1 - a_3 b_2 c_1$$

Using, say, the first column elements,

$$|A| = a_1 \begin{vmatrix} b_2 & c_2 \\ b_3 & c_3 \end{vmatrix} - a_2 \begin{vmatrix} b_1 & c_1 \\ b_3 & c_3 \end{vmatrix} + a_3 \begin{vmatrix} b_1 & c_1 \\ b_2 & c_2 \end{vmatrix}$$

$$= a_1 (b_2 c_3 - b_3 c_2) - a_2 (b_1 c_3 - b_3 c_1) + a_3 (b_1 c_2 - b_2 c_1)$$
$$= a_1 b_2 c_3 - a_1 b_3 c_2 - a_2 b_1 c_3 + a_2 b_3 c_1 + a_3 b_1 c_2 - a_3 b_2 c_1.$$

This result is the same as that obtained by using the first row elements, and any of the six expressions for the value of a determinant given above will give the same result. When finding the value of a determinant containing one or more 0's as elements, it is usually easier to select the row or column containing the most 0's as the basis of the expansion, since the cofactor of a zero element is zero and need not be calculated.

Worked problems on the determinant of a 3 by 3 matrix

Problem 1. Determine the value of (a) the minors of elements -1, 0 and 5, and (b) the cofactors of elements 7, 3 and -2 for the matrix

$$\begin{pmatrix} 2 & -1 & 7 \\ 3 & 4 & -2 \\ 0 & 8 & 5 \end{pmatrix}$$

(a) The minor of the element -1 is obtained by writing the 2 by 2 determinant left when the row and column containing the element -1 are covered up (i.e. the first row and the second column).

Thus the value of the minor of the element $-1 = \begin{vmatrix} 3 & -2 \\ 0 & 5 \end{vmatrix} = 3 \times 5 - 0 \times (-2)$
$$= 15$$

Similarly, the value of the minor of element $0 = \begin{vmatrix} -1 & 7 \\ 4 & -2 \end{vmatrix} = (-1) \times (-2) - 4 \times 7$
$$= -26$$

The value of the minor of element $5 = \begin{vmatrix} 2 & -1 \\ 3 & 4 \end{vmatrix} = 2 \times 4 - 3 \times (-1)$
$$= 11$$

(b) Cofactors are signed $-$ minors, the signs following the pattern

$$\begin{pmatrix} + & - & + \\ - & + & - \\ + & - & + \end{pmatrix}$$. For the element 7, the cofactor $= + \begin{vmatrix} 3 & 4 \\ 0 & 8 \end{vmatrix} = 3 \times 8 - 0 \times 4$

$$= 24$$

For the element 3, the cofactor $= - \begin{vmatrix} -1 & 7 \\ 8 & 5 \end{vmatrix} = -((-1) \times 5 - 8 \times 7)$
$$= 61$$

For the element -2, the cofactor $= - \begin{vmatrix} 2 & -1 \\ 0 & 8 \end{vmatrix} = -(2 \times 8 - 0 \times (-1))$
$$= -16$$

Problem 2. Find the value of the determinant $|A|$ of the matrix

$$A = \begin{pmatrix} 3 & 2 & -5 \\ 0 & 1 & 7 \\ 4 & 5 & 2 \end{pmatrix}$$

The value of the determinant is the sum of the products of the elements and their cofactors of any row or any column. Since element a_{21} is 0, the arithmetic is kept to a minimum by selecting either the second row or the first column as the basis for expansion. Selecting, say, the first column, i.e. $|A| = a_1 A_1 + a_2 A_2 + a_3 A_3$, gives

$$|A| = \begin{vmatrix} 3 & 2 & -5 \\ 0 & 1 & 7 \\ 4 & 5 & 2 \end{vmatrix} = 3 \begin{vmatrix} 1 & 7 \\ 5 & 2 \end{vmatrix} - 0 \begin{vmatrix} 2 & -5 \\ 5 & 2 \end{vmatrix} + 4 \begin{vmatrix} 2 & -5 \\ 1 & 7 \end{vmatrix}$$

$$= 3(1 \times 2 - 5 \times 7) - (0) + 4(2 \times 7 - 1 \times (-5))$$

$$= 3(-33) + 4(19) = -23$$

Further problems on the determinant of a 3 by 3 matrix may be found in Section 5 (Problems 9 to 16) page 36.

4 The inverse or reciprocal of a matrix

To determine the inverse or reciprocal of a matrix, two new operations associated with matrices are introduced below. These refer to finding the transpose of a matrix and to finding the adjoint of a matrix.

The **transpose** of a matrix A, usually denoted by A^T, is obtained by writing the rows of matrix A as the columns of matrix A^T. Thus if

$$A = \begin{pmatrix} a_1 & b_1 & c_1 \\ a_2 & b_2 & c_2 \\ a_3 & b_3 & c_3 \end{pmatrix}, \text{ then } A^T = \begin{pmatrix} a_1 & a_2 & a_3 \\ b_1 & b_2 & b_3 \\ c_1 & c_2 & c_3 \end{pmatrix}$$

It can be seen that the elements on the leading diagonal, $(\setminus)$, of A and A^T are the same.

The algebraic expression for the determinant of A, using the first row as the basis for the expansion, is:

$$|A| = a_1 \begin{vmatrix} b_2 & c_2 \\ b_3 & c_3 \end{vmatrix} - b_1 \begin{vmatrix} a_2 & c_2 \\ a_3 & c_3 \end{vmatrix} + c_1 \begin{vmatrix} a_2 & b_2 \\ a_2 & b_3 \end{vmatrix}$$

$$= a_1 b_2 c_3 - a_1 b_3 c_2 - a_2 b_1 c_3 + a_3 b_1 c_2 + a_2 b_3 c_1 - a_3 b_2 c_1$$

and for A^T, using the first column as the basis for the expansion, is:

$$|A^T| = a_1 \begin{vmatrix} b_2 & b_3 \\ c_2 & c_3 \end{vmatrix} - b_1 \begin{vmatrix} a_2 & a_3 \\ c_2 & c_3 \end{vmatrix} + c_1 \begin{vmatrix} a_2 & a_3 \\ b_2 & b_3 \end{vmatrix}$$

$$= a_1 \, b_2 \, c_3 - a_1 \, b_3 \, c_2 - a_2 \, b_1 \, c_3 + a_3 \, b_1 \, c_2 + a_2 \, b_3 \, c_1 - a_3 \, b_2 \, c_1$$

Since the algebraic equations for $|A|$ and $|A^T|$ are the same, it follows that:

$$|A| \; = \; |A^T|$$

The **adjoint** matrix of matrix A, usually abbreviated to 'adj A', is obtained by:

(i)   determining the nine cofactors of matrix A,
(ii)   forming a matrix of these cofactors, and
(iii)   transposing the matrix formed in (ii) above.

Thus, to determine the adjoint matrix of matrix A, where

$$A = \begin{pmatrix} 1 & 2 & 1 \\ 2 & 0 & 2 \\ 1 & 1 & 0 \end{pmatrix}, \text{ and following the steps outlined above, gives:}$$

(i)   the cofactors of the first row are:

$$+ \begin{vmatrix} 0 & 2 \\ 1 & 0 \end{vmatrix} = -2, - \begin{vmatrix} 2 & 2 \\ 1 & 0 \end{vmatrix} = 2 \text{ and } + \begin{vmatrix} 2 & 0 \\ 1 & 1 \end{vmatrix} = 2,$$

the cofactors of the second row are:

$$- \begin{vmatrix} 2 & 1 \\ 1 & 0 \end{vmatrix} = 1, + \begin{vmatrix} 1 & 1 \\ 1 & 0 \end{vmatrix} = -1 \text{ and } - \begin{vmatrix} 1 & 2 \\ 1 & 1 \end{vmatrix} = 1$$

and the cofactors of the third row are:

$$+ \begin{vmatrix} 2 & 1 \\ 0 & 2 \end{vmatrix} = 4, - \begin{vmatrix} 1 & 1 \\ 2 & 2 \end{vmatrix} = 0 \text{ and } + \begin{vmatrix} 1 & 2 \\ 2 & 0 \end{vmatrix} = -4.$$

(ii)   the matrix of these cofactors is $\begin{pmatrix} -2 & 2 & 2 \\ 1 & -1 & 1 \\ 4 & 0 & -4 \end{pmatrix}$

(iii)   transposing the matrix of cofactors, gives

$$\text{adj } A = \begin{pmatrix} -2 & 1 & 4 \\ 2 & -1 & 0 \\ 2 & 1 & -4 \end{pmatrix}$$

The **unit matrix** I for a 2 by 2 matrix has been previously defined in *Mathematics for Electrical Technicians, level 3* by J. O. Bird and A. J. C. May and is the matrix having elements of value 1 on its leading diagonal, all other elements being 0. It is analogous to the number 1 in ordinary algebra. For a 3 by 3 matrix, the unit matrix is

$$I = \begin{pmatrix} 1 & 0 & 0 \\ 0 & 1 & 0 \\ 0 & 0 & 1 \end{pmatrix}$$

It has also been shown previously that the inverse or reciprocal of matrix A

is the matrix A^{-1}, such that

$$A \cdot A^{-1} = I$$

$$\text{i.e. } A^{-1} = \frac{I}{A} \tag{4}$$

The inverse of matrix A can also be expressed in terms of the adjoint of matrix A and the determinant of matrix A, the relationship being:

$$A^{-1} = \frac{\text{adj } A}{|A|} \tag{5}$$

Equation (5) may be verified for matrix A, where A is, say,

$\begin{pmatrix} 1 & 2 & 1 \\ 2 & 0 & 2 \\ 1 & 1 & 0 \end{pmatrix}$, as follows.

It is shown above that adj $A = \begin{pmatrix} -2 & 1 & 4 \\ 2 & -1 & 0 \\ 2 & 1 & -4 \end{pmatrix}$

The product of A and adj A is

$$A \times \text{adj } A = \begin{pmatrix} 1 & 2 & 1 \\ 2 & 0 & 2 \\ 1 & 1 & 0 \end{pmatrix} \times \begin{pmatrix} -2 & 1 & 4 \\ 2 & -1 & 0 \\ 2 & 1 & -4 \end{pmatrix}$$

$$= \begin{pmatrix} 4 & 0 & 0 \\ 0 & 4 & 0 \\ 0 & 0 & 4 \end{pmatrix} = 4 \begin{pmatrix} 1 & 0 & 0 \\ 0 & 1 & 0 \\ 0 & 0 & 1 \end{pmatrix}$$

(Extracting a common factor is dealt with in Chapter 4 following.)

i.e. $A \times \text{adj } A = 4I$.

Also, the determinant of A, $|A| = 1 \begin{vmatrix} 0 & 2 \\ 1 & 0 \end{vmatrix} - 2 \begin{vmatrix} 2 & 2 \\ 1 & 0 \end{vmatrix} + 1 \begin{vmatrix} 2 & 0 \\ 1 & 1 \end{vmatrix}$

$$= 4$$

i.e. $A \times \text{adj } A = |A| \times I$.
But from equation (4), $A \times A^{-1} = I$
Hence, $A \times \text{adj } A = |A| \times A \times A^{-1}$

i.e. $A^{-1} = \dfrac{\text{adj } A}{A}$

Although this relationship has been verified for a specific matrix, the proof can be done, following the same steps, for the general matrix

$\begin{pmatrix} a_1 & b_1 & c_1 \\ a_2 & b_2 & c_2 \\ a_3 & b_3 & c_3 \end{pmatrix}$

In this case, the algebraic expressions produced become numerous and contain a large number of factors and hence become difficult to follow.

It is now possible, using equation (5), to find the reciprocal of a matrix and since dividing by, say a, is the same as multiplying by $\frac{1}{a}$, the inverse matrix may be used for the operation of matrix division,

i.e. $\dfrac{A}{B} = A \times \dfrac{1}{B} = AB^{-1}$

Worked problems on the inverse or reciprocal of a matrix

Problem 1. Find the adjoint matrix of matrix A, where

$$A = \begin{pmatrix} 4 & 2 & 0 \\ 0 & 1 & 1 \\ 3 & 2 & 0 \end{pmatrix}$$

Using the procedure given in the text for determining the adjoint of a matrix:

(i) determine the nine cofactors of A.

First row: $+\begin{vmatrix} 1 & 1 \\ 2 & 0 \end{vmatrix} = -2, -\begin{vmatrix} 0 & 1 \\ 3 & 0 \end{vmatrix} = 3, +\begin{vmatrix} 0 & 1 \\ 3 & 2 \end{vmatrix} = -3.$

Second row: $-\begin{vmatrix} 2 & 0 \\ 2 & 0 \end{vmatrix} = 0, +\begin{vmatrix} 4 & 0 \\ 3 & 0 \end{vmatrix} = 0, -\begin{vmatrix} 4 & 2 \\ 3 & 2 \end{vmatrix} = -2.$

Third row: $+\begin{vmatrix} 2 & 0 \\ 1 & 1 \end{vmatrix} = 2, -\begin{vmatrix} 4 & 0 \\ 0 & 1 \end{vmatrix} = -4, +\begin{vmatrix} 4 & 2 \\ 0 & 1 \end{vmatrix} = 4.$

(ii) form the matrix of these cofactors, i.e. $\begin{pmatrix} -2 & 3 & -3 \\ 0 & 0 & -2 \\ 2 & -4 & 4 \end{pmatrix}$

(iii) transpose the matrix of cofactors, giving

$$\text{adj } A = \begin{pmatrix} -2 & 0 & 2 \\ 3 & 0 & -4 \\ -3 & -2 & 4 \end{pmatrix}$$

Problem 2. Determine the inverse or reciprocal matrix, A^{-1}, for the matrix A given in Problem 1 above.

From equation (5), $A^{-1} = \dfrac{\text{adj } A}{|A|}$

It is shown in Problem 1 above that the adjoint of matrix A is:

$$\text{adj } A = \begin{pmatrix} -2 & 0 & 2 \\ 3 & 0 & -4 \\ -3 & -2 & 4 \end{pmatrix}$$

The determinant of matrix A, $|A|$, using the third column as the basis of the expansion, since it contains two 0's, is: $0 - 1 \begin{vmatrix} 4 & 2 \\ 3 & 2 \end{vmatrix} + 0$

i.e. $|A| = -2$

Hence, $A^{-1} = -\dfrac{1}{2}\begin{pmatrix} -2 & 0 & 2 \\ 3 & 0 & -4 \\ -3 & -2 & 4 \end{pmatrix}$

Further problems on the inverse or reciprocal of a matrix may be found in the following Section (5) (Problems 17 to 25), page 36.

5 Further problems

Addition, subtraction and multiplication of third-order matrices

In problems 1 to 4, matrix $M = \begin{pmatrix} 1 & 2 & 1 \\ 3 & 2 & 1 \\ 2 & 3 & 2 \end{pmatrix}$ and

$$\text{matrix } N = \begin{pmatrix} 3 & 2 & 1 \\ 1 & 1 & 2 \\ 1 & 2 & 1 \end{pmatrix}$$

1. Determine matrix L, where $L = M + N$. $\left[\begin{pmatrix} 4 & 4 & 2 \\ 4 & 3 & 3 \\ 3 & 5 & 3 \end{pmatrix}\right]$

2. Prove that $M + N = N + M$.

3. Determine the matrix $M \times N$. $\left[\begin{pmatrix} 6 & 6 & 6 \\ 12 & 10 & 8 \\ 11 & 11 & 10 \end{pmatrix}\right]$

4. Prove that $M \times N \neq N \times M$. $\left[N \times M = \begin{pmatrix} 11 & 13 & 7 \\ 8 & 10 & 6 \\ 9 & 9 & 5 \end{pmatrix}\right]$

In Problems 5 to 8, $A = \begin{pmatrix} 2 & 5 & 4 & 3 \\ 3 & 1 & 4 & 5 \\ 1 & 8 & 3 & 5 \end{pmatrix}, B = \begin{pmatrix} 1 & 2 & 3 & 1 \\ 0 & 1 & 2 & -2 \\ 1 & -3 & 1 & 0 \end{pmatrix}$

$C = \begin{pmatrix} 2 & 2 & 1 \\ 4 & 5 & 0 \\ 3 & 5 & 7 \\ 5 & 3 & 5 \end{pmatrix}$ and $Q = \begin{pmatrix} 2 \\ 4 \\ 3 \\ 5 \end{pmatrix}$

5. Determine $A + B$. $\left[\begin{pmatrix} 3 & 7 & 7 & 4 \\ 3 & 2 & 6 & 3 \\ 2 & 5 & 4 & 5 \end{pmatrix}\right]$

6. Determine $A \times Q$.
$$\left[\begin{pmatrix} 51 \\ 47 \\ 68 \end{pmatrix} \right]$$

7. Determine $A \times C$.
$$\left[\begin{pmatrix} 51 & 58 & 45 \\ 47 & 46 & 56 \\ 68 & 72 & 47 \end{pmatrix} \right]$$

8. Determine $C \times A$.
$$\left[\begin{pmatrix} 11 & 20 & 19 & 21 \\ 23 & 25 & 36 & 37 \\ 28 & 76 & 53 & 69 \\ 24 & 68 & 47 & 55 \end{pmatrix} \right]$$

The determinant of a 3 by 3 matrix

Problems 9 to 12 refer to the determinant $\begin{vmatrix} a & b & c \\ m & n & p \\ x & y & z \end{vmatrix}$

9. Write down the minors for elements a, p and y.
$$[nz - yp, \ ay - xb, \ ap - mc]$$

10. Write down the cofactors for elements c, m and x.
$$[my - xn, \ yc - bz, \ bp - nc]$$

11. Find the algebraic expression for the determinant based on the first row elements. $\quad [a\,(nz - yp) - b\,(mz - xp) + c\,(my - xn)]$

12. Find the algebraic expression for the determinant based on the second column elements. $\quad [b\,(mz - xp) - n\,(az - xc) + y\,(ap - mc)]$

In problems 13 to 16, determine the values of the determinants for the matrices given.

13. $\begin{pmatrix} 2 & -1 & 3 \\ -2 & 3 & -2 \\ 2 & 1 & 5 \end{pmatrix}$ [4]

14. $\begin{pmatrix} 3 & 5 & 1 \\ -2 & 3 & 1 \\ 4 & -2 & 1 \end{pmatrix}$ [37]

15. $\begin{pmatrix} 1 & 2 & -3 \\ 2 & -3 & 4 \\ 3 & 4 & 5 \end{pmatrix}$ [−78]

16. $\begin{pmatrix} -13 & 3 & 2 \\ 32 & -6 & 3 \\ 12 & -4 & 1 \end{pmatrix}$ [−178]

The inverse or reciprocal of a matrix

17. Write down the transpose matrix of the matrix $\begin{pmatrix} 2 & -7 & 4 \\ 3 & 2 & 5 \\ -4 & 1 & 0 \end{pmatrix}$

$$\left[\begin{pmatrix} 2 & 3 & -4 \\ -7 & 2 & 1 \\ 4 & 5 & 0 \end{pmatrix}\right]$$

18. Write down the transpose matrix of the matrix $\begin{pmatrix} 4 & 8 & 3 \\ 1 & 4 & 7 \\ -3 & -2 & 6 \end{pmatrix}$

$$\left[\begin{pmatrix} 4 & 1 & -3 \\ 8 & 4 & -2 \\ 3 & 7 & 6 \end{pmatrix}\right]$$

In Problems 19 to 21, determine the adjoint matrices for the matrices given.

19. $\begin{pmatrix} 3 & 2 & 1 \\ 4 & 2 & 2 \\ 1 & 3 & 1 \end{pmatrix}$ $\left[\begin{pmatrix} -4 & 1 & 2 \\ -2 & 2 & -2 \\ 10 & -7 & -2 \end{pmatrix}\right]$

20. $\begin{pmatrix} 1 & -3 & 0 \\ 2 & 0 & 1 \\ 4 & 1 & 3 \end{pmatrix}$ $\left[\begin{pmatrix} -1 & 9 & -3 \\ -2 & 3 & -1 \\ 2 & -13 & 6 \end{pmatrix}\right]$

21. $\begin{pmatrix} 1 & 2 & 3 \\ 2 & -1 & 4 \\ 0 & -1 & 1 \end{pmatrix}$ $\left[\begin{pmatrix} 3 & -5 & 11 \\ -2 & 1 & 2 \\ -2 & 1 & -5 \end{pmatrix}\right]$

In Problems 22 to 25, determine the reciprocals of the matrices given. Show, in each case, that the product of the matrix and its reciprocal is the unit matrix.

22. $\begin{pmatrix} -1 & 2 & -3 \\ 2 & -1 & 4 \\ 3 & 4 & 1 \end{pmatrix}$ $\left[\frac{1}{4}\begin{pmatrix} -17 & -14 & 5 \\ 10 & 8 & -2 \\ 11 & 10 & -3 \end{pmatrix}\right]$

23. $\begin{pmatrix} 0 & 0 & 1 \\ 0 & 1 & 0 \\ 1 & 0 & 0 \end{pmatrix}$ $\left[-1\begin{pmatrix} 0 & 0 & -1 \\ 0 & -1 & 0 \\ -1 & 0 & 0 \end{pmatrix}\right]$

24. $\begin{pmatrix} 2 & 1 & 1 \\ 1 & 2 & 2 \\ 1 & 3 & 2 \end{pmatrix}$ $\left[-\frac{1}{3}\begin{pmatrix} -2 & 1 & 0 \\ 0 & 3 & -3 \\ 1 & -5 & 3 \end{pmatrix}\right]$

25. $\begin{pmatrix} 1 & 2 & 1 \\ 2 & 4 & 6 \\ 3 & 1 & 2 \end{pmatrix}$ $\left[\frac{1}{20}\begin{pmatrix} 2 & -3 & 8 \\ 14 & -1 & -4 \\ -10 & 5 & 0 \end{pmatrix}\right]$

Chapter 4

The general properties of 3 by 3 determinants and the solution of simultaneous equations

1   The properties of third-order determinants

The amount of arithmetic used to determine the value of a determinant can often be reduced by simplifying the determinant before evaluation. The principal properties of a determinant used in the simplifying process are given below.

(a) Changing rows or columns

If all the elements in a row or a column of a determinant are interchanged with the corresponding elements in another row or column, the value of the determinant obtained by doing this is (-1) multiplied by the value of the original determinant. It is shown in Chapter 3 that, using the first-row elements:

$$\begin{vmatrix} a_1 & b_1 & c_1 \\ a_2 & b_2 & c_2 \\ a_3 & b_3 & c_3 \end{vmatrix} = a_2\,(b_2c_3 - b_3c_2) - b_1\,(a_2c_3 - a_3c_2) + c_1\,(a_2b_3 - a_3b_2)$$

$$= a_1b_2c_3 - a_1b_3c_2 - a_2b_1c_3 + a_3b_1c_2 + a_2b_3c_1 - a_3b_2c_1$$

Suppose now that the first and third columns are interchanged. The determinant becomes:

$$\begin{vmatrix} c_1 & b_1 & a_1 \\ c_2 & b_2 & a_2 \\ c_3 & b_3 & a_3 \end{vmatrix}$$

Evaluating, using the first row elements, gives:

$$c_1(b_2a_3 - a_2b_3) - b_1(c_2a_3 - c_3a_2) + a_1(c_2b_3 - c_3b_2)$$
$$= a_3b_2c_1 - a_2b_3c_1 - a_3b_1c_2 + a_2b_1c_3 + a_1b_3c_2 - a_1b_2c_3$$

Comparing these terms with those obtained for the determinant $\begin{vmatrix} a_1 & b_1 & c_1 \\ a_2 & b_2 & c_2 \\ a_3 & b_3 & c_3 \end{vmatrix}$

shows that each term in the second expression is (-1) multiplied by its corresponding term in the first expression. Thus

$$\begin{vmatrix} a_1 & b_1 & c_1 \\ a_2 & b_2 & c_2 \\ a_3 & b_3 & c_3 \end{vmatrix} = (-1) \times \begin{vmatrix} c_1 & b_1 & a_1 \\ c_2 & b_2 & a_2 \\ c_3 & b_3 & a_3 \end{vmatrix}$$

Although this has been proved by interchanging the first and third columns, it may be shown in the same way that it applies when interchanging any row with any other row and when interchanging any column with any other column.

(b) The value of a determinant if two rows or two columns are equal

Let $|A|$ be a determinant in which, say, the first and second columns are equal,

i.e. $|A| = \begin{vmatrix} a_1 & a_1 & c_1 \\ a_2 & a_2 & c_2 \\ a_3 & a_3 & c_3 \end{vmatrix}$

Let $|B|$ be the determinant $|A|$ in which the first and second columns have been interchanged, then $|B| = \begin{vmatrix} a_1 & a_1 & c_1 \\ a_2 & a_2 & c_2 \\ a_3 & a_3 & c_3 \end{vmatrix}$

Since the first and second columns of $|A|$ are interchanged in $|B|$, then from (a) above

$$|A| = -|B|$$

But it can be seen that $|A| = |B|$ from the elements obtained. These two statements can only be true if $|A| = 0$. It follows that the value of a determinant in which two rows or two columns are equal is 0.

This statement can also be verified by obtaining the algebraic expression for a determinant in which, say, the first and second columns are equal. For the determinant $|A|$ given above, using the first-row elements as the basis for the expansion:

$$|A| = a_1 \begin{vmatrix} a_2 & c_2 \\ a_3 & c_3 \end{vmatrix} - a_1 \begin{vmatrix} a_2 & c_2 \\ a_3 & c_3 \end{vmatrix} + c_1 \begin{vmatrix} a_2 & a_2 \\ a_3 & a_3 \end{vmatrix}$$

i.e. $|A| = 0$.

(c) Common factors

All elements in a row or a column having a common factor can be divided by this factor. This factor then becomes a factor of the determinant. Thus

$$\begin{vmatrix} a_1 & b_1 & c_1 \\ \lambda a_2 & \lambda b_2 & \lambda c_2 \\ a_3 & b_3 & c_3 \end{vmatrix} = \lambda \begin{vmatrix} a_1 & b_1 & c_1 \\ a_2 & b_2 & c_2 \\ a_2 & b_3 & c_3 \end{vmatrix}$$

This property can be verified by obtaining the algebraic expression for the determinant. Using the first-row elements to obtain this, gives:

$$\begin{vmatrix} a_1 & b_1 & c_1 \\ \lambda a_2 & \lambda b_2 & \lambda c_2 \\ a_3 & b_3 & c_3 \end{vmatrix} = a_1 \begin{vmatrix} \lambda b_2 & \lambda c_2 \\ b_3 & c_3 \end{vmatrix} - b_1 \begin{vmatrix} \lambda a_2 & \lambda c_2 \\ a_3 & c_3 \end{vmatrix} + c_1 \begin{vmatrix} \lambda a_2 & \lambda b_2 \\ a_3 & b_3 \end{vmatrix}$$

$$= a_1(\lambda b_2 c_3 - \lambda b_3 c_2) - b_1(\lambda a_2 c_3 - \lambda a_3 c_2)$$
$$+ c_1(\lambda a_2 b_3 - \lambda a_3 b_2)$$

$$= \lambda[a_1(b_2 c_3 - b_3 c_2) - b_1(a_2 c_3 - a_3 c_2)$$
$$+ c_1(a_2 b_3 - a_3 b_2)]$$

$$= \lambda \begin{vmatrix} a_1 & b_1 & c_1 \\ a_2 & b_2 & c_2 \\ a_3 & b_3 & c_3 \end{vmatrix}$$

(d) Addition or subtraction of rows or columns

If a multiple of the elements of any row or column are added to the corresponding elements of any other row or column, then the value of the determinant so obtained is equal to the value of the original determinant. Thus multiplying the third-row elements by λ and adding them to the corresponding elements in the first row, gives:

$$\begin{vmatrix} a_1 & b_1 & c_1 \\ a_2 & b_2 & c_2 \\ a_3 & b_3 & c_3 \end{vmatrix} = \begin{vmatrix} a_1 + \lambda c_1 & b_1 & c_1 \\ a_2 + \lambda c_2 & b_2 & c_2 \\ a_3 + \lambda c_3 & b_3 & c_3 \end{vmatrix}$$

This property may be verified as follows. Expanding the determinant $\begin{vmatrix} a_1 + \lambda c_1 & b_1 & c_1 \\ a_2 + \lambda c_2 & b_2 & c_2 \\ a_3 + \lambda c_3 & b_3 & c_3 \end{vmatrix}$, using the first-row elements, gives:

$$(a_1 + \lambda c_1)(b_2 c_3 - b_3 c_2) - b_1[(a_2 + \lambda c_2)c_3 - (a_3 + \lambda c_3)c_2]$$
$$+ c_1[(a_2 + \lambda c_2)b_3 - (a_3 + \lambda c_3)b_2]$$

i.e. $a_1(b_2 c_3 - b_3 c_2) - b_1(a_2 c_3 - a_3 c_2) + c_1(a_2 b_3 - a_3 b_2)$

$$+ \lambda(c_1 b_2 c_3 - c_1 b_3 c_2 - b_1 c_2 c_3 + b_1 c_3 c_2 + c_1 c_2 b_3 - c_1 c_3 b_2)$$

All the terms containing λ cancel out and the remaining terms are equal to $\begin{vmatrix} a_1 & b_1 & c_1 \\ a_2 & b_2 & c_2 \\ a_3 & b_3 & c_3 \end{vmatrix}$. Thus

$$\begin{vmatrix} a_1 + \lambda c_1 & b_1 & c_1 \\ a_2 + \lambda c_2 & b_2 & c_2 \\ a_3 + \lambda c_3 & b_3 & c_3 \end{vmatrix} = \begin{vmatrix} a_1 & b_1 & c_1 \\ a_2 & b_2 & c_2 \\ a_3 & b_3 & c_3 \end{vmatrix}$$

When λ is equal to 1, then the elements of a row or column are being added to the corresponding elements of another row or column. When $\lambda = -1$, then the elements of a row or column are being subtracted from the corresponding elements of another row or column.

The four properties introduced above are used mainly in two ways. These are (i) to reduce the magnitude of the elements in a row or column, making subsequent arithmetic easier, and (ii) to introduce as many 0's as practical into a determinant before evaluating it. The way in which this is done is shown in the worked problems following.

Worked problems on the properties of third order determinants

---

Problem 1. Use the properties of determinants to simplify the determinant
$$\begin{vmatrix} 3 & 4 & 5 \\ 6 & 7 & 8 \\ 1 & 11 & 6 \end{vmatrix}, \text{ and hence find its value.}$$

There are many ways of simplifying this determinant and one method is shown below.

Taking the first row elements from the corresponding second-row elements (property (d) above) and then taking out a common factor of 3 from the second row elements (property (c) above) gives:

$$\begin{vmatrix} 3 & 4 & 5 \\ 6 & 7 & 8 \\ 1 & 11 & 6 \end{vmatrix} = \begin{vmatrix} 3 & 4 & 5 \\ 3 & 3 & 3 \\ 1 & 11 & 6 \end{vmatrix} = 3\begin{vmatrix} 3 & 4 & 5 \\ 1 & 1 & 1 \\ 1 & 11 & 6 \end{vmatrix}$$

To introduce some elements whose values are 0, property (d) is applied again. The second-row elements are taken from the corresponding third-row elements, then the first-column elements are taken from the third-column elements and finally the first-column elements from the corresponding second-column elements. These three steps are shown below.

$$3\begin{vmatrix} 3 & 4 & 5 \\ 1 & 1 & 1 \\ 1 & 11 & 6 \end{vmatrix} = 3\begin{vmatrix} 3 & 4 & 5 \\ 1 & 1 & 1 \\ 0 & 10 & 5 \end{vmatrix} = 3\begin{vmatrix} 3 & 4 & 2 \\ 1 & 1 & 0 \\ 0 & 10 & 5 \end{vmatrix} = 3\begin{vmatrix} 3 & 1 & 2 \\ 1 & 0 & 0 \\ 0 & 10 & 5 \end{vmatrix}$$

It is now fairly easy to evaluate this determinant by using the second row elements, since this row contains two 0's.

Hence $3\begin{vmatrix} 3 & 1 & 2 \\ 1 & 0 & 0 \\ 0 & 10 & 5 \end{vmatrix} = 3\left\{-1\begin{vmatrix} 1 & 2 \\ 10 & 5 \end{vmatrix} + 0 + 0\right\} = -3(1 \times 5 - 10 \times 2)$

i.e $\begin{vmatrix} 3 & 4 & 5 \\ 6 & 7 & 8 \\ 1 & 11 & 6 \end{vmatrix} = 45$

Problem 2. Evaluate the determinant $\begin{vmatrix} 3 & 5 & 7 \\ 11 & 9 & 13 \\ 15 & 17 & 19 \end{vmatrix}$

To reduce the magnitude of the elements in the second and third rows, the second-row elements are taken from the corresponding third-row elements, then the first-row elements are taken from the corresponding second-row elements, giving:

$$\begin{vmatrix} 3 & 5 & 7 \\ 11 & 9 & 13 \\ 15 & 17 & 19 \end{vmatrix} = \begin{vmatrix} 3 & 5 & 7 \\ 11 & 9 & 13 \\ 4 & 8 & 6 \end{vmatrix} = \begin{vmatrix} 3 & 5 & 7 \\ 8 & 4 & 6 \\ 4 & 8 & 6 \end{vmatrix}$$

To introduce 0's, the second-row elements are taken from the corresponding third-row elements and then the first-column elements are added to the corresponding second-column elements. This gives:

$$\begin{vmatrix} 3 & 5 & 7 \\ 8 & 4 & 6 \\ 4 & 8 & 6 \end{vmatrix} = \begin{vmatrix} 3 & 5 & 7 \\ 8 & 4 & 6 \\ -4 & 4 & 0 \end{vmatrix} = \begin{vmatrix} 3 & 8 & 7 \\ 8 & 12 & 6 \\ -4 & 0 & 0 \end{vmatrix}$$

Evaluating, using the third-row elements, gives

$$\begin{vmatrix} 3 & 8 & 7 \\ 8 & 12 & 6 \\ -4 & 0 & 0 \end{vmatrix} = -4 \begin{vmatrix} 8 & 7 \\ 12 & 6 \end{vmatrix} = -4(48 - 84)$$

i.e. $\begin{vmatrix} 3 & 5 & 7 \\ 11 & 9 & 13 \\ 15 & 17 & 19 \end{vmatrix} = \mathbf{144}$

Further problems on the properties of third order determinants may be found in Section 3 (Problems 1 to 7), page 47.

2 The solution of simultaneous equations having three unknowns

In engineering and science, the principal use of determinants is for the solution of simultaneous equations. Solving simultaneous equations having two unknowns by determinants is discussed in *Mathematics for Electrical Technicians, level III* by J.O. Bird and A.J.C. May. This section deals with the solution of simultaneous equations having three unknowns by determinants.

Let the simultaneous equations to be solved be:

$$a_1 x + b_1 y + c_1 z + d_1 = 0 \tag{1}$$
$$a_2 x + b_2 y + c_2 z + d_2 = 0 \tag{2}$$
$$a_3 x + b_3 y + c_3 z + d_3 = 0 \tag{3}$$

where a, b, c and d are constants.

The variables y and z may be eliminated from equations (1), (2) and (3) by the following procedure:

    (i) multiplying equation (1) by $(b_2 c_3 - b_3 c_2)$,

(ii) multiplying equation (2) by $-(b_1c_3 - b_3c_1)$,
(iii) multiplying equation (3) by $(b_1c_2 - b_2c_1)$ and
(iv) adding the equations obtained in (i), (ii) and (iii) above.

(i) Equation (1) multiplied by $(b_2c_3 - b_3c_2)$ gives:

$$a_1(b_2c_3 - b_3c_2)x + b_1(b_2c_3 - b_3c_2)y + c_1(b_2c_3 - b_3c_2)z$$
$$+ d_1(b_2c_3 - b_3c_2) \quad = 0$$

(ii) Equation (2) multiplied by $-(b_1c_3 - b_3c_1)$ gives:

$$-a_2(b_1c_3 - b_3c_1)x - b_2(b_1c_3 - b_3c_1)y - c_2(b_1c_3 - b_3c_1)z$$
$$- d_2(b_1c_3 - b_3c_1) \quad = 0$$

(iii) Equation (3) multiplied by $(b_1c_2 - b_2c_1)$ gives:

$$a_3(b_1c_2 - b_2c_1)x + b_3(b_1c_2 - b_2c_1)y + c_3(b_1c_2 - b_2c_1)z$$
$$+ d_3(b_1c_2 - b_2c_1) \quad = 0$$

(iv) Adding (i), (ii) and (iii) above gives:

$$x\{a_1(b_2c_3 - b_3c_2) - a_2(b_1c_3 - b_3c_1) + a_3(b_1c_2 - b_2c_1)\}$$
$$+ d_1(b_2c_3 - b_3c_2) - d_2(b_1c_3 - b_3c_1) + d_3(b_1c_2 - b_2c_1) \quad = 0.$$

Writing these algebraic expressions in determinant form gives:

$$x \begin{vmatrix} a_1 & b_1 & c_1 \\ a_2 & b_2 & c_2 \\ a_3 & b_3 & c_3 \end{vmatrix} + \begin{vmatrix} b_1 & c_1 & d_1 \\ b_2 & c_2 & d_2 \\ b_3 & c_3 & d_3 \end{vmatrix} = 0$$

$$\text{Thus } x = \frac{- \begin{vmatrix} b_1 & c_1 & d_1 \\ b_2 & c_2 & d_2 \\ b_3 & c_3 & d_3 \end{vmatrix}}{\begin{vmatrix} a_1 & b_1 & c_1 \\ a_2 & b_2 & c_2 \\ a_3 & b_3 & c_3 \end{vmatrix}}$$

Equations (1), (2) and (3) may be written as:

$$b_1y + a_1x + c_1z + d_1 = 0$$
$$b_2y + a_2x + c_2z + d_2 = 0$$
$$b_3y + a_3x + c_3z + d_3 = 0$$

By comparison with the determinant obtained for x:

$$y = \frac{- \begin{vmatrix} a_1 & c_1 & d_1 \\ a_2 & c_2 & d_2 \\ a_3 & c_3 & d_3 \end{vmatrix}}{\begin{vmatrix} b_1 & a_1 & c_1 \\ b_2 & a_2 & c_2 \\ b_3 & a_3 & c_3 \end{vmatrix}}$$

Using property (a) of determinants introduced in Section 1, the first and second columns of the denominator can be interchanged, giving

$$y = \dfrac{+ \begin{vmatrix} a_1 & c_1 & d_1 \\ a_2 & c_2 & d_2 \\ a_3 & c_3 & d_3 \end{vmatrix}}{\begin{vmatrix} a_1 & b_1 & c_1 \\ a_2 & b_2 & c_2 \\ a_3 & b_3 & c_3 \end{vmatrix}}$$

Similarly, to find the determinant for z, equations (1), (2) and (3) may be written as:

$$c_1 z + b_1 y + a_1 x + d_1 = 0$$
$$c_2 z + b_2 y + a_2 x + d_2 = 0$$
$$c_3 z + b_3 y + a_3 x + d_3 = 0$$

By comparison with the determinant obtained for x:

$$z = \dfrac{- \begin{vmatrix} b_1 & a_1 & d_1 \\ b_2 & a_2 & d_2 \\ b_3 & a_3 & d_3 \end{vmatrix}}{\begin{vmatrix} c_1 & b_1 & a_1 \\ c_2 & b_2 & a_2 \\ c_3 & b_3 & a_3 \end{vmatrix}}$$

Using property (a) of determinants introduced in Section 1, the first and second columns of the numerator and the first and third columns of the denominator may be interchanged, giving

$$z = \dfrac{+ \begin{vmatrix} a_1 & b_1 & d_1 \\ a_2 & b_2 & d_2 \\ a_3 & b_3 & d_3 \end{vmatrix}}{- \begin{vmatrix} a_1 & b_1 & c_1 \\ a_2 & b_2 & c_2 \\ a_3 & b_3 & c_3 \end{vmatrix}} = \dfrac{- \begin{vmatrix} a_1 & b_1 & d_1 \\ a_2 & b_2 & d_2 \\ a_3 & b_3 & d_3 \end{vmatrix}}{\begin{vmatrix} a_1 & b_1 & c_1 \\ a_2 & b_2 & c_2 \\ a_3 & b_3 & c_3 \end{vmatrix}}$$

These results can be combined, showing that

$$\frac{x}{|D_1|} = \frac{-y}{|D_2|} = \frac{z}{|D_3|} = \frac{-1}{|D|}$$

where $|D_1|$ is the determinant remaining when the x-column is 'covered up',

$|D_2|$ is the determinant remaining when the y-column is 'covered up',

$|D_3|$ is the determinant remaining when the z-column is 'covered up',

and $|D|$ is the determinant remaining when the constants-column is 'covered up'.

Thus, for the simultaneous equations:

$$a_1x + b_1y + c_1z + d_1 = 0$$
$$a_2x + b_2y + c_2z + d_2 = 0$$
$$a_3x + b_3y + c_3z + d_3 = 0$$

the solution is given by:

$$\frac{x}{\begin{vmatrix} b_1 & c_1 & d_1 \\ b_2 & c_2 & d_2 \\ b_3 & c_3 & d_3 \end{vmatrix}} = \frac{-y}{\begin{vmatrix} a_1 & c_1 & d_1 \\ a_2 & c_2 & d_2 \\ a_3 & c_3 & d_3 \end{vmatrix}} = \frac{z}{\begin{vmatrix} a_1 & b_1 & d_1 \\ a_2 & b_2 & d_2 \\ a_3 & b_3 & d_3 \end{vmatrix}} = \frac{-1}{\begin{vmatrix} a_1 & b_1 & c_1 \\ a_2 & b_2 & c_2 \\ a_3 & b_3 & c_3 \end{vmatrix}}$$

This relationship is used to solve simultaneous equations having three unknown values, as shown in the worked problems following.

Worked problems on the solution of simultaneous equations having three unknowns

Problem 1. Use determinants to solve the simultaneous equations:

$$2x + y = 2 \tag{1}$$
$$-4y + z = 0 \tag{2}$$
$$4x + z = 6 \tag{3}$$

Writing the equations in the form $a_1x + b_1y + c_1z + d_1 = 0$, gives

$$2x + y + 0z - 2 = 0$$
$$0x - 4y + z + 0 = 0$$
$$4x + 0y + z - 6 = 0$$

The solution can be obtained from

$$\frac{x}{|D_1|} = \frac{-y}{|D_2|} = \frac{z}{|D_3|} = \frac{-1}{|D|}$$

$|D_1|$ is the third-order determinant obtained by covering up the x-column and writing down the remaining coefficients.

i.e. $|D_1| = \begin{vmatrix} 1 & 0 & -2 \\ -4 & 1 & 0 \\ 0 & 1 & -6 \end{vmatrix} = 1 \begin{vmatrix} 1 & 0 \\ 1 & -6 \end{vmatrix} - 0 \begin{vmatrix} -4 & 0 \\ 0 & -6 \end{vmatrix} + (-2) \begin{vmatrix} -4 & 1 \\ 0 & 1 \end{vmatrix}$

$$= 2$$

$|D_2|$ is the third-order determinant remaining when the y-column is covered up, i.e.,

$$|D_2| = \begin{vmatrix} 2 & 0 & -2 \\ 0 & 1 & 0 \\ 4 & 1 & -6 \end{vmatrix} = 0 + 1 \begin{vmatrix} 2 & -2 \\ 4 & -6 \end{vmatrix} + 0 = -4$$

$|D_3|$ is the third-order determinant remaining when the z-column is covered up, i.e.

$$|D_3| = \begin{vmatrix} 2 & 1 & -2 \\ 0 & -4 & 0 \\ 4 & 0 & -6 \end{vmatrix} = -4 \begin{vmatrix} 2 & -2 \\ 4 & -6 \end{vmatrix} = 16$$

$|D|$ is the third-order determinant remaining when the constants-column is covered up, i.e.

$$|D| = \begin{vmatrix} 2 & 1 & 0 \\ 0 & -4 & 1 \\ 4 & 0 & 1 \end{vmatrix} = 2 \begin{vmatrix} -4 & 1 \\ 0 & 1 \end{vmatrix} - 1 \begin{vmatrix} 0 & 1 \\ 4 & 1 \end{vmatrix} + 0 = -4$$

Thus, $\dfrac{x}{2} = \dfrac{-y}{-4} = \dfrac{z}{16} = \dfrac{-1}{-4}$

i.e. $x = \frac{1}{2}, y = 1, z = 4$.

Problem 2. Solve the simultaneous equations:

$$\frac{3}{x} - \frac{4}{y} - \frac{2}{z} = 1 \tag{1}$$

$$\frac{2}{x} + \frac{5}{y} - \frac{2}{z} = 3 \tag{2}$$

$$\frac{1}{x} + \frac{2}{y} + \frac{1}{z} = 2 \tag{3}$$

Letting $p = \dfrac{1}{x}$, $q = \dfrac{1}{y}$ and $r = \dfrac{1}{z}$ and writing the equations in the form $a_1x + b_1y + c_1z + d_1 = 0$, gives

$$3p - 4q - 2r - 1 = 0$$
$$2p + 5q - 2r - 3 = 0$$
$$p + 2q + r - 2 = 0$$

The solution can be obtained from

$$\frac{p}{|D_1|} = \frac{-q}{|D_2|} = \frac{r}{|D_3|} = \frac{-1}{|D|}$$

$$|D_1| = \begin{vmatrix} -4 & -2 & -1 \\ 5 & -2 & -3 \\ 2 & 1 & -2 \end{vmatrix}$$

The arithmetic of evaluating this determinant can be simplified by taking twice the second-column elements from the first.

Thus: $\begin{vmatrix} -4 & -2 & -1 \\ 5 & -2 & -3 \\ 2 & 1 & -2 \end{vmatrix} = \begin{vmatrix} 0 & -2 & -1 \\ 9 & -2 & -3 \\ 0 & 1 & -2 \end{vmatrix}$

$$= -9 \begin{vmatrix} -2 & -1 \\ 1 & -2 \end{vmatrix} = -45$$

$$|D_2| = \begin{vmatrix} 3 & -2 & -1 \\ 2 & -2 & -3 \\ 1 & 1 & -2 \end{vmatrix}$$

Taking the second-row elements from the first and then twice the first-column elements from the third, gives:

$$|D_2| = \begin{vmatrix} 1 & 0 & 2 \\ 2 & -2 & -3 \\ 1 & 1 & -2 \end{vmatrix} = \begin{vmatrix} 1 & 0 & 0 \\ 2 & -2 & -7 \\ 1 & 1 & -4 \end{vmatrix} = 1 \begin{vmatrix} -2 & -7 \\ 1 & -4 \end{vmatrix} = 15$$

$$|D_3| = \begin{vmatrix} 3 & -4 & -1 \\ 2 & 5 & -3 \\ 1 & 2 & -2 \end{vmatrix}$$

Adding the third-column elements to the second, and twice the first-column elements to the third gives:

$$|D_3| = \begin{vmatrix} 3 & -5 & -1 \\ 2 & 2 & -3 \\ 1 & 0 & -2 \end{vmatrix} = \begin{vmatrix} 3 & -5 & 5 \\ 2 & 2 & 1 \\ 1 & 0 & 0 \end{vmatrix} = 1 \begin{vmatrix} -5 & 5 \\ 2 & 1 \end{vmatrix} = -15$$

$$|D| = \begin{vmatrix} 3 & -4 & -2 \\ 2 & 5 & -2 \\ 1 & 2 & 1 \end{vmatrix}$$

Adding twice the third-row elements to the first, gives:

$$|D| = \begin{vmatrix} 5 & 0 & 0 \\ 2 & 5 & -2 \\ 1 & 2 & 1 \end{vmatrix} = 5 \begin{vmatrix} 5 & -2 \\ 2 & 1 \end{vmatrix} = 45$$

Thus, $\dfrac{p}{-45} = \dfrac{-q}{15} = \dfrac{r}{-15} = \dfrac{-1}{45}$

i.e. $p = 1, q = \dfrac{1}{3}, r = \dfrac{1}{3}$.

Thus $x = 1, y = 3$ and $z = 3$.

Further problems on the solution of simultaneous equations having three unknowns may be found in the following Section (3) (Problems 8 to 18), page 48.

3 Further problems

The properties of third-order determinants

In Problems 1 to 7, use the properties of determinants to simplify the determinants given and then evaluate them.

1. $\begin{vmatrix} 1 & 3 & 5 \\ 2 & 4 & 6 \\ 3 & 5 & 7 \end{vmatrix}$ [0]

2. $\begin{vmatrix} 13 & 2 & 23 \\ 30 & 7 & 53 \\ 39 & 9 & 70 \end{vmatrix}$ [34]

3. $\begin{vmatrix} 14 & 9 & 33 \\ 13 & 11 & 36 \\ 17 & 2 & 22 \end{vmatrix}$ [1]

4. $\begin{vmatrix} 1 & 11 & 16 \\ 23 & 37 & 58 \\ 16 & 9 & 1 \end{vmatrix}$ [3 310]

5. $\begin{vmatrix} 6 & 10 & 14 \\ -2 & 1 & 1 \\ -9 & -15 & -12 \end{vmatrix}$ [234]

6. $\begin{vmatrix} 22 & 25 & 28 \\ 26 & 28 & 31 \\ 24 & 27 & 29 \end{vmatrix}$ [40]

7. $\begin{vmatrix} 5.2 & 7.5 & 8.6 \\ 5.4 & 7.2 & 8.7 \\ 5.6 & 6.9 & 8.3 \end{vmatrix}$ [1.53]

The solution of simultaneous equations

In Problems 8 to 11, solve the simultaneous equations given, using determinants.

8. $11p + 7q + 2r = 31$
$p + q + r = 4$
$31p + 15q + 13r = 90$ $[p = 2, q = 1, r = 1]$

9. $4l + 9m + 2n = 21$
$13l + 5m + 7n = 1$
$17l + 19m + 8n = 26$ $[l = -7, m = 3, n = 11]$

10. $\dfrac{1}{x} + \dfrac{2}{y} + \dfrac{2}{z} = 4$

$\dfrac{3}{x} - \dfrac{1}{y} + \dfrac{4}{z} = 25$

$\dfrac{3}{x} + \dfrac{2}{y} - \dfrac{1}{z} = -4$ $[x = \frac{1}{2}, y = -\frac{1}{3}, z = \frac{1}{4}]$

11. $\dfrac{2x}{3} - y + \dfrac{2z}{3} = 2$

$x + 8y + 3z = -31$

$\dfrac{6x}{5} - \dfrac{4y}{5} + \dfrac{2z}{5} = -2$ $[x = -5, y = -4, z = 2]$

12. Kirchhoff's laws are applied to an electrical network and the following simultaneous equations for the current flowing in amperes in various closed circuits are obtained.

$$2i_1 \quad + \; i_2 \quad + \; i_3 \quad = \; 1.67$$
$$3i_1 \quad + \; 4.5i_2 - \; 1.5i_3 = \; 0$$
$$2.25i_1 - \; 1.5i_2 + \; 5.25i_3 = \; 0$$

Use determinants to find the values of i_1, i_2 and i_3, correct to three significant figures. $\quad [i_1 = 3.97\text{A}, i_2 = -3.55\text{A}, i_3 = -2.71\text{A}]$

13. The rate of working, $\dot{W}$, is given by:

$$\dot{W} = \Sigma\, F \cdot \dot{r},$$

where F is the force in newtons, r is the distance in metres from some reference point, and $\dot{r}$ is the velocity in metres per second. The following simultaneous equations arise from experiments carried out on a system.

$$\dot{r}_1 + 3\dot{r}_2 + 2\dot{r}_3 = -13$$
$$2\dot{r}_1 - 6\dot{r}_2 + 3\dot{r}_3 = 32$$
$$3\dot{r}_1 - 4\dot{r}_2 - \; \dot{r}_3 = 12$$

Use determinants to find the values of $\dot{r}_1$, $\dot{r}_2$ and $\dot{r}_3$.
$$[\dot{r}_1 = -2 \text{ m s}^{-1}, \dot{r}_2 = -5 \text{ m s}^{-1}, \dot{r}_3 = 2 \text{ m s}^{-1}]$$

14. The simultaneous equations representing the currents flowing in an unbalanced, three-phase, star-connected, electrical network are as follows:

$$3.6i_1 + 2.4i_2 + 4.8i_3 = 1.2$$
$$1.3i_1 - 3.9i_2 - 6.5i_3 = 2.6$$
$$11.9i_1 + 1.7i_2 + 8.5i_3 = 0$$

Use determinants to find the values of i_1, i_2 and i_3.
$$[i_1 = 1, i_2 = 3, i_3 = -2]$$

15. In a mass-spring-damper system, the acceleration $\ddot{x}$ m s^{-2}, velocity $\dot{x}$ m s^{-1} and displacement x m are related by the following simultaneous equations:

$$5.8\ddot{x} + 8.7\dot{x} + 14.5x = \quad 23.2$$
$$8.1\ddot{x} + 5.4\dot{x} + \quad 5.4x = \quad\; 8.1$$
$$14.8\ddot{x} + 3.7\dot{x} - 14.8x = -22.2$$

By using determinants, determine the acceleration, velocity and displacement for the system, correct to two decimal places.
$$[\ddot{x} = -0.33 \text{ m s}^{-2}, \dot{x} = 0.67 \text{ m s}^{-1}, x = 1.33 \text{ m}]$$

16. The currents in a network are represented by I_1, I_2 and I_3. Application of Kirchhoff's laws to the circuit leads to the following three equations:

$$14I_1 - \; 5I_2 - \; 6I_3 = 10$$
$$-5I_1 + 14I_2 - \; 2I_3 = \; 3$$
$$-6I_1 - \; 2I_2 + 18I_3 = \; 5$$

50

Determine I_1, I_2 and I_3 correct to four significant figures.

$$[1.358, 0.816\ 5, 0.821\ 1]$$

17. The tensions in a simple framework, T_1, T_2 and T_3, are given by the equations:

$$6T_1 + 6T_2 + 6T_3 = 8.4$$
$$T_1 + 2T_2 + 4T_3 = 2.4$$
$$4T_1 + 2T_2 \qquad = 4.0$$

Determine T_1, T_2 and T_3. $\qquad\qquad\qquad\qquad$ [0.8, 0.4, 0.2]

18. When a number of mass/spring systems are connected together and have a mode of oscillation, and all masses oscillate with a frequency $\dfrac{n}{2\pi}$ but having different amplitudes, n can be given in terms of eigenvalues λ (where $\lambda = n^2$) by the determinant:

$$\begin{vmatrix} 1 - \lambda & -\dfrac{1}{2} & 0 \\[2mm] -\dfrac{3}{4} & \dfrac{6}{4} - \lambda & -\dfrac{3}{4} \\[2mm] 0 & -\dfrac{3}{4} & 1 - \lambda \end{vmatrix} = 0$$

Show that $16\lambda^3 - 56\lambda^2 + 49\lambda - 9 = 0$ and verify that $\lambda = 1$, $\dfrac{9}{4}$ and $\dfrac{1}{4}$.

Chapter 5

Maclaurin's and Taylor's Series

1 Maclaurin's series

Conditions

Certain mathematical functions may be represented as a power series, containing terms in ascending powers of the variable. This enables mixed functions containing, say, algebraic, trigonometric and exponential functions to be expressed solely as algebraic functions. In this form, subsequent operations such as differentiation or integration often can be more readily performed. Maclaurin's theorem may be used to express any function, say $f(x)$, as a power series, provided that, at $x = 0$, the three conditions given below are met:

(i) The function is finite, i.e. $f(0) \neq \infty$. For the two functions $f(x) = \sin x$ and $g(x) = \ln x$, $f(0) = \sin 0 = 0$ and $g(0) = \ln 0 = -\infty$, hence $f(x) = \sin x$ meets this condition but $g(x) = \ln x$ does not.

(ii) The derivatives of the function are finite, i.e. $f'(0), f''(0), f'''(0) \dots \neq \infty$. For the function $f(x) = \sin x$, $f'(0) = \cos 0 = 1$, $f''(0) = -\sin 0 = 0$, $f'''(0) = -\cos(0) = -1$ and so on. For the function $g(x) = \ln x$, $g'(x) = \dfrac{1}{x}$ and $g'(0) = \dfrac{1}{0} = \infty$. Hence the function $f(x) = \sin x$ meets this condition, but again $g(x) = \ln x$ does not meet this condition, and therefore cannot be expressed as a power series by using Maclaurin's theorem.

(iii) The power series arising from the use of Maclaurin's theorem must be convergent, that is, in general, the values of the terms (or groups of terms) must get progressively smaller and the sum of the term reach a limiting value. For example, the series $1 + \dfrac{1}{2} + \dfrac{1}{4} + \dfrac{1}{8} + \dots$ is convergent, since the values of the terms are getting smaller and the sum of the terms is approaching a limiting value of 2. However, for the series $1 + \frac{1}{2} + \frac{1}{3} + \frac{1}{4} + \frac{1}{5} + \ldots$, groups of terms are getting successively

larger, i.e. $(\frac{1}{3} + \frac{1}{4}) > \frac{1}{2}$, $(\frac{1}{5} + \frac{1}{6} + \frac{1}{7} + \frac{1}{8}) > \frac{1}{2}$ and so on,

so although the values of the terms are getting smaller, groups of terms are not and this series diverges to an infinite number. A full study of the conditions for a series to be convergent is a topic in its own right but one test which applies to most series is as follows. If two consecutive terms in a series are u_n and u_{n+1}, then for the series to be convergent,

the limiting value of $\frac{u_{n+1}}{u_n}$ as $n \to \infty$ must be less than 1, i.e.

$$\lim_{n \to \infty} \left| \frac{u_{n+1}}{u_n} \right| < 1, \text{ for the series to converge,}$$

where the vertical lines indicate 'the modulus of' or 'the positive value' of the quantity. For example, in the series $1 + \frac{1}{2!} + \frac{1}{3!} + \frac{1}{4!} + \ldots$,

the third term is $\frac{1}{3!}$, the fourth term is $\frac{1}{4!}$ and the nth term is $\frac{1}{n!}$,

i.e. $u_n = \frac{1}{n!}$. The term after the nth term is the $(n + 1)^{th}$ term, hence,

$$u_{n+1} = \frac{1}{(n+1)!}. \text{ Thus } \frac{u_{n+1}}{u_n} = \frac{\frac{1}{(n+1)!}}{\frac{1}{n!}} = \frac{n!}{(n+1)!} = \frac{1}{n+1}.$$

As $n \to \infty$, $\frac{1}{n+1} \to 0$, hence this series is convergent.

Derivation of Maclaurin's theorem

Let the power series for $f(x)$ be:

$$f(x) = a_0 + a_1 x + a_2 x^2 + a_3 x^3 + a_4 x^4 + a_5 x^5 + \ldots, \tag{1}$$

where $a_0, a_1, a_2, \ldots$ are constants.
When $x = 0$, $f(0) = a_0$, since all terms having a factor of x^n in them become zero. Thus, $a_0 = f(0)$.

Differentiating equation (1) with respect to x, gives:

$$f'(x) = a_1 + 2a_2 x + 3a_3 x^2 + 4a_4 x^3 + 5a_5 x^4 + \ldots \tag{2}$$

When $x = 0$, $f'(0) = a_1$, i.e. $a_1 = f'(0)$.

Differentiating equation (2) with respect to x, gives:

$$f''(x) = 2a_2 + 3 \cdot 2a_3 x + 4 \cdot 3a_4 x^2 + 5 \cdot 4a_5 x^3 + \ldots \tag{3}$$

When $x = 0$, $f''(0) = 2a_2 = 2!a_2$, i.e. $a_2 = \frac{f''(0)}{2!}$

Differentiating equation (3) with respect to x, gives:

$$f'''(x) = 3 \cdot 2a_3 + 4 \cdot 3 \cdot 2a_4 x + 5 \cdot 4 \cdot 3a_5 x^2 + \ldots \tag{4}$$

When $x = 0, f'''(0) = 3 \cdot 2a_3 = 3!a_3$, i.e. $a_3 = \dfrac{f'''(0)}{3!}$

The same procedure may be repeated to find $f^{iv}(x), f^{v}(x), \ldots$ giving $f^{iv}(0) = 4!a_4, f^{v}(0) = 5!a_5$ and so on. Thus

$$a_4 = \frac{f^{iv}(0)}{4!}, a_5 = \frac{f^{v}(0)}{5!} \ldots$$

Substituting for $a_0, a_1, a_2, a_3, \ldots$ in equation (1), gives

$$f(x) = f(0) + f'(0)x + \frac{f''(0)}{2!} x^2 + \frac{f'''(0)}{3!} x^3 + \ldots$$

i.e. $f(x) = f(0) + xf'(0) + \dfrac{x^2}{2!} f''(0) + \dfrac{x^3}{3!} f'''(0) + \ldots$ \hfill (5)

This mathematical statement is called **Maclaurin's theorem** or **Maclaurin's series**. The uses of Maclaurin's series include expressing mixed functions, containing some or all of trigonometric, exponential, algebraic and logarithmic terms, in a form where such techniques as Newton's method (see Chapter 1), or approximate integration (see Section 3 of this chapter), can be employed. At a more advanced level, such theories as Webb's buckling of struts are based on an understanding of Maclaurin's series.

Worked problems on Maclaurin's series

Problem 1. Determine the first four terms of the power series for $\sin x$. Hence determine the value of $\sin 0.5$ radians correct to three decimal places. What will be the power series of $\sin 2x$?

The values of $f(0), f'(0), f''(0) \ldots$ in Maclaurin's series are determined as follows:

$f(x)$	$= \sin x$	$f(0)$	$= \sin 0 = 0$
$f'(x)$	$= \cos x$	$f'(0)$	$= \cos 0 = 1$
$f''(x)$	$= -\sin x$	$f''(0)$	$= -\sin 0 = 0$
$f'''(x)$	$= -\cos x$	$f'''(0)$	$= -\cos 0 = -1$
$f^{iv}(x)$	$= \sin x$	$f^{iv}(0)$	$= \sin 0 = 0$

Since $f^{iv}(x) = f(x)$, then $f^{v}(x) = f'(x)$, $f^{vi}(x) = f''(x)$ and so on. Thus the coefficients are $0, 1, 0, -1, 0, 1, 0, -1, \ldots$ Substituting the values of $f(0), f'(0), f''(0), \ldots$ in equation (5) gives:

$$\sin x = 0 + (1)x + 0 + (-1)\frac{x^3}{3!} + 0 + (1)\frac{x^5}{5!} + 0 + (-1)\frac{x^7}{7!} + \ldots$$

i.e. $\sin x = x - \dfrac{x^3}{3!} + \dfrac{x^5}{5!} - \dfrac{x^7}{7!} + \ldots$

When $x = 0.5$ radians

$$\textbf{Sin } 0.5 = 0.5 - \frac{0.5^3}{6} + \frac{0.5^5}{120} - \frac{0.5^7}{5040} + \ldots$$

$$= 0.500\,00 - 0.020\,83 + 0.000\,26 - 0.000\,00$$
$$= \textbf{0.479 correct to three decimal places.}$$

Since $\sin x = x - \dfrac{x^3}{3!} + \dfrac{x^5}{5!} - \dfrac{x^7}{7!} + \ldots$

then $\sin u = u - \dfrac{u^3}{3!} + \dfrac{u^5}{5!} - \dfrac{u^7}{7!} + \ldots$

Replacing u by $2x$, gives

$$\sin 2x = 2x - \frac{(2x)^3}{3!} + \frac{(2x)^5}{5!} - \frac{(2x)^7}{7!} + \ldots$$

i.e. $\sin 2x = 2x - \dfrac{4x^3}{3} + \dfrac{4x^5}{15} - \dfrac{8x^7}{315} + \ldots$

Problem 2. Determine the first five terms of the series for $\ln(1+x)$.

The values of $f(0), f'(0), f''(0), \ldots$ in Maclaurin's series are defined as follows:

$f(x) = \ln(1+x)$ $f(0) = \ln(1+0) = 0$

$f'(x) = \dfrac{1}{1+x}$ $f'(0) = \dfrac{1}{1+0} = 1$

$f''(x) = -\dfrac{1}{(1+x)^2}$ $f''(0) = -\dfrac{1}{(1+0)^2} = -1$

$f'''(x) = \dfrac{2}{(1+x)^3}$ $f'''(0) = \dfrac{2}{(1+0)^3} = 2!$

$f^{iv}(x) = -\dfrac{3 \cdot 2}{(1+x)^4}$ $f^{iv}(0) = -\dfrac{3!}{(1+0)^4} = (-1)(3!)$

$f^{v}(x) = \dfrac{4 \cdot 3 \cdot 2}{(1+x)^5}$ $f^{v}(0) = \dfrac{4!}{(1+0)^5} = 4!$

Substituting the values of $f(0), f'(0), f''(0), \ldots$ in equation (5), gives:

$$\ln(1+x) = 0 + (1)x + (-1)\frac{x^2}{2!} + \frac{(2!)x^3}{3!} + \frac{(-1)(3!)x^4}{4!} + \frac{4!x^5}{5!} + \ldots$$

i.e. $\ln(1+x) = x - \dfrac{x^2}{2} + \dfrac{x^3}{3} - \dfrac{x^4}{4} + \dfrac{x^5}{5} - \ldots$

In this series, $|U_n| = \dfrac{x^n}{n}$, i.e. when n is, say, 4 the term in x^4 is $\dfrac{x^4}{4}$, and so on.

Also $|U_{n+1}| = \dfrac{x^{n+1}}{n+1}$.

Testing for convergence (see page 52),

$$\lim_{n \to \infty} \left| \frac{U_{n+1}}{U_n} \right| = \lim_{n \to \infty} \left| \frac{x^{n+1}}{n+1} \times \frac{n}{x^n} \right|$$

$$= \lim_{n \to \infty} \left| \frac{nx}{n+1} \right|$$

As $n \to \infty$, $n \to n+1$ and $\dfrac{n}{n+1} \to 1$

Hence, $\displaystyle\lim_{n \to \infty} \left| \frac{nx}{n+1} \right| = |x|$

The modulus of x must be less than 1 for the series to converge, hence this series only converges for values of x such that $-1 < x < 1$ and hence is only valid for these values of x.

Problem 3. Use the power series for $\ln\left(\dfrac{1+x}{1-x}\right)$ to determine the value of $\ln 3$ correct to three decimal places.

The power series for $\ln(1+x)$ is derived in Problem 2 above, and is

$$\ln(1+x) = x - \frac{x^2}{2} + \frac{x^3}{3} - \frac{x^4}{4} + \frac{x^5}{5} - \dots$$

Replacing x by $(-x)$ in this power series, gives:

$$\ln(1+(-x)) = (-x) - \frac{(-x)^2}{2} + \frac{(-x)^3}{3} - \frac{(-x)^4}{4} + \frac{(-x)^5}{5} - \dots$$

i.e. $\ln(1-x) = -x - \dfrac{x^2}{2} - \dfrac{x^3}{3} - \dfrac{x^4}{4} - \dfrac{x^5}{5} - \dots$

A law of logarithms states that $\ln\left(\dfrac{A}{B}\right) = \ln A - \ln B$.

Applying this law to $\ln\left(\dfrac{1+x}{1-x}\right)$ gives:

$$\ln\left(\frac{1+x}{1-x}\right) = \ln(1+x) - \ln(1-x) = \left(x - \frac{x^2}{2} + \frac{x^3}{3} - \frac{x^4}{4} + \frac{x^5}{5} - \dots\right)$$

$$- \left(-x - \frac{x^2}{2} - \frac{x^3}{3} - \frac{x^4}{4} - \frac{x^5}{5} - \dots\right)$$

$$= 2x + \frac{2x^3}{3} + \frac{2x^5}{5} + \frac{2x^7}{7} + \dots$$

i.e. $\ln\left(\dfrac{1+x}{1-x}\right) = 2\left(x + \dfrac{x^3}{3} + \dfrac{x^5}{5} + \dfrac{x^7}{7} + \dots\right)$

The series for $\ln\left(\dfrac{1+x}{1-x}\right)$ is convergent provided $-1<x<1$, and is far more suitable than the power series for $\ln(1+x)$ or $\ln(1-x)$ when determining the values of Naperian logarithms of numbers. This is because it converges far more rapidly since its terms contain only odd powers of x and it reaches the term containing, say, x^7 in four terms instead of the seven terms in the expansion of $\ln(1\pm x)$. Hence less terms have to be calculated to give the value of a Naperian logarithm to a required degree of accuracy; also it may be used to determine the values of Naperian logarithms for numbers equal to or greater than 2.

Since the series for $\ln\left(\dfrac{1+x}{1-x}\right)$ is being used to find the value of $\ln 3$, then

$$\frac{1+x}{1-x} = 3, \therefore 1+x = 3-3x$$

i.e. $x = \dfrac{1}{2}$

Thus $\ln\left(\dfrac{1+\frac{1}{2}}{1-\frac{1}{2}}\right) = \ln 3 = 2\left(\dfrac{1}{2} + \dfrac{(\frac{1}{2})^3}{3} + \dfrac{(\frac{1}{2})^5}{5} + \dfrac{(\frac{1}{2})^7}{7} + \dfrac{(\frac{1}{2})^9}{9} + \dots\right)$

$$= 2\,(0.500\,0 + 0.041\,7 + 0.006\,3 + 0.001\,1$$
$$+ 0.000\,2 + \dots)$$

$$= 2\,(0.549\,3) = 1.098\,6$$

i.e. $\qquad\qquad$ **$\ln 3 = 1.099$ correct to three decimal places.**

Problem 4. Use Maclaurin's series to find the expansion of $(3+x)^4$.

The values of $f(0), f'(0), f''(0), \dots$ in Maclaurin's series are determined as follows:

$$\begin{array}{llll}
f(x) & = (3+x)^4 & f(0) & = 3^4 \\
f'(x) & = 4\,(3+x)^3 & f'(0) & = 4\cdot 3^3 \\
f''(x) & = 4\cdot 3\,(3+x)^2 & f''(0) & = 4\cdot 3\cdot 3^2 \\
f'''(x) & = 4\cdot 3\cdot 2\,(3+x) & f'''(0) & = 4\cdot 3\cdot 2\cdot 3^1 \\
f^{iv}(x) & = 4\cdot 3\cdot 2\cdot 1\,(3+x)^0 & f^{iv}(0) & = 4\cdot 3\cdot 2\cdot 1\cdot 3^0
\end{array}$$

Substituting the values of $f(0), f'(0), f''(0), \dots$ in equation (5) gives:

$$(3+x)^4 = 3^4 + 4\cdot 3^3 x + \frac{4\cdot 3\cdot 3^2 \cdot x^2}{2!} + \frac{4\cdot 3\cdot 2\cdot 3^1 \cdot x^3}{3!} + \frac{4\cdot 3\cdot 2\cdot 1\cdot x^4}{4!}$$

i.e. the expansion which would have been obtained by applying the binomial series. Thus

$$(3+x)^4 = 81 + 108x + 54x^2 + 12x^3 + x^4$$

Problem 5. Use Maclaurin's series to determine the terms of the power series for $\ln(1+e^x)$ as far as the term in x^3.

$$f(x) = \ln(1 + e^x) \qquad\qquad f(0) = \ln(1 + 1) = \ln 2$$

$$f'(x) = \frac{e^x}{1 + e^x} \qquad\qquad f'(0) = \frac{1}{1 + 1} = \frac{1}{2}$$

$$f''(x) = \frac{(1 + e^x)\, e^x - e^x\,(e^x)}{(1 + e^x)^2}$$

$$= \frac{e^x\,(1 + e^x - e^x)}{(1 + e^x)^2} = \frac{e^x}{(1 + e^x)^2} \cdot \qquad f''(0) = \frac{1}{2^2} = \frac{1}{4}$$

$$f'''(x) = \frac{(1 + e^x)^2 e^x - e^x \cdot 2\,(1 + e^x)\,e^x}{(1 + e^x)^4} \cdot \qquad f'''(0) = \frac{2^2 - 2(2)}{2^4} = 0$$

Substituting the values of $f(0), f'(0), f''(0), \ldots$ in equation (5) gives:

$$\ln(1 + e^x) = \ln 2 + \frac{x}{2} + \frac{x^2}{4(2!)} + 0 \cdot x^3$$

i.e. $\ln(1 + e^x) = \ln 2 + \dfrac{x}{2} + \dfrac{x^2}{8}$, as far as the term in x^3

Power series

Many mathematical functions may be determined along similar lines to those shown in the worked problems above. Some of the results which may be obtained by applying Maclaurin's theorem to various functions are listed below.

$$\sin x = x - \frac{x^3}{3!} + \frac{x^5}{5!} - \frac{x^7}{7!} + \ldots$$

$$\cos x = 1 - \frac{x^2}{2!} + \frac{x^4}{4!} - \frac{x^6}{6!} + \ldots$$

$$e^x = 1 + x + \frac{x^2}{2!} + \frac{x^3}{3!} + \ldots$$

$$\sinh x = x + \frac{x^3}{3!} + \frac{x^5}{5!} + \frac{x^7}{7!} + \ldots$$

$$\cosh x = 1 + \frac{x^2}{2!} + \frac{x^4}{4!} + \frac{x^6}{6!} + \ldots$$

$$\ln(1 + x) = x - \frac{x^2}{2} + \frac{x^3}{3} - \frac{x^4}{4} + \ldots \text{(Only valid for } -1 < x < 1)$$

$$(a + x)^n = a^n + na^{n-1}x + \frac{n(n - 1)}{2!} a^{n-2}x^2 + \frac{n(n - 1)(n - 2)}{3!} a^{n-3}x^3 + \ldots$$

(This last power series is called the **binomial series**.)

These results can be used to determine the power series for more complicated functions. For example, since the series for e^x is $1 + x + \dfrac{x^2}{2!} + \dfrac{x^3}{3!} + \ldots$

the series for e^{2x} is $1 + (2x) + \dfrac{(2x)^2}{2!} + \dfrac{(2x)^3}{3!} + \ldots$

    A power series for the product of two functions can be determined by multiplying the corresponding power series. For example, the power series for $e^x \cos x$ is $\left(1 + x + \dfrac{x^2}{2!}\right)\left(1 - \dfrac{x^2}{2!}\right)$, that is, $1 - \dfrac{x^2}{2!} + x - \dfrac{x^3}{2!} + \dfrac{x^2}{2!}$,

i.e. $1 + x - \dfrac{x^3}{2!}$, as far as the term in x^3. Functions involving quotients may be determined in a similar way.

Further problems on Maclaurin's series may be found in Section 4 (Problems 1 to 10), page 64.

2 Taylor's series

Maclaurin's series, introduced in section 1, expresses the height of a curve at any point x in terms of its height at the origin. With reference to Fig. 1(a),

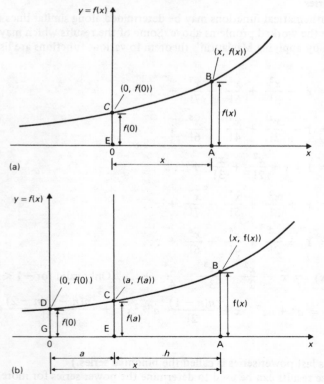

(a)

(b)

Figure 1

the height of the curve at B is AB, that is, $f(x) = $ AB and the height at the origin is EC, that is, $f(0) = $ EC. Maclaurin's series, introduced in equation (5), expresses $f(x)$ in terms of $f(0)$ and is

$$f(x) = f(0) + x f'(0) + \frac{x^2}{2!} f''(0) + \ldots$$

Taylor's series expresses the height of a curve at distance $(a + h)$ from the origin in terms of its height at distance a from the origin. With reference to Fig. 1(b), the origin of the vertical axis is moved until it lies along GD, that is, it is moved distance a to the left. Since the shape of ECBA in Fig. 1(b) is the same as in Fig. 1(a), then Maclaurin's series must still relate the height of AB to the height of EC. However, since the origin has been moved, EC is now of height $f(a)$ and distance x in Fig. 1(a) now becomes $(x - a)$ with reference to the new axis. Writing $f(a)$ for $f(0)$ and $(x - a)$ for x in Maclaurin's series gives the height of the curve at AB as:

$$f(x) = f(a) + (x - a)f'(a) + \frac{(x - a)^2}{2!} f''(a) + \frac{(x - a)^3}{3!} f'''(a) + \ldots \qquad (6)$$

This equation is called **Taylor's theorem** or **Taylor's series**.

If a is made zero in Taylor's series given in equation (6), then:

$$f(x) = f(0) + x f'(0) + \frac{x^2}{2!} f''(0) + \ldots,$$

i.e. Maclaurin's series is a special case of Taylor's series when a is equal to 0. Some of the applications of Taylor's series are to such techniques as numerical differentiation, limits, small errors and the numerical solution of certain differential equations.

An alternative way of writing Taylor's series is obtained by writing the series in terms of h rather than x. Since, from Fig. 1(b), $x = (a + h)$, and $(x - a) = h$, then substituting for x in equation (6) gives:

$$f(a + h) = f(a) + h f'(a) + \frac{h^2}{2!} f''(a) + \frac{h^3}{3!} f'''(a) + \ldots \qquad (7)$$

Taylor's series may be used for the following two applications:

(i) To determine the approximate values of functions which are close to known values of that function. For example, the value of cos 60° is known (i.e. cos 60° = 0.5). Taylor's series may be used to determine the value of, say, cos 61°. In the power series for $f(a + h)$, given in equation (7), a is taken as 60°, i.e. $\pi/3$ radians and h is taken as 1°, i.e. $\frac{1 \times \pi}{180}$ radians. These values are then substituted in the power series for cos $(a + h)$.

(ii) Newton's method of finding the approximate value of the roots of an equation is introduced in Chapter 1, and may be derived as follows. Let the equation $f(x) = 0$ have a root which is approximately equal to a. Also, let the true value of the root be $(a + $ a small quantity), say, $(a + h)$. Then $f(a + h) = 0$. By Taylor's series:

$$f(a + h) = f(a) + h f'(a) + \frac{h^2}{2!} f''(a) + \frac{h^3}{3!} f'''(a) + \ldots$$

Since $f(a + h) = 0$, then

$$f(a) + h f'(a) + \frac{h^2}{2!} f''(a) + \ldots = 0$$

Because h is small, terms containing h^2 and higher powers of h are very small and may be neglected, hence

$$f(a) + h f'(a) \doteq 0$$

i.e. $\quad h \doteq \dfrac{-f(a)}{f'(a)}$

Since the true value of the root is at $(a + h)$ and the value obtained for h is only an approximate value, then

$$a + h \doteq a - \frac{f(a)}{f'(a)}$$

Let this latter approximation to the value of the root be a_1 then

$$a_1 = a - \frac{f(a)}{f'(a)} \tag{8}$$

This is the basis of Newton's method of determining the approximate value of roots of equations.

Worked problems on Taylor's series

Problem 1. Determine the power series for $\sin (a + h)$ and hence determine the value of $\sin 30° 30'$ correct to five decimal places.

Taylor's series for $f(a + h)$ is given in equation (7) and is:

$$f(a + h) = f(a) + h f'(a) + \frac{h^2}{2!} f''(a) + \frac{h^3}{3!} f'''(a) + \ldots$$

The function f is the sine function in this problem, hence

$f(a + h) = \sin (a + h)$ and $f(a) = \sin a$.

By repeated differentiation, $f'(a) = \cos a$, $f''(a) = - \sin a$, $f'''(a) = - \cos a$, $f^{iv}(a) = \sin a$, and so on.

Substituting for $f(a), f'(a), f''(a), \ldots$ in Taylor's series, gives:

$$\sin (a + h) = \sin a + h \cos a - \frac{h^2}{2!} \sin a - \frac{h^3}{3!} \cos a + \ldots$$

To evaluate $\sin 30° 30'$, radian measure must be used and a is taken as $30°$, i.e. $\pi/6$ radians, and h as $30'$ or $0.5°$, i.e. $\dfrac{0.5 \times \pi}{180}$ radians.

Thus, $\sin 30° 30' = \sin \dfrac{\pi}{6} + \dfrac{0.5\pi}{180} \cos \dfrac{\pi}{6} - \dfrac{1}{2!} \left(\dfrac{0.5\pi}{180} \right)^2 \sin \dfrac{\pi}{6}$

$$- \dfrac{1}{3!} \left(\dfrac{0.5\pi}{180} \right)^3 \cos \pi/6 + \ldots$$

$$= 0.500\,000 + 0.008\,727 \cdot \dfrac{\sqrt{3}}{2} - \dfrac{1}{2} \cdot 0.000\,076 \cdot \dfrac{1}{2}$$

$$- \dfrac{1}{6} \cdot 0.000\,0007 \dfrac{\sqrt{3}}{2}$$

$$= 0.500\,000 + 0.007\,558 - 0.000\,019 - 0.000\,0001,$$

i.e. **$\sin 30° 30'$ = 0.507 54, correct to five decimal places.**

Problem 2. Given that $\cos 60° = 0.5$, determine the value of $\cos 70°$ by using Taylor's series, correct to four decimal places.

Taylor's series for $f(x)$ is given in equation (6) and is:

$$f(x) = f(a) + (x - a)f'(a) + \dfrac{(x-a)^2}{2!}f''(a) + \dfrac{(x-a)^3}{3!}f'''(a) + \ldots$$

With reference to Fig. 1(b), x corresponds to $70°$, i.e. $\dfrac{7\pi}{18}$ radians, a corresponds to $60°$, i.e. $\dfrac{\pi}{3}$ radians, and $(x - a)$ corresponds to $10°$, i.e. $\dfrac{\pi}{18}$ radians. Also, $f(a) = \cos \dfrac{\pi}{3}$, and by repeated differentiation,

$f'(a) = - \sin \dfrac{\pi}{3}$, $f''(a) = - \cos \dfrac{\pi}{3}$, $f'''(a) = \sin \dfrac{\pi}{3}$ and so on.

Thus, substituting these values in Taylor's series, gives:

$$f(x) = \cos 70° = \cos \dfrac{\pi}{3} - \left(\dfrac{\pi}{18} \right) \sin \dfrac{\pi}{3} - \dfrac{1}{2!} \left(\dfrac{\pi}{18} \right)^2 \cos \dfrac{\pi}{3}$$

$$+ \dfrac{1}{3!} \left(\dfrac{\pi}{18} \right)^3 \sin \dfrac{\pi}{3} + \dfrac{1}{4!} \left(\dfrac{\pi}{18} \right)^4 \cos \dfrac{\pi}{3} - \ldots$$

$$= 0.500\,00 - 0.151\,15 - 0.007\,62 + 0.000\,77 + 0.000\,02$$

$$= 0.342\,02$$

i.e. **$\cos 70°$ = 0.342 0, correct to four decimal places.**

Problem 3. Determine the value of the positive root of the equation $x^3 - 2x - 5 = 0$, correct to three decimal places, using Newton's method.

The first approximation to the value of the root is found by determining the approximate point where the graph cuts the x-axis. Using functional notation:

$$f(x) = x^3 - 2x - 5$$

$$f(0) = (0)^3 - 2(0) - 5 = -5$$
$$f(1) = (1)^3 - 2(1) - 5 = -6$$
$$f(2) = (2)^3 - 2(2) - 5 = -1$$
$$f(3) = (3)^3 - 2(3) - 5 = +16$$

This shows that $x^3 - 2x - 5$ is very nearly equal to 0 at $x = 2$, i.e. the first approximation is $a = 2$.

Since $f(x) = x^3 - 2x - 5$, $f'(x) = 3x^2 - 2$.

From equation (8), a second approximation is given by

$$a_1 = a - \frac{f(a)}{f'(a)}$$

When $a = 2$, $a_1 = 2 - \frac{(-1)}{3(2)^2 - 2} = 2.1$

The third approximation is given by:

$$a_2 = a_1 - \frac{f(a_1)}{f'(a_1)}$$

When $a_1 = 2.1$, $a_2 = 2.1 - \left(\frac{(2.1)^3 - 2(2.1) - 5}{3(2.1)^2 - 2} \right)$

$$= 2.1 - \frac{0.06}{11.23} = 2.095$$

The fourth approximation is given by:

$$a_3 = a_2 - \frac{f(a_2)}{f'(a_2)}$$

When $a_2 = 2.095$, $a_3 = 2.095 - \left(\frac{(2.095)^3 - 2(2.095) - 5}{3(2.095)^2 - 2} \right)$

$$= 2.095 - \frac{0.005}{11.17}$$

$$= 2.095 - 0.000\,4 = 2.094\,6$$

Since the values of the third and fourth approximations are the same, when expressed to the required degree of accuracy (i.e. three decimal places), then

the positive root of $x^3 - 2x - 5 = 0$ is 2.095, correct to three decimal places.

Further problems on Taylor's series may be found in Section 4 (Problems 11 to 17), page 65.

3 Approximate values of definite integrals by using series expansions

The value of many definite integrals cannot be determined by the various analytical methods introduced at Level 3 and in Chapters 15 and 16. Several methods exist of finding the approximate value of definite integrals. One of these is to express the function as a power series and then to integrate the terms of the power series. This method is particularly suitable for use with functions whose power series converge fairly rapidly, where evaluation of a few terms gives the required degree of accuracy. This method is shown in the worked problems following.

Worked problems on approximate values of definite integrals by using series expansions

Problem 1. Determine the value of $\int_{0.1}^{0.5} e^{\sin x} dx$, correct to three significant figures.

The power series for $e^{\sin x}$ is obtained by applying Maclaurin's theorem given in equation (5), i.e.

$$f(x) = f(0) + f'(0)x + f''(0)\frac{x^2}{2!} + f'''(0)\frac{x^3}{3!} + \dots$$

$f(x) = e^{\sin x}$. Hence, $f(0) = e^{\sin 0} = 1$

$f'(x) = \cos x e^{\sin x}$. Hence, $f'(0) = 1 \cdot e^0 = 1$

$f''(x) = \cos^2 x e^{\sin x} + (-\sin x) e^{\sin x}$
$\qquad = e^{\sin x}(\cos^2 x - \sin x)$. Hence, $f''(0) = 1(1-0) = 1$

$f'''(x) = e^{\sin x}(2\cos x \sin x - \cos x) + \cos x e^{\sin x}(\cos^2 x - \sin x)$
$\qquad = \cos x e^{\sin x}(\sin x - 1 + \cos^2 x)$. Hence, $f'''(0) = 1(0 - 1 + 1) = 0$

Thus, the power series for $e^{\sin x}$ is $1 + x + \dfrac{x^2}{2} + 0 + \dots$

It follows that

$$\int_{0.1}^{0.5} e^{\sin x} dx \simeq \int_{0.1}^{0.5} \left(1 + x + \frac{x^2}{2}\right) dx$$

$$\simeq \left[x + \frac{x^2}{2} + \frac{x^3}{6}\right]_{0.1}^{0.5}$$

$$\simeq 0.645\,8 - 0.105\,2$$

$$\simeq 0.540\,6$$

i.e. $\int_{0.1}^{0.5} e^{\sin x} dx = 0.541$, correct to three significant figures.

Problem 2. Determine the value of $\int_0^1 \dfrac{\cos 2x}{x^{\frac{1}{3}}} dx$, correct to two decimal places.

Maclaurin's theorem is used to express $\cos 2x$ as a power series.

$f(x) = \cos 2x$ $\qquad\qquad f(0) = \cos 0 = 1$

$f'(x) = -2 \sin 2x$ $\qquad\quad f'(0) = -2(0) = 0$

$f''(x) = -4 \cos 2x$ $\qquad\quad f''(0) = -4(1) = -4$

$f'''(x) = 8 \sin 2x$ $\qquad\qquad f'''(0) = 8(0) = 0$

$f^{iv}(x) = 16 \cos 2x$ $\qquad\quad f^{iv}(0) = 16(1) = 16$

64

$$f^v(x) = -32 \sin 2x \qquad f^v(0) = -32\,(0) = 0$$
$$f^{vi}(x) = -64 \cos 2x \qquad f^{vi}(0) = -64\,(1) = -64$$

Hence, $\cos 2x = 1 - \dfrac{4x^2}{2!} + \dfrac{16x^4}{4!} - \dfrac{64x^6}{6!} + \dots$

$$= 1 - 2x^2 + \dfrac{2x^4}{3} - \dfrac{4x^6}{45} + \dots$$

Thus, $\displaystyle\int_0^1 \dfrac{\cos 2x}{x^{\frac{1}{3}}}\, dx = \int_0^1 x^{-\frac{1}{3}}(\cos 2x)\, dx \doteq \int_0^1 \left\{ x^{-\frac{1}{3}}\left(1 - 2x^2 + \dfrac{2x^4}{3} - \dfrac{4x^6}{45}\right)\right\} dx$

$$\doteq \int_0^1 \left(x^{-\frac{1}{3}} - 2x^{\frac{5}{3}} + \dfrac{2x^{\frac{11}{3}}}{3} - \dfrac{4x^{\frac{17}{3}}}{45}\right) dx$$

$$\doteq \left[\dfrac{3x^{\frac{2}{3}}}{2} - \dfrac{(2)3x^{\frac{8}{3}}}{8} + \dfrac{(2)3x^{\frac{14}{3}}}{(3)(14)} - \dfrac{(4)3x^{\frac{20}{3}}}{(45)(20)}\right]_0^1$$

$$\doteq \left[\dfrac{3x^{\frac{2}{3}}}{2} - \dfrac{3x^{\frac{8}{3}}}{4} + \dfrac{x^{\frac{14}{3}}}{7} - \dfrac{x^{\frac{20}{3}}}{75}\right]_0^1$$

$$\doteq \left(\dfrac{3}{2} - \dfrac{3}{4} + \dfrac{1}{7} - \dfrac{1}{75}\right)$$

$$\doteq 0.879\,5$$

i.e. $\displaystyle\int_0^1 x^{-\frac{1}{3}}(\cos 2x)\, dx = 0.88$, **correct to two decimal places**

Further problems on approximate values of definite integrals by series expansions may be found in the following Section (4) (Problems 18 to 27).

4 Further problems

Maclaurin's series

In Problems 1 to 9, use Maclaurin's theorem or the series given on page 57, to determine the power series stated.

1. $\cos x$, as far as the term in x^6. Hence determine the value of $\cos 0.38$

   radians correct to six decimal places. $\left[1 - \dfrac{x^2}{2!} + \dfrac{x^4}{4!} - \dfrac{x^6}{6!} \ ; 0.928\,665\right]$

2. $e^{\frac{x}{2}}$, as far as the term in x^5. $\left[1 + \dfrac{x}{2} + \dfrac{x^2}{8} + \dfrac{x^3}{48} + \dfrac{x^4}{384} + \dfrac{x^5}{3\,840}\right]$

3. $\tan ax$, as far as the term in x^5. $\left[ax + \dfrac{a^3 x^3}{3} + \dfrac{2a^5 x^5}{15}\right]$

4. $\sec \frac{x}{2}$, as far as the term in x^4. $\left[1 + \frac{x^2}{8} + \frac{5x^4}{384}\right]$

5. $\cos^2 2x$, as far as the term in x^6. $\left[1 - 4x^2 + \frac{16x^4}{3} - \frac{128x^6}{45}\right]$

6. $(1 - x)^{-3}$, as far as the term in x^4. $[1 + 3x + 6x^2 + 10x^3 + 15x^4]$

7. $e^{2x} \cos 3x$, as far as the term in x^2. $\left[1 + 2x - \frac{5}{2} x^2\right]$

8. $\frac{1 + x}{(1 - x)^3}$, as far as the term in x^3. $[1 + 4x + 9x^2 + 16x^3]$

9. $\ln (1 + 3x) \sin \frac{3x}{2}$, as far as the term in x^4. $\left[\frac{9x^2}{2} - \frac{27x^3}{4} + \frac{189x^4}{16}\right]$

10. By expressing $\ln\left(\frac{1 + x}{1 - x}\right)$ as a power series, determine the value of $\ln 0.8$, correct to five decimal places. $[-0.223\,14]$

Taylor's series

11. Determine the power series for $\cos (a + h)$ and hence determine the value of $\cos 31°$, correct to five decimal places.

    $[\cos (a + h) = \cos a - h \sin a - \frac{h^2}{2!} \cos a + \frac{h^3}{3!} \sin a + \ldots; 0.857\,17]$

12. Find the value of $\tan 31°$ correct to five decimal places by determining the power series for $\tan (a + h)$.

    $[\tan (a + h) = \tan a + h \sec^2 a + h^2 \sec^2 a \tan a$
    $+ \frac{h}{3} \sec^2 a \; (\sec^2 a + 2 \tan^2 a) + \ldots; 0.600\,86]$

13. Given that $\ln 10 = 2.302\,585$, determine the value of $\ln 12$, correct to five decimal places. $[2.484\,91]$

14. Use Taylor's series to determine the value of $\tan 50°$, correct to six significant figures, given that $\tan 40° = 0.839\,100$. $[1.191\,75]$

15. Determine the three roots of $x^3 = 7x + 1$, correct to four significant figures, by Newton's method. $[2.714, -0.143\,3, -2.571]$

16. Find the value of the real root of the equation

    $$9 - 9\left(\frac{7}{6}x - 1\right)^{\frac{8}{7}} - 10x = 0, \text{ correct to three significant figures, by}$$

    using Newton's method. $[0.883]$

17. Determine the value of the root of the equation

    $$1 - \frac{x}{6} + \frac{x^2}{180} - \frac{x^3}{12\,960} = 0,$$

    which lies near to 8, correct to four significant figures. $[7.815]$

Approximate values of definite integrals by using series expansions

In Problems 18 to 22, determine the values of the definite integrals given,

66

correct to two significant figures.

18. $\int_0^1 x^{\frac{1}{2}} \cos x \, dx$ [0.53]

19. $\int_0^{\frac{1}{2}} \sqrt{(x)} \ln (x+1) \, dx$ [0.061]

20. $\int_0^{\pi/3} (\sec x)^{\frac{1}{2}} \, dx$ [1.2]

21. $\int_1^2 \frac{1}{2} \cos 3\theta \, d\theta$ [−0.070]

22. $\int_{\frac{1}{3}}^1 \sec^2 (3x-1) \, dx$ [−0.73]

23. Determine the value of $\int_0^{\frac{1}{2}} x^2 \ln (1 + \sin x) \, dx$, correct to two significant figures.   [0.013]

24. By determining the power series for $\sec \theta$ as far as the term in θ^4, determine the approximate value of

$$\int_0^{\pi/6} \sec \theta \, d\theta$$

$$\left[\int_0^{\pi/6} \left(1 + \frac{\theta^2}{2} + \frac{5}{24} \theta^4 \right) d\theta \simeq 0.667 \text{ correct to three decimal places.} \right]$$

25. Use Maclaurin's theorem to determine the power series for e^{-x^2} as far as the term in x^4 and hence determine the value of $\int_0^1 e^{-x^2} \, dx$, correct to three significant figures.   [0.767]

26. Determine the value of $\int_0^{0.3} \frac{(1-y^2)^{\frac{1}{2}}}{4\sqrt{y}} \, dy$, correct to three significant figures.   [0.534]

27. Use the power series for $\sin \phi$ to determine the value of $\int_0^{\pi/6} \phi^{\frac{1}{2}} \sin \phi \, d\phi$, correct to three decimal places.   [0.077]

Chapter 6

De Moivre's theorem

1 The exponential form of a complex number

By using Maclaurin's theorem, the power series for e^θ, $\sin \theta$ and $\cos \theta$ may be derived, and are:

$$e^\theta \quad = 1 + \theta + \frac{\theta^2}{2!} + \frac{\theta^3}{3!} + \frac{\theta^4}{4!} + \frac{\theta^5}{5!} + \ldots \qquad (1)$$

$$\sin \theta \quad = \theta - \frac{\theta^3}{3!} + \frac{\theta^5}{5!} - \frac{\theta^7}{7!} + \ldots \qquad (2)$$

$$\text{and } \cos \theta = 1 - \frac{\theta^2}{2!} + \frac{\theta^4}{4!} - \frac{\theta^6}{6!} + \ldots \qquad (3)$$

Replacing θ by the imaginary number $j\theta$ in equation (1) gives:

$$e^{j\theta} = 1 + j\theta + \frac{(j\theta)^2}{2!} + \frac{(j\theta)^3}{3!} + \frac{(j\theta)^4}{4!} + \frac{(j\theta)^5}{5!} + \ldots$$

$$= 1 + j\theta + j^2 \frac{\theta^2}{2!} + j^3 \frac{\theta^3}{3!} + j^4 \frac{\theta^4}{4!} + j^5 \frac{\theta^5}{5!} + \ldots$$

Since $j^2 = -1, j^3 = -j, j^4 = 1$ and $j^5 = j$,

$$\text{then } e^{j\theta} = 1 + j\theta - \frac{\theta^2}{2!} - j \frac{\theta^3}{3!} + \frac{\theta^4}{4!} + j \frac{\theta^5}{5!} - \ldots$$

Grouping real and imaginary terms gives:

$$e^{j\theta} = \left(1 - \frac{\theta^2}{2!} + \frac{\theta^4}{4!} - \ldots \right) + j \left(\theta - \frac{\theta^3}{3!} + \frac{\theta^5}{5!} - \ldots \right)$$

$= \cos\theta + j\sin\theta$, from equations (2) and (3) above.

Thus $e^{j\theta} = \cos\theta + j\sin\theta$ (4)

If each side of equation (4) is multiplied by r, then

$r\,e^{j\theta} = r(\cos\theta + j\sin\theta)$, often abbreviated to $r\angle\theta$.

Thus $z = r\,e^{j\theta}$ is a complex number having a modulus, $|z| = r$ and argument, $\arg z = \theta$.

Writing $-\theta$ for θ in equation (4) gives:

$e^{j(-\theta)} = \cos(-\theta) + j\sin(-\theta)$.

Since $\cos(-\theta) = \cos\theta$ and $\sin(-\theta) = -\sin\theta$, then

$$e^{-j\theta} = \cos\theta - j\sin\theta \qquad (5)$$

By adding equations (4) and (5)

$e^{j\theta} + e^{-j\theta} = 2\cos\theta$

i.e. $\cos\theta = \dfrac{e^{j\theta} + e^{-j\theta}}{2}$ (6)

By subtracting equation (5) from equation (4)

$e^{j\theta} - e^{-j\theta} = 2j\sin\theta$

i.e. $\sin\theta = \dfrac{e^{j\theta} - e^{-j\theta}}{j2}$ (7)

The identities given in equations (6) and (7) are used when establishing the relationship between trigonometric and hyperbolic functions (see Chapter 8).

2 De Moivre's theorem

De Moivre's theorem is deduced from the polar form of a complex number in *Mathematics for Electrical Technicians level 3* by J. O. Bird and A. J. C. May. In this section, it is deduced from the exponential form of a complex number.

For a complex number, z, expressed in exponential form

$$z = r\,e^{j\theta}$$

Then, $z^2 = z \times z = r\,e^{j\theta} \times r\,e^{j\theta}$

$\qquad\qquad = r^2 e^{(j\theta + j\theta)}$ by the laws of indices

i.e. $\qquad\qquad z^2 = r^2\,e^{j2\theta}$

Similarly, $\qquad z^3 = r\,e^{j\theta} \times r\,e^{j\theta} \times r\,e^{j\theta}$

$\qquad\qquad = r^3\,e^{(j\theta + j\theta + j\theta)}$

$\qquad\qquad = r^3\,e^{j3\theta}$

A general relationship is that:

$$z^n = r^n e^{jn\theta}$$

But from equation (4) in section 1, $e^{jn\theta} = \cos n\,\theta + j \sin n\theta$

Thus, $z^n = r^n (\cos n\theta + j \sin n\theta) = r^n \angle n\theta$ \hfill (8)

This way of expressing the power of a complex number is known as De Moivre's theorem. The same theorem may be used when n is a fraction for finding the roots of a complex number, and also applies when n is negative.

The uses of De Moivre's theorem include the solution of problems associated with electrical circuit theory, simplifying certain types of integrals and verifying and determining certain types of trigonometric identities.

Worked problems on De Moivre's theorem

Problem 1. Use De Moivre's theorem to find the value of $(2 - j\sqrt{5})^7$ in the form $r\angle\theta$ and in the form $a + jb$.

The general relationship between a complex number in $a + jb$ and polar forms is that if $z = x + jy$, then the modulus of z, i.e. $|z| = \sqrt{(x^2 + y^2)}$ and the argument of z, i.e. $\arg z = \arctan \dfrac{y}{x}$. Thus, since $z = 2 - j\sqrt{5}$,

$|z| = \sqrt{(2^2 + \sqrt{5}^2)} = \sqrt{9} = 3$, since a modulus cannot have a negative value.

$$\arg z = \arctan - \frac{\sqrt{5}}{2} = \arctan - 1.118$$

An angle having a negative value for its tangent can lie in either the second or fourth quadrants. However, visualising the Argand diagram, $+2$ is to the right ($\rightarrow$) and $-j\sqrt{5}$ is downwards, ($\downarrow$) showing that only the fourth quadrant value for the argument is possible. Thus:

$\arg z = \arctan - 1.118 = -48° \ 11'$
Then, $(2 - j\sqrt{5}) = 3 \angle -48° \ 11'$

Applying De Moivre's theorem (equation (8) above) gives:

$$(2 - j\sqrt{5})^7 = 3^7 \angle [7 \times (-48° \ 11')]$$
$$= 2\,187 \angle -337° \ 17'$$

It is usual to express angles in terms of their principal values, that is, so that $-180° \leqslant \theta \leqslant 180°$, hence

$(2 - j\sqrt{5})^7 = 2\,187 \angle 22° \ 43'$ **in polar form**

The general relationship between a complex number in polar form and the $a + jb$ form is that if

$r (\cos \theta + j \sin \theta) \quad = a + jb$, then $a = r \cos \theta$ and $b = r \sin \theta$
Thus, $2\,187 \angle 22° \ 43' = (2\,187 \cos 22° \ 43') + j(2\,187 \sin 22° \ 43')$
giving $(2 - j\sqrt{5})^7 \quad = 2\,017 + j\,844.6$ in $a + jb$ form, correct to four **significant figures**.

Problem 2. Determine the four roots of $(-3 - j7)^{\frac{1}{4}}$ in polar form.

Let $z = -3 - j7$, then $|z| = \sqrt{[(-3)^2 + (-7)^2]}$

$\qquad\qquad\qquad\qquad = 7.616$

$\arg z \qquad\qquad = \arctan \dfrac{-7}{-3} = \arctan 2.333$

$\qquad\qquad\qquad\qquad = 66° \ 48'$ or $246° \ 48'$

Visualising the Argand diagram, -3 is to the left ($\leftarrow$) and $-j7$ is downward ($\downarrow$), showing that only the third quadrant value for the argument is possible. Expressing the argument of $246° \ 48'$ as a principal value gives

$-3 - j7 \qquad = 7.616 \angle -113° \ 12'$

By De Moivre's theorem,

$$(-3 - j7)^{\frac{1}{4}} = 7.616^{\frac{1}{4}} \angle \left[\frac{1}{4} \times (-113° \ 12') \right]$$

$$= 1.661 \angle -28° \ 18'$$

The remaining three roots have the same modulus and are calculated by adding $360°$, $720°$ and $1\,080°$ to the argument of the complex number in turn. This is done because when dividing these arguments by 4, angles of less than $360°$ are still obtained.

Thus the arguments of the other three roots are

$\dfrac{1}{4} \times (-113° \ 12' + 360°) \ = 61° \ 42'$

$\dfrac{1}{4} \times (-113° \ 12' + 720°) \ = 151° \ 42'$ and

$\dfrac{1}{4} \times (-113° \ 12' + 1\,080°) = 241° \ 42'$ or $-118° \ 18'$

(The argument of the root obtained by adding $4 \times 360°$, that is, $1\,440°$, is $\dfrac{1}{4} \times (-113° \ 12' + 1\,440°) \ = 331° \ 42'$ or $-28° \ 18'$, which is the same value as the first of the arguments shown above. Hence all the roots have been calculated.) The values of the arguments can be easily checked, since they are symmetrically placed on an Argand diagram $\left(\dfrac{360}{n} \right)°$ apart where n is the number of roots required. Thus for a fourth root problem, the arguments are $\left(\dfrac{360}{4} \right)°$, i.e. $90°$ apart, and it can be seen that the four arguments obtained in this problem are displaced from one another by $90°$.

Thus the four roots of $(-3 - j7)^{\frac{1}{4}}$ are:

$1.661 \angle -28° \ 18'$, $\qquad$ **$1.661 \angle 61° \ 42'$**
$1.661 \angle 151° \ 42'$ and $\quad$ **$1.661 \angle -118° \ 18'$.**

Problem 3. Calculate the five roots of $(4 \angle 21°)^{-\frac{2}{5}}$ in the $a + jb$ form, correct to four decimal places.

For fractional powers of a complex number, the relationship between rectangular and polar forms is:

$$(a + jb)^{\frac{p}{q}} = [r \angle \theta]^{\frac{p}{q}} = r^{\frac{p}{q}} \angle \left[\frac{p}{q} (\theta + 360\,n) \right]$$

where $r = \sqrt{(a^2 + b^2)}$, $\theta = \arctan \dfrac{b}{a}$ and $n = 0, 1, 2, \ldots q - 1$, since the three values of a cube root are given by $n = 0, 1, 2$, the four values of a fourth root by $n = 0, 1, 2$ and 3 and q values of a qth root by $n = 0, 1, 2, \ldots, q - 1$.

By De Moivre's theorem, $(4 \angle 21°)^{-\frac{2}{5}} = 4^{-\frac{2}{5}} \angle \left[-\frac{2}{5}(21° + 360\,n°) \right]$

where $n = 0, 1, 2, 3$ and 4, since there are 5 roots.

When $n = 0$, $(4 \angle 21°)^{-\frac{2}{5}} = 0.5743 \angle \left[-\frac{2}{5}(21°) \right] = 0.5743 \angle 351° \, 36'$

When $n = 1$, $(4 \angle 21°)^{-\frac{2}{5}} = 0.5743 \angle \left[-\frac{2}{5}(21 + 360)° \right]$
$$= 0.5743 \angle 207° \, 36'$$

When $n = 2$, $(4 \angle 21°)^{-\frac{2}{5}} = 0.5743 \angle \left[-\frac{2}{5}(21 + 360 \times 2)° \right]$
$$= 0.5743 \angle 63° \, 36'$$

When $n = 3$, $(4 \angle 21°)^{-\frac{2}{5}} = 0.5743 \angle \left[-\frac{2}{5}(21 + 360 \times 3)° \right]$
$$= 0.5743 \angle 279° \, 36'$$

When $n = 4$, $(4 \angle 21°)^{-\frac{2}{5}} = 0.5743 \angle \left[-\frac{2}{5}(21 + 360 \times 4)° \right]$
$$= 0.5743 \angle 135° \, 36'$$

Checking: the 5 roots should be $\dfrac{360}{5}$, i.e. $72°$ apart. The values of the arguments obtained are spaced $72°$ apart, so no error has been made. (The arguments have been expressed in $0 \leqslant \theta \leqslant 360°$ form to enable this check to be easily carried out.)

Since $a = r \cos \theta$ and $b = r \sin \theta$, the five roots expressed in the $a + jb$ form are:

$0.5743 \cos (351° \, 36') + j\, 0.5743 \sin (351° \, 36')$
$$= 0.5681 - j\, 0.0839$$
$0.5743 \cos (207° \, 36') + j\, 0.5743 \sin (207° \, 36')$
$$= -0.5089 - j\, 0.2661$$

$$0.5743 \cos (63° 36') + j 0.5743 \sin (63° 36')$$
$$= 0.2554 + j 0.5144$$
$$0.5743 \cos (279° 36') + j 0.5743 \sin (279° 36')$$
$$= 0.0958 - j 0.5663 \text{ and}$$
$$0.5743 \cos (135° 36') + j 0.5743 \sin (135° 36')$$
$$= -0.4103 + j 0.4018$$

Further problems on De Moivre's theorem may be found in Section 5 (Problems 1 to 10), page 76.

3 Expressing $\cos n\theta$ and $\sin n\theta$ in terms of powers of $\cos \theta$ and $\sin \theta$

The mathematical statement of De Moivre's theorem is

$$z^n = r^n (\cos n\theta + j \sin n\theta)$$
But $z^n = [r (\cos \theta + j \sin \theta)]^n$
Hence, $r^n (\cos n\theta + j \sin n\theta) = [r (\cos \theta + j \sin \theta)]^n$
When $r = 1$, this equation becomes

$$\cos n\theta + j \sin n\theta = (\cos \theta + j \sin \theta)^n$$

The right-hand side of this equation can be expanded using the binomial series. (The general binomial expansion is given by:

$$(a + b)^n = a^n + na^{n-1}b + \frac{n(n-1)}{2!} a^{n-2}b^2 + \ldots)$$

Since the terms produced when doing this are long, abbreviations are used below, where

    (i) $\cos^n \theta$ is written as C^n
    (ii) $\sin^n \theta$ is written as S^n

and (iii) binomial coefficients such as $\dfrac{n(n-1)(n-2)}{3!}$ are written as $\binom{n}{3}$.

Using this notation:

$$(\cos n\theta + j \sin n\theta) = C^n + n C^{n-1} \cdot jS + \binom{n}{2} C^{n-2} \cdot j^2 S^2$$
$$+ \binom{n}{3} C^{n-3} \cdot j^3 S^3 + \binom{n}{4} C^{n-4} \cdot j^4 S^4 + \ldots$$

But $j^2 = -1, j^3 = -j$ and $j^4 = 1$, hence:

$$(\cos n\theta + j \sin n\theta) = C^n + jn C^{n-1} S - \binom{n}{2} C^{n-2} S^2 - j \binom{n}{3} C^{n-3} S^3$$
$$+ \binom{n}{4} C^{n-4} S^4 + \ldots$$

Grouping real and imaginary terms gives:

$$(\cos n\theta + j \sin n\theta) = [C^n - \binom{n}{2} C^{n-2} S^2 + \binom{n}{4} C^{n-4} S^4 - \ldots]$$
$$+ j [n C^{n-1} S - \binom{n}{3} C^{n-3} S^3 + \binom{n}{5} C^{n-5} S^5 - \ldots]$$

Equating real parts gives:

$$\cos n\theta = C^n - \binom{n}{2} C^{n-2} S^2 + \binom{n}{4} C^{n-4} S^4 - \ldots$$

That is, $\cos n\theta = \cos^n\theta - \dfrac{n\,(n-1)}{2!}\cos^{n-2}\theta\,\sin^2\theta$

$$+ \dfrac{n\,(n-1)(n-2)(n-3)}{4!}\cos^{n-4}\theta\,\sin^4\theta - \ldots \text{(9)}$$

Equating imaginary parts gives.

$\sin n\theta = n\,C^{n-1}S - \binom{n}{3}C^{n-3}S^3 + \binom{n}{5}C^{n-5}S^5 - \ldots$

That is, $\sin n\theta = n\cos^{n-1}\theta\,\sin\theta - \dfrac{n\,(n-1)(n-2)}{3!}\cos^{n-3}\theta\,\sin^3\theta$

$$+ \dfrac{n\,(n-1)(n-2)(n-3)(n-4)}{5!}\cos^{n-5}\theta\,\sin^5\theta - \ldots \qquad \text{(10)}$$

Worked problems on expressing $\cos n\theta$ and $\sin n\theta$ in terms of powers of $\cos\theta$ and $\sin\theta$

Problem 1. Express $\cos 3\,\theta$ in terms of powers of $\cos\theta$.

Substituting $n = 3$ in equation (9) gives:

$$\cos 3\,\theta = \cos^3\theta - \dfrac{3(3-1)}{2!}\cos^{3-2}\theta\,\sin^2\theta$$

(when $n = 3$, $\dfrac{n\,(n-1)(n-2)(n-3)}{4!} = \dfrac{3\times 2\times 1\times 0}{4\times 3\times 2\times 1} = 0$.)

(Hence all terms after the second in equation (9) are zero when $n = 3$.)

Thus $\cos 3\,\theta = \cos^3\theta - \dfrac{3\times 2}{2\times 1}\cos\theta\,\sin^2\theta$

$$= \cos^2\theta - 3\cos\theta\,\sin^2\theta$$

But $\quad \sin^2\theta = 1 - \cos^2\theta$,

Hence, $\cos 3\,\theta = \cos^3\theta - 3\cos\theta\,(1 - \cos^2\theta)$

$$= \cos^3\theta - 3\cos\theta + 3\cos^3\theta$$

i.e. $\quad \cos 3\theta = 4\cos^3\theta - 3\cos\theta$ **in terms of powers of $\cos\theta$.**

Problem 2. Express $\sin 5\,\theta$ in terms of powers of $\cos\theta$ and $\sin\theta$.

Substituting $n = 5$ in equation (10) gives:

$$\sin 5\,\theta = 5\cos^{5-1}\theta\,\sin\theta - \dfrac{5\times 4\times 3}{3\times 2\times 1}\cos^{5-3}\theta\,\sin^3\theta$$

$$+ \dfrac{5\times 4\times 3\times 2\times 1}{5\times 4\times 3\times 2\times 1}\cos^{5-5}\theta\,\sin^5\theta.$$

i.e. $\sin 5\,\theta = 5\cos^4\theta\,\sin\theta - 10\cos^2\theta\,\sin^3\theta + \sin^5\theta$.

Further problems on expressing $\cos n\,\theta$ and $\sin n\,\theta$ in terms of powers of $\cos\theta$ and $\sin\theta$ may be found in Section 5 (Problems 11 to 16), page 77.

4 Expressing $\cos^n \theta$ and $\sin^n \theta$ in terms of sines and cosines of multiples of θ

A complex number, z, may be written as

$z = re^{j\theta} = r(\cos \theta + j \sin \theta)$
When $r = 1$, then $z = \cos \theta + j \sin \theta = e^{j\theta}$ (11)
Also, from section 1, equation (5), $e^{-j\theta} = \cos \theta - j \sin \theta$

But, if $e^{j\theta} = z$, then $e^{-j\theta} = \dfrac{1}{z}$

Hence $\dfrac{1}{z} = \cos \theta - j \sin \theta$ (12)

Adding equations (11) and (12) gives:

$$z + \frac{1}{z} = 2 \cos \theta \tag{13}$$

Subtracting equation (12) from equation (11) gives:

$$z - \frac{1}{z} = j2 \sin \theta \tag{14}$$

Raising equation (13) to a power of n gives:

$$\left(z + \frac{1}{z}\right)^n = 2^n \cos^n \theta \tag{15}$$

where $z = \cos \theta + j \sin \theta$ and $\dfrac{1}{z} = \cos \theta - j \sin \theta$

Raising equation (14) to a power of n gives:

$$\left(z - \frac{1}{z}\right)^n = j^n 2^n \sin^n \theta \tag{16}$$

By expanding $\left(z + \dfrac{1}{z}\right)^n$ by the binomial theorem, equation (15) can be used to express $\cos^n \theta$ in terms of $\cos n\theta$ and $\sin n\theta$. Similarly, equation (16) can be used to express $\sin^n \theta$ in terms of $\cos n\theta$ and $\sin n\theta$. This technique is used to express integrals of $\cos^n x$ and $\sin^n x$ in a form which enables integration to be achieved.

Worked problems on expressing $\cos^n \theta$ and $\sin^n \theta$ in terms of sines and cosines of multiples of θ

Problem 1. Express $\cos^2 \theta$ in terms of cosines of multiples of θ.

From equation (15): $2^n \cos^n \theta = \left(z + \dfrac{1}{z}\right)^n$, where

$z = \cos \theta + j \sin \theta$ and $\dfrac{1}{z} = \cos \theta - j \sin \theta$

When $n = 2$, $2^2 \cos^2 \theta = \left(z + \dfrac{1}{z} \right)^2 = z^2 + 2 + \dfrac{1}{z^2}$

$$= \left(z^2 + \dfrac{1}{z^2} \right) + 2$$

However, $z^n + \dfrac{1}{z^n} = (\cos n\theta + j \sin n\theta) + (\cos n\theta - j \sin n\theta)$

$$= 2 \cos n\theta$$

Hence, $\left(z^2 + \dfrac{1}{z^2} \right) + 2 = 2 \cos 2\theta + 2$

Thus, $2^2 \cos^2 \theta = 2 \cos 2\theta + 2$

$\cos^2 \theta = \dfrac{1}{2} (\cos 2\theta + 1)$

Problem 2. Express $\sin^4 \theta$ in terms of $\cos n\theta$.

From equation (16): $j^n 2^n \sin^n \theta = \left(z - \dfrac{1}{z} \right)^n$

When $n = 4$: $\qquad j^4 2^4 \sin^4 \theta = \left(z - \dfrac{1}{z} \right)^4$

Using the binomial theorem to expand $\left(z - \dfrac{1}{z} \right)^4$ gives:

$\left(z - \dfrac{1}{z} \right)^4 = z^4 - 4z^3 \cdot \dfrac{1}{z} + 6z^2 \cdot \dfrac{1}{z^2} - 4z \cdot \dfrac{1}{z^3} + \dfrac{1}{z^4}$

$$= \left(z^4 + \dfrac{1}{z^4} \right) - 4 \left(z^2 + \dfrac{1}{z^2} \right) + 6$$

But it is shown in problem 1 above that $\left(z^n + \dfrac{1}{z^n} \right) = 2 \cos n\theta$.

Hence, $j^4 2^4 \sin^4 \theta = 2 \cos 4\theta - 8 \cos 2\theta + 6$

But, $j^4 = 1$, hence $\sin^4 \theta = \dfrac{1}{8} \cos 4\theta - \dfrac{1}{2} \cos 2\theta + \dfrac{3}{8}$.

Problem 3. Express $\sin^7 \theta$ in terms of $\sin n\theta$.

From equation (16): $j^n 2^n \sin^n \theta = \left(z - \dfrac{1}{z} \right)^n$

When $n = 7$, $\qquad j^7 2^7 \sin^7 \theta = \left(z - \dfrac{1}{z} \right)^7$

Using the binomial theorem to expand the right-hand side of this equation gives:

$$\left(z - \frac{1}{z}\right)^7 = z^7 - 7z^6 \cdot \frac{1}{z} + 21z^5 \cdot \frac{1}{z^2} - 35z^4 \cdot \frac{1}{z^3} + 35z^3 \cdot \frac{1}{z^4}$$

$$- 21z^2 \cdot \frac{1}{z^5} + 7z \cdot \frac{1}{z^6} - \frac{1}{z^7}$$

$$= \left(z^7 - \frac{1}{z^7}\right) - 7\left(z^5 - \frac{1}{z^5}\right) + 21\left(z^3 - \frac{1}{z^3}\right) - 35\left(z - \frac{1}{z}\right)$$

However, $\left(z^n - \frac{1}{z^n}\right) = (\cos n\,\theta + j \sin n\,\theta) - (\cos n\,\theta - j \sin n\,\theta)$

$$= 2j \sin n\,\theta.$$

Hence, $j^7 \, 2^7 \sin^7 \theta = 2j \,(\sin 7\,\theta - 7 \sin 5\,\theta + 21 \sin 3\,\theta - 35 \sin \theta)$
But $j^7 = -j$, hence

$$-j\, 2^7 \sin^7 \theta = 2j \,(\sin 7\,\theta - 7 \sin 5\,\theta + 21 \sin 3\,\theta - 35 \sin \theta)$$

i.e. $\sin^7 \theta = \dfrac{1}{2^6} \,(-\sin 7\,\theta + 7 \sin 5\,\theta - 21 \sin 3\,\theta + 35 \sin \theta).$

Further problems on expressing $\cos^n \theta$ and $\sin^n \theta$ in terms of sines and cosines of multiples of θ may be found in the following Section (5) (Problems 17 to 22).

5 Further problems

De Moivre's theorem

1. Determine the value of $(3 + j4)^3$, expressing the result in the $a + jb$ form. $[-117 + j44]$
2. Find the value of $(\sqrt{3} - j)^9$, expressing the result in both polar and $a + jb$ forms. $[512 \angle 90°; j\,512]$
3. Determine the values of (a) $(3 - j4)^5$ and (b) $(3 + j4)^5$, giving the answers in polar form.   (a) $[3125 \angle 94°\,21']$ (b) $[3125 \angle -94°\,21']$
4. Find the roots of $(-2 + j2)^{\frac{1}{2}}$ in polar form, correct to four significant figures. $[1.682 \angle 67°\,30', 1.682 \angle -112°\,30']$
5. Determine the three cube roots of $8\,(1 - j)$ in polar form.
$$[2.245 \angle -15°, 2.245 \angle 105°, 2.245 \angle -135°]$$
6. Find the four fourth roots of $3 - j4$, expressing them in the $a + jb$ form, correct to four significant figures.
$$\begin{bmatrix} 1.455 - j\,0.3436, & -0.3436 - j\,1.455 \\ -1.455 + j\,0.3436, & 0.3436 + j\,1.455 \end{bmatrix}$$
7. Express the roots of $(-2 + j3)^{-\frac{2}{3}}$ in the $a + jb$ form, correct to four decimal places. $\begin{bmatrix} 0.055\,8 - j\,0.421\,6, 0.337\,2 + j\,0.259\,1 \\ -0.393\,0 + j\,0.162\,5 \end{bmatrix}$
8. Evaluate $(-2 - j)^{-\frac{1}{5}}$, expressing the answers in both polar and $a + jb$ forms, correct to four decimal places.

$$\begin{bmatrix} 0.851\,3 \angle\,30°\,41',\,0.851\,3 \angle\,102°\,41',\,0.851\,3 \angle\,174°\,41', \\ 0.851\,3 \angle -41°\,19',\,0.851\,3 \angle -113°\,19';\,0.732\,1 + j\,0.434\,4, \\ -0.186\,9 + j\,0.830\,5,\,-0.847\,6 + j\,0.078\,9,\,0.639\,4 - j\,0.562\,0, \\ -0.337\,0 - j\,0.781\,8 \end{bmatrix}$$

9. The characteristic impedance (Z_0) and the propagation constant (P) of a transmission line are given by the equations:

$$Z_0 = (0.3 + j\,0.5)^{\frac{1}{2}} (2 + j\,500)^{-\frac{1}{2}} \times 10^4$$

$$P = (0.3 + j\,0.5)^{\frac{1}{2}} (2 + j\,500)^{\frac{1}{2}} \times 10^{-4}$$

Evaluate Z_0 and P.

$$\begin{bmatrix} Z_0 = (329 - j\,90.5) \\ P = (0.000\,46 + j\,0.001\,64) \end{bmatrix}$$

10. When displaced electrons oscillate about an equilibrium position the displacement x is given by the equation:

$$x = A\,e^{\left\{ -\frac{ht}{2m} + j\,\frac{\sqrt{(4\,mf - h^2)}}{2m - a} \right\}}$$

Determine the real part of x assuming $(4\,mf - h^2)$ is positive.

$$\left[A\,e^{-\frac{ht}{2m}}\ \cos\left(\frac{\sqrt{(4\,mf - h^2)}}{2\,m - a} \right) t \right]$$

Expressing $\cos n\,\theta$ and $\sin n\,\theta$ in terms of powers of $\cos \theta$ and $\sin \theta$

11. Express $\cos 4\,\theta$ in terms of powers of $\cos \theta$ and $\sin \theta$.
$$[\cos^4 \theta - 6 \cos^2 \theta \sin^2 \theta + \sin^4 \theta]$$

12. Express $\sin 4\,\theta$ in terms of powers of $\cos \theta$ and $\sin \theta$.
$$[4 \cos^3 \theta \sin \theta - 4 \cos \theta \sin^3 \theta]$$

13. Express $\cos 6\,\theta$ in terms of powers of $\cos \theta$.
$$[32 \cos^6 \theta - 48 \cos^4 \theta + 18 \cos^2 \theta - 1]$$

14. Express $\dfrac{\sin 5\,\theta}{\sin \theta}$ in terms of powers of $\sin \theta$.
$$[16 \sin^4 \theta - 20 \sin^2 \theta + 5]$$

15. Express $\dfrac{\sin 7\,\theta}{\sin \theta}$ in terms of powers of $\cos \theta$ and $\sin \theta$.
$$[7 \cos^6 \theta - 35 \cos^4 \theta \sin^2 \theta + 21 \cos^2 \theta \sin^4 \theta - \sin^6 \theta.]$$

16. Express $\cos 7\,\theta$ in terms of powers of $\cos \theta$.
$$[64 \cos^7 \theta - 112 \cos^5 \theta + 56 \cos^3 \theta - 7 \cos \theta]$$

Expressing $\cos^n \theta$ and $\sin^n \theta$ in terms of sines and cosines of multiples of θ

17. Express $\cos^6 \theta$ in terms of cosines of multiples of θ.
$$\left[\frac{1}{2^5} (\cos 6\,\theta + 6 \cos 4\,\theta + 15 \cos 2\,\theta + 10) \right]$$

18. Express $\sin^6 \theta$ in terms of cosines of multiples of θ.

$$\left[\frac{1}{2^5} \left(- \cos 6\theta + 6 \cos 4\theta - 15 \cos 2\theta + 10\right)\right]$$

19. Express $\cos^8 \theta$ in terms of cosines of multiples of θ.

$$\left[\frac{1}{2^7} \left(\cos 8\theta + 8 \cos 6\theta + 28 \cos 4\theta + 56 \cos 2\theta + 35\right)\right]$$

20. Show that $\sin^2 \theta \cos^5 \theta = \frac{1}{2^6} \left(- \cos 7\theta - 3 \cos 5\theta - \cos 3\theta + 5 \cos \theta\right)$.

21. Show that $\sin^4 \theta \cos^3 \theta = \frac{1}{2^6} \left(\cos 7\theta - \cos 5\theta - 3 \cos 3\theta + 3 \cos \theta\right)$.

22. Express $\sin^5 \theta$ in terms of sines of multiples of θ.

$$\left[\frac{1}{2^4} \left(\sin 5\theta - 5 \sin 3\theta + 10 \sin \theta\right)\right]$$

Chapter 7

Hyperbolic functions

1 Definitions of hyperbolic functions

Trigonmetric functions are called **'circular functions'** since they arise naturally in connection with the geometry of the circle. There are other functions which are associated with the geometry of the conic section called a **hyperbola** (see page 122) and are thus classified as **hyperbolic functions**. Such functions have several applications, in particular with transmission line theory and with catenary problems, a catenary being a curve formed by a chain or rope of uniform density hanging freely from two fixed points not in the same vertical line.

Six hyperbolic functions are defined below, each of them being closely connected with the six trigonometrical ratios.

By definition:

(a) Hyperbolic sine of x, $\sinh x = \dfrac{e^x - e^{-x}}{2}$ (1)

'sinh x' is often abbreviated to 'sh x' and is pronounced as 'shine x'.

(b) Hyperbolic cosine of x, $\cosh x = \dfrac{e^x + e^{-x}}{2}$ (2)

'cosh x' is often abbreviated to 'ch x' and is pronounced as 'kosh x'.

(c) Hyperbolic tangent of x, $\tanh x = \dfrac{\sinh x}{\cosh x} = \dfrac{e^x - e^{-x}}{e^x + e^{-x}}$ (3)

'tanh x' is often abbreviated to 'th x' and is pronounced as 'than x'.

(d) Hyperbolic cosecant of x, **cosech** $x = \dfrac{1}{\sinh x} = \dfrac{2}{e^x - e^{-x}}$ $\qquad$ (4)

'cosech x' is pronounced as 'coshec x'.

(e) Hyperbolic secant of x, **sech** $x = \dfrac{1}{\cosh x} = \dfrac{2}{e^x + e^{-x}}$ $\qquad$ (5)

'sech x' is pronounced as 'shec x'.

(f) Hyperbolic contangent of x, **coth** $x = \dfrac{1}{\tanh x} = \dfrac{e^x + e^{-x}}{e^x - e^{-x}}$ $\qquad$ (6)

'coth x' is pronounced as 'koth x'.

2   Some properties of hyperbolic functions

(a) If for a function of x, $f(-x) = f(x)$, i.e. $f(x)$ is unchanged when x is replaced by $-x$, then $f(x)$ is called an **even function** of x.
Cos $(-x) = \cos x$, thus the cosine function is an even function.
Replacing x by $-x$ in the formula for cosh x gives:

$$\cosh(-x) = \frac{e^{-x} + e^{-(-x)}}{2} = \frac{e^{-x} + e^x}{2} = \cosh x$$

Thus the hyperbolic cosine, **cosh** x, **is an even function**. The graph of an even function is symmetrical about the y-axis (see section 4). Since cosh x is an even function then sech $x = \left(\dfrac{1}{\cosh x}\right)$ is an even function also.

(b) If for a function of x, $f(-x) = -f(x)$ then $f(x)$ is called an **odd function** of x. Sin $(-x) = -\sin x$, thus the sine function is an odd function. Replacing x by $-x$ in the formula for sinh x gives:

$$\sinh(-x) = \frac{e^{-x} - e^{-(-x)}}{2} = \frac{e^{-x} - e^x}{2} = -\left(\frac{e^x - e^{-x}}{2}\right) = -\sinh x$$

Thus the hyperbolic sine, **sinh** x, **is an odd function**. The graph of an odd function is symmetrical about the origin (see Section 4).
Since sinh x is an odd function then cosech $x = \left(\dfrac{1}{\sinh x}\right)$ is an odd function also.

(c) Replacing x by $-x$ in the formula for tanh x gives:

$$\tanh(-x) = \frac{e^{-x} - e^{-(-x)}}{e^{-x} + e^{-(-x)}} = \frac{e^{-x} - e^x}{e^{-x} + e^x} = -\left(\frac{e^x - e^{-x}}{e^x + e^{-x}}\right) = -\tanh x.$$

Thus the hyperbolic tangent, **tanh** x, **is an odd function**. It follows that coth $x = \left(\dfrac{1}{\tanh x}\right)$ is also an odd function.

(d) replacing x by 0 in the formula for sinh x gives:

$$\sinh 0 = \frac{e^0 - e^{-0}}{2} = \frac{1-1}{2} = 0$$

(e) Replacing x by 0 in the formula for cosh gives:

$$\cosh 0 = \frac{e^0 + e^{-0}}{2} = \frac{1+1}{2} = 1$$

3 Evaluation of hyperbolic functions

Tables of exponential and hyperbolic functions are readily available and one such table is shown in Table 1, where it is seen that values of sinh x and cosh x may be read directly from an argument of $x = 0$ to $x = 6.0$. (Note that 'argument' means 'the value on which calculation of another quantity depends'.) For example,

sh 0.32 = 0.325 5,
ch 0.80 = 1.337 4,
sh 4.6 = 49.737 and
ch 5.9 = 182.52

Values of hyperbolic functions for values of x greater than 6.0 and for values of x of greater accuracy than two significant figures are evaluated as shown in the following worked problems.

Worked problems on evaluating hyperbolic functions

Problem 1. Evaluate, using tables, sinh 6.3, correct to four significant figures.

$$\sinh 6.3 = \frac{1}{2}(e^{6.3} - e^{-6.3}) = \frac{1}{2}\left[(e^{6.0})(e^{0.3}) - (e^{-6.0})(e^{-0.3})\right]$$

$$= \frac{1}{2}\left[(403.43)(1.349\,9) - (0.002\,48)(0.740\,8)\right]$$

$$= \frac{1}{2}\left[544.59 - 0.001\,837\right]$$

Hence sinh 6.3 = **272.3 correct to four significant figures.**

Problem 2. Using tables, find the value of cosh 2.54 correct to four significant figures.

$$\cosh 2.54 = \frac{1}{2}(e^{2.54} + e^{-2.54}) = \frac{1}{2}\left[(e^{2.5})(e^{0.04}) + (e^{-2.5})(e^{-0.04})\right]$$

$$= \frac{1}{2}\left[(12.182)(1.040\,8) + (0.082\,1)(0.960\,8)\right]$$

Table 1 Exponential and hyperbolic functions

x	e^x	e^{-x}	$\sinh x$	$\cosh x$	x	e^x	e^{-x}	$\sinh x$	$\cosh x$
0.02	1.0202	0.9802	0.0200	1.0002	1.0	2.7183	0.3679	1.1752	1.5431
0.04	1.0408	0.9608	0.0400	1.0008	1.1	3.0042	0.3329	1.3356	1.6685
0.06	1.0618	0.9418	0.0600	1.0018	1.2	3.3201	0.3012	1.5095	1.8107
0.08	1.0833	0.9231	0.0801	1.0032	1.3	3.6693	0.2725	1.6984	1.9709
0.10	1.1052	0.9048	0.1002	1.0050	1.4	4.0552	0.2466	1.9043	1.1509
0.11	1.1163	0.8958	0.1102	1.0061	1.5	4.4817	0.2231	2.1293	1.3524
0.12	1.1275	0.8869	0.1203	1.0072	1.6	4.9530	0.2109	2.3756	2.5775
0.13	1.1388	0.8781	0.1304	1.0085	1.7	5.4739	0.1827	2.6456	2.8283
0.14	1.1503	0.8694	0.1405	1.0098	1.8	6.0497	0.1653	2.9422	3.1075
0.15	1.1618	0.8607	0.1506	1.0113	1.9	6.6859	0.1496	3.2682	3.5177
0.16	1.1735	0.8521	0.1607	1.0128	2.0	7.3891	0.1353	3.6269	3.7622
0.17	1.1853	0.8437	0.1708	1.0145	2.1	8.1662	0.1225	4.0219	4.1443
0.18	1.1972	0.8353	0.1810	1.0162	2.2	9.0250	0.1108	4.4571	4.5679
0.19	1.2092	0.8270	0.1911	1.0181	2.3	9.9742	0.1003	4.9370	5.0372
0.20	1.2214	0.8187	0.2013	1.0201	2.4	11.023	0.0907	5.4662	5.5569
0.21	1.2337	0.8106	0.2115	1.0221	2.5	12.182	0.0821	6.0502	6.1323
0.22	1.2461	0.8025	0.2218	1.0243	2.6	13.464	0.0743	6.6947	6.7690
0.23	1.2586	0.7945	0.2320	1.0266	2.7	14.880	0.0672	7.4063	7.4735
0.24	1.2712	0.7866	0.2423	1.0289	2.8	16.445	0.0608	8.1919	8.2527
0.25	1.2840	0.7788	0.2526	1.0314	3.0	20.085	0.0498	10.018	10.068
0.26	1.2969	0.7711	0.2629	1.0340	3.0	20.085	0.0498	10.018	10.068
0.27	1.3100	0.7634	0.2733	1.0367	3.1	22.198	0.0450	11.076	11.121
0.28	1.3231	0.7558	0.2837	1.0395	3.2	24.532	0.0408	12.246	12.287
0.29	1.3364	0.7483	0.2941	1.0423	3.3	27.113	0.0369	13.538	13.575
0.30	1.3499	0.7408	0.3045	1.0453	3.4	29.964	0.0334	14.965	14.999
0.31	1.3634	0.7335	0.3150	1.0484	3.5	33.115	0.0302	16.543	16.573
0.32	1.3771	0.7261	0.3255	1.0516	3.6	36.598	0.0273	18.285	18.313
0.33	1.3910	0.7189	0.3360	1.0550	3.7	40.447	0.0247	20.211	20.236
0.34	1.4050	0.7118	0.3466	1.0584	3.8	44.701	0.0224	22.339	22.362
0.35	1.4191	0.7047	0.3572	1.0619	3.9	49.402	0.0202	24.691	24.711
0.36	1.4333	0.6977	0.3678	1.0655	4.0	54.598	0.0183	27.290	27.308
0.37	1.4477	0.6907	0.3785	1.0692	4.1	60.340	0.0166	30.162	30.178
0.38	1.4623	0.6839	0.3892	1.0731	4.2	66.686	0.0150	33.336	33.351
0.39	1.4770	0.6771	0.4000	1.0770	4.3	73.700	0.0136	36.843	36.857
0.40	1.4918	0.6703	0.4107	1.0811	4.4	81.451	0.0123	40.719	40.732
0.41	1.5068	0.6636	0.4216	1.0852	4.5	90.017	0.0111	45.003	45.014
0.42	1.5220	0.6570	0.4325	1.0895	4.6	99.484	0.0100	49.737	49.747
0.43	1.5373	0.6505	0.4434	1.0939	4.7	109.95	0.00910	54.969	54.978
0.44	1.5527	0.6440	0.4543	1.0984	4.8	121.51	0.00823	60.751	60.759
0.45	1.5683	0.6376	0.4653	1.1033	4.9	134.29	0.00745	67.141	67.149
0.46	1.5841	0.6313	0.4764	1.1077	5.0	148.41	0.00674	74.203	74.210
0.47	1.6000	0.6250	0.4875	1.1125	5.1	164.02	0.00610	82.008	82.014
0.48	1.6161	0.6188	0.4986	1.1174	5.2	181.27	0.00552	90.633	90.639
0.49	1.6323	0.6126	0.5098	1.1225	5.3	200.34	0.00499	100.17	100.17
0.50	1.6487	0.6065	0.5211	1.1276	5.4	221.41	0.00452	110.70	110.71
0.6	1.8221	0.5488	0.6367	1.1855	5.5	244.69	0.00409	122.34	122.35
0.7	2.0138	0.4966	0.7586	1.2552	5.6	270.43	0.00370	135.21	135.21
0.8	2.2255	0.4493	0.8881	1.3374	5.7	298.87	0.00335	149.43	149.43
0.9	2.4596	0.4066	1.0265	1.4331	5.8	330.30	0.00303	165.15	165.15
					5.9	365.04	0.00274	182.52	182.52
					6.0	403.43	0.00248	201.71	201.72

$$= \frac{1}{2}\left[12.679 + 0.078\,88 \right]$$

Hence, cosh 2.54   **= 6.379 correct to four significant figures.**

Problem 3. Evaluate sinh 1.487 2 correct to four significant figures.

$$\sinh 1.487\,2 = \frac{1}{2}\,(e^{1.487\,2} - e^{-1.487\,2})$$

$e^{1.487\,2}$ may be evaluated using a calculator or by the following method (which was explained in *Technician Mathematics level* 3 by J.O. Bird and A.J.C. May, Chapter 3):

Let $e^{1.487\,2} = y$

Taking logarithms to base e of both sides gives:

$$\ln e^{1.487\,2} = \ln y$$
$$1.487\,2 \ln e = \ln y$$
i.e. $1.487\,2 = \ln y$ (since $\ln e = 1$)

Taking antilogarithms, using four figure tables of hyperbolic or Naperian logarithms,

$$y = 4.425$$
Hence $e^{1.487\,2} = 4.425$

Similarly, let $e^{-1.487\,2} = y$
then $-1.487\,2 = \ln y$
$$-1.487\,2 = -2 + 0.512\,8$$
Hence $\bar{2}.512\,8 = \ln y$
$$\bar{2}.512\,8 = \bar{3}.697\,4 + (\bar{2}.512\,8 - \bar{3}.697\,4)$$
Therefore $\ln y = \bar{3}.697\,4 + 0.815\,4$
$$y = (10^{-1})\,(2.26)$$
$$y = 0.226$$

Hence $e^{-1.487\,2} = 0.226$

$$\sinh 1.487\,2 = \frac{1}{2}\,(4.425 - 0.226) = 2.099\,5$$

        = 2.100 correct to four significant figures.

Problem 4. Evaluate correct to four significant figures:

(a) th 0.62   (b) cosech 1.6   (c) sech 0.96   (d) coth 0.44

(a) $\tanh 0.62 = \dfrac{e^{0.62} - e^{-0.62}}{e^{0.62} + e^{-0.62}}$

84

$$e^{0.62} = (e^{0.6})(e^{0.02}) = (1.822\ 1)(1.020\ 2) = 1.858\ 9$$
$$e^{-0.62} = (e^{-0.6})(e^{-0.02}) = (0.548\ 8)(0.980\ 2) = 0.537\ 9$$

Hence $\tanh 0.62 = \dfrac{1.858\ 9 - 0.537\ 9}{1.858\ 9 + 0.537\ 9} = \dfrac{1.321\ 0}{2.396\ 8} = \mathbf{0.551\ 2}$

(b) $\operatorname{cosech} 1.6 = \dfrac{1}{\sinh 1.6} = \dfrac{1}{2.375\ 6}$ (from Table 1) $= \mathbf{0.420\ 9}$

(c) $\operatorname{sech} 0.96 = \dfrac{1}{\cosh 0.96} = \dfrac{2}{e^{0.96} + e^{-0.96}}$

$$e^{0.96} = (e^{0.9})(e^{0.06}) = (2.459\ 6)(1.061\ 8) = 2.611\ 6$$
$$e^{-0.96} = (e^{-0.9})(e^{-0.06}) = (0.406\ 6)(0.941\ 8) = 0.382\ 9$$

Hence $\operatorname{sech} 0.96 = \dfrac{2}{2.611\ 6 + 0.382\ 9} = \dfrac{2}{2.994\ 5} = \mathbf{0.667\ 9}$

(d) $\coth 0.44 = \dfrac{1}{\operatorname{th} 0.44} = \dfrac{\operatorname{ch} 0.44}{\operatorname{sh} 0.44} = \dfrac{1.098\ 4}{0.454\ 3}$ (from Table 1) $= \mathbf{2.418}$

Further problems on evaluating hyperbolic functions may be found in Section 9 (Problems 1 to 14), page 98.

4 Graphs of hyperbolic functions

(a) $y = \cosh x$

A graph of $y = \cosh x$ may be plotted using the values in Table 1, or by taking the average of the graphs of $y = e^x$ and $y = e^{-x}$. Either method will produce the curve shown in Fig. 1.

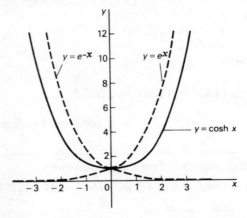

Figure 1   Graphs of $y = e^x$, $y = e^{-x}$ and $y = \cosh x$

It is noted that: (i) ch $0 = 1$
(ii) The graph is symmetrical about the y-axis, thus, cosh x is an even function.
(iii) As $x \to \infty$, ch $x \to \frac{1}{2} e^x$ (since $e^{-x} \to 0$).
and as $x \to -\infty$, ch $x \to \frac{1}{2} e^{-x}$ (since $e^x \to 0$)

The shape of the curve $y = \cosh x$ is that of a heavy rope or chain hanging freely under gravity and is called a **catenary**. Examples include a telegraph wire, a fisherman's line or a transmission line.

(b) $y = \sinh x$

A graph of $y = \sinh x$ may be plotted using the values in Table 1 or by taking half the difference of the graphs $y = e^x$ and $y = e^{-x}$. Either of these methods will produce the curve shown in Fig. 2.

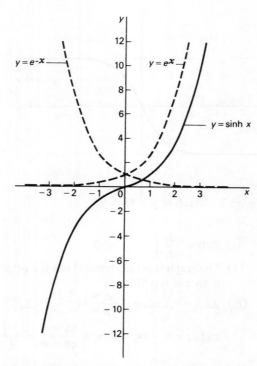

Figure 2 Graphs of $y = e^x$, $y = e^{-x}$ and $y = \sinh x$

It is noted that: (i) sh $0 = 0$
(ii) The graph is symmetrical about the origin, thus sinh x is an odd function.
(iii) As $x \to \infty$, sh $x \to \frac{1}{2} e^x$ (since $e^{-x} \to 0$)
and as $x \to -\infty$, sh $x \to -\frac{1}{2} e^{-x}$ (since $e^x \to 0$)

(c) $y = \tanh x$

Since $\tanh x = \dfrac{\operatorname{sh} x}{\operatorname{ch} x}$, Table 1 may be used to draw up the table of values shown below.

x		-4	-3	-2	-1	0	1	2	3	4
$\operatorname{sh} x$		-27.29	-10.02	-3.63	-1.18	0	1.18	3.63	10.02	27.29
$\operatorname{ch} x$		27.31	10.07	3.76	1.54	1	1.54	3.76	10.07	27.31
$y = \operatorname{th} x = \dfrac{\operatorname{sh} x}{\operatorname{ch} x}$		-0.999	-0.995	-0.97	-0.77	0	0.77	0.97	0.995	0.999

The graph $y = \tanh x$ is shown in Fig. 3.

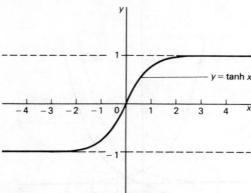

Figure 3 Graph of $y = \tanh x$

It is noted that:
(i) $\operatorname{th} 0 = \dfrac{\operatorname{sh} 0}{\operatorname{ch} 0} = \dfrac{0}{1} = 0$

(ii) The graph is symmetrical about the origin, thus $\tanh x$ is an odd function.

(iii) As $x \to \infty$, $\tanh x \to \dfrac{\operatorname{sh} \infty}{\operatorname{ch} \infty} \to \dfrac{\frac{1}{2} e^x}{\frac{1}{2} e^x} \to 1$,

and as $x \to -\infty$, $\tanh x \to \dfrac{\operatorname{sh} -\infty}{\operatorname{ch} -\infty} \to \dfrac{-\frac{1}{2} e^{-x}}{\frac{1}{2} e^{-x}} \to -1$

(d) $y = \operatorname{cosech} x$

Since $\operatorname{cosech} x = \dfrac{1}{\sinh x}$, Table 1 may be used to draw up the table of values shown below

x		-4	-3	-2	-1	0	1	2	3	4
$\operatorname{sh} x$		-27.29	-10.02	-3.63	-1.18	0	1.18	3.63	10.02	27.29
$\operatorname{cosech} x = \dfrac{1}{\operatorname{sh} x}$		-0.04	-0.10	-0.28	-0.85	$\pm\infty$	0.85	0.28	0.10	0.04

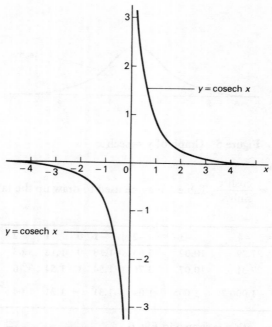

Figure 4 Graph of $y = \text{cosech } x$

The graph $y = \text{cosech } x$ is shown in Fig. 4.

It is noted that: (i) $\text{cosech } 0 = \pm \infty$
                     (ii) The graph is symmetrical about the origin, thus cosech x is an odd function.
                    (iii) As $x \to \infty$, cosech $x \to 0$ and as $x \to -\infty$, cosech $x \to 0$.

(e) $y = \text{sech } x$

Since $\text{sech } x = \dfrac{1}{\cosh x}$, Table 1 may be used to draw up the table of values shown below.

x	−4	−3	−2	−1	0	1	2	3	4
ch x	27.31	10.07	3.76	1.54	1	1.54	3.76	10.07	27.31
sech $x = \dfrac{1}{\text{ch } x}$	0.04	0.10	0.27	0.65	1	0.65	0.27	0.10	0.04

The graph $y = \text{sech } x$ is shown in Fig. 5.

It is noted that: (i) $\text{sech } 0 = 1$.
                     (ii) The graph is symmetrical about the y-axis, thus sech x is an even function.
                    (iii) As $x \to \infty$, sech $x \to 0$ and as $x \to -\infty$, sech $x \to 0$.

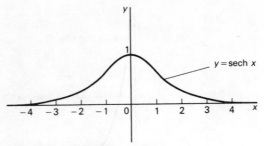

Figure 5 Graph of $y = \text{sech } x$

(f) $y = \coth x.$

Since $\coth x = \dfrac{\cosh x}{\sinh x}$, Table 1 may be used to draw up the table of values shown below.

x	−4	−3	−2	−1	0	1	2	3	4
sh x	−27.29	−10.02	−3.63	−1.18	0	1.18	3.63	10.02	27.29
ch x	27.31	−10.07	3.76	1.54	1	1.54	3.76	10.07	27.31
$\coth x = \dfrac{\text{ch } x}{\text{sh } x}$	−1.000 7	−1.005	−1.04	−1.31	±∞	1.31	1.04	1.005	1.000 7

The graph $y = \coth x$ is shown in Fig. 6.

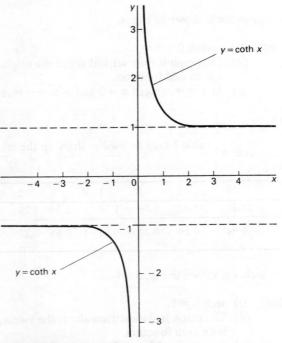

Figure 6 Graph of $y = \coth x$

It is noted that: (i) $\coth 0 = \pm \infty$
                (ii) The graph is symmetrical about the origin, thus $\coth x$ is an odd function.
              (iii) As $x \to \infty$, $\coth x \to 1$ and as $x \to -\infty$, $\coth x \to -1$.

5 Hyperbolic identities — Osborne's rule

$$\cosh x + \sinh x = \left(\frac{e^x + e^{-x}}{2}\right) + \left(\frac{e^x - e^{-x}}{2}\right) = e^x$$

$$\cosh x - \sinh x = \left(\frac{e^x + e^{-x}}{2}\right) - \left(\frac{e^x - e^{-x}}{2}\right) = e^{-x}$$

$$(\text{ch } x + \text{sh } x)(\text{ch } x - \text{sh } x) = (e^x)(e^{-x}) = e^0 = 1$$

i.e. $\text{ch}^2 x - \text{sh}^2 x = 1$ (1)

Dividing each term of equation (1) by $\text{ch}^2 x$ gives:

$$\frac{\text{ch}^2 x}{\text{ch}^2 x} - \frac{\text{sh}^2 x}{\text{ch}^2 x} = \frac{1}{\text{ch}^2 x}$$

i.e. $1 - \text{th}^2 x = \text{sech}^2 x$ (2)

Dividing each term of equation (1) by $\text{sh}^2 x$ gives:

$$\frac{\text{ch}^2 x}{\text{sh}^2 x} - \frac{\text{sh}^2 x}{\text{sh}^2 x} = \frac{1}{\text{sh}^2 x}$$

i.e. $\coth^2 x - 1 = \text{cosech}^2 x$ (3)

    Osborne's rule *states that the six trigonometrical ratios used in trigonometrical identities relating general angles may be replaced by their corresponding hyperbolic functions, but the sign of any direct or implied product of two sines must be changed.*
    Hence, since $\cos^2 x + \sin^2 x = 1$, by Osborne's rule $\text{ch}^2 x - \text{sh}^2 x = 1$, i.e. the trigonometrical functions have been changed to their corresponding hyperbolic functions and since $\sin^2 x$ is a direct product of two sines the sign is changed from $+$ to $-$.
    Similarly, since $1 + \tan^2 x = \sec^2 x$, then $1 - \text{th}^2 x = \text{sech}^2 x$.
In this case $\tan^2 x = \left(\dfrac{\sin^2 x}{\cos^2 x}\right)$ is considered as an implied product of two sines, hence the sign change.
Also, since $\cot^2 x + 1 = \text{cosec}^2 x$, then, from Osborne's rule,
          $-\coth^2 x + 1 = -\text{cosech}^2 x$
      i.e. $\coth^2 x - 1 = \text{cosech}^2 x$.
In this case, $\cot^2 x = \dfrac{\cos^2 x}{\sin^2 x}$ and $\text{cosec}^2 x = \dfrac{1}{\sin^2 x}$ are both implied products of two sines, hence the sign changes.
    Below are listed some trigonometrical identities, with their corresponding

hyperbolic identities which may be proved either by replacing sh A by $\left(\dfrac{e^A - e^{-A}}{2}\right)$ and ch A by $\left(\dfrac{e^A + e^{-A}}{2}\right)$ or by using Osborne's rule. Those identities marked with an asterisk involve sign changes due to a product of two sines.

Trigonometrical identities	Corresponding hyperbolic identities	
$\cos^2 A + \sin^2 A = 1$	$\text{ch}^2 A - \text{sh}^2 A = 1$	*
$1 + \tan^2 A = \sec^2 A$	$1 - \tanh^2 A = \text{sech}^2 A$	*
$\cot^2 A + 1 = \text{cosec}^2 A$	$\coth^2 A - 1 = \text{cosech}^2 A$	*

Compound angle addition and subtraction formulae

$\sin (A + B)$	$\text{sh} (A + B)$	
$= \sin A \cos B + \cos A \sin B$	$= \text{sh} A \,\text{ch} B + \text{ch} A \,\text{sh} B$	
$\sin (A - B)$	$\text{sh} (A - B)$	
$= \sin A \cos B - \cos A \sin B$	$= \text{sh} A \,\text{ch} B - \text{ch} A \,\text{sh} B$	
$\cos (A + B)$	$\text{ch} (A + B)$	
$= \cos A \cos B - \sin A \sin B$	$= \text{ch} A \,\text{ch} B + \text{sh} A \,\text{sh} B$	*
$\cos (A - B)$	$\text{ch} (A - B)$	
$= \cos A \cos B + \sin A \sin B$	$= \text{ch} A \,\text{ch} B - \text{sh} A \,\text{sh} B$	*

Dougle angles

$\sin 2A = 2 \sin A \cos A$	$\text{sh} 2A = 2 \,\text{sh} A \,\text{ch} A$	
$\cos 2A = \cos^2 A - \sin^2 A$	$\text{ch} 2A = \text{ch}^2 A + \text{sh}^2 A$	*
$= 2 \cos^2 A - 1$	$= 2 \,\text{ch}^2 A - 1$	
$= 1 - 2 \sin^2 A$	$= 1 + 2 \,\text{sh}^2 A$	*

Worked problems on hyperbolic identities

Problem 1. Prove that (a) $2 \,\text{ch}^2 \theta - 1 \equiv \text{ch} 2\theta$.

and (b) $\dfrac{\text{sech} x - \text{cosech} x}{\text{th} x - \coth x} \equiv \text{ch} x - \text{sh} x$

(a) Left hand side (L.H.S.) $\equiv 2 \,\text{ch}^2 \theta - 1$

$$\equiv 2\left(\frac{e^\theta + e^{-\theta}}{2}\right)^2 - 1$$

$$\equiv 2\left(\frac{e^{2\theta} + 2e^\theta e^{-\theta} + e^{-2\theta}}{4}\right) - 1$$

$$\equiv \frac{e^{2\theta} + 2 + e^{-2\theta}}{2} - 1$$

$$\equiv \frac{e^{2\theta} + e^{-2\theta}}{2} + \frac{2}{2} - 1$$

$$\equiv \frac{e^{2\theta} + e^{-2\theta}}{2} \equiv \text{ch} 2\theta = \text{R.H.S.}$$

(b) L.H.S. $\quad\equiv\quad \dfrac{\text{sech } x - \text{cosech } x}{\text{th } x - \text{coth } x}$

$$\equiv\quad \dfrac{\dfrac{1}{\text{ch } x} - \dfrac{1}{\text{sh } x}}{\dfrac{\text{sh } x}{\text{ch } x} - \dfrac{\text{ch } x}{\text{sh } x}} \quad\equiv\quad \dfrac{\dfrac{\text{sh } x - \text{ch } x}{\text{ch } x \ \text{sh } x}}{\dfrac{\text{sh}^2 x - \text{ch}^2 x}{\text{ch } x \ \text{sh } x}}$$

$$\equiv\quad \dfrac{\text{sh } x - \text{ch } x}{\text{sh}^2 x - \text{ch}^2 x} \quad\equiv\quad \dfrac{\text{sh } x - \text{ch } x}{-1}$$

$$\left(\begin{array}{l} \text{since } \text{ch}^2\, x - \text{sh}^2\, x = 1 \\ \text{then } \text{sh}^2\, x - \text{ch}^2\, x = -1 \end{array} \right)$$

$$\equiv\quad \text{ch } x - \text{sh } x = \text{R.H.S.}$$

Problem 2. Prove that $\text{ch } A \ \text{ch } B - \text{sh } A \ \text{sh } B \equiv \text{ch } (A - B)$

L.H.S. $\quad\equiv\quad \text{ch } A \ \text{ch } B - \text{sh } A \ \text{sh } B$

$$\equiv\quad \left(\dfrac{e^A + e^{-A}}{2} \right) \left(\dfrac{e^B + e^{-B}}{2} \right) - \left(\dfrac{e^A - e^{-A}}{2} \right) \left(\dfrac{e^B - e^{-B}}{2} \right)$$

$$\equiv\quad \tfrac{1}{4} \ (e^A e^B + e^A e^{-B} + e^{-A} e^B + e^{-A} e^{-B})$$

$$- \tfrac{1}{4} \ (e^A e^B - e^A e^{-B} - e^{-A} e^B + e^{-A} e^{-B})$$

$$\equiv\quad \tfrac{1}{4} \ (e^{A+B} + e^{A-B} + e^{-A+B} + e^{-A-B}$$

$$- e^{A+B} + e^{A-B} + e^{-A+B} - e^{-A-B})$$

$$\equiv\quad \tfrac{1}{2} \ e^{A-B} + \tfrac{1}{2} \ e^{-A+B}$$

$$\equiv\quad \tfrac{1}{2} \ (e^{A-B} + e^{-(A-B)}) \quad\equiv\quad \text{ch } (A - B) \quad=\quad \text{R.H.S.}$$

Problem 3.   (a) If $Pe^x + Qe^{-x} \equiv 3 \ \text{ch } x - 4 \ \text{sh } x$ find the values of P and Q.
           (b) If $5e^x - 2e^{-x} \equiv A \ \text{sh } x + B \ \text{ch } x$ find the values of A and B.

(a) $Pe^x + Qe^{-x} \equiv 3 \ \text{ch } x - 4 \ \text{sh } x$

$$\equiv 3 \left(\dfrac{e^x + e^{-x}}{2} \right) - 4 \left(\dfrac{e^x - e^{-x}}{2} \right)$$

$$Pe^x + Qe^{-x} \equiv \tfrac{3}{2} \ e^x + \tfrac{3}{2} \ e^{-x} - 2e^x + 2e^{-x}$$

$$Pe^x + Qe^{-x} \equiv - \tfrac{1}{2} \ e^x + \tfrac{7}{2} \ e^{-x}$$

Equating coefficients gives $P = -\tfrac{1}{2}$ and $Q = 3\tfrac{1}{2}$

(b) $5e^x - 2e^{-x} \equiv A \ \text{sh } x + B \ \text{ch } x$

$$\equiv A \left(\dfrac{e^x - e^{-x}}{2} \right) + B \left(\dfrac{e^x + e^{-x}}{2} \right)$$

$$\equiv \tfrac{A}{2} \ e^x - \tfrac{A}{2} \ e^{-x} + \tfrac{B}{2} \ e^x + \tfrac{B}{2} \ e^{-x}$$

$$5e^x - 2e^{-x} = \left(\frac{A+B}{2}\right)e^x + \left(\frac{B-A}{2}\right)e^{-x}$$

Equating coefficients gives: $5 = \dfrac{A+B}{2}$ and $-2 = \dfrac{B-A}{2}$

i.e. $A + B = 10$ (1)

and $-A + B = -4$ (2)

Adding equations (1) and (2) gives: $2B = 6$, i.e. $B = 3$

Substituting in equation (1) gives: $A = 7$

Further problems on hyperbolic identities may be found in Section 9 (Problems 15 to 21), page 99.

6. Differentiation of hyperbolic functions

(a) $\dfrac{d}{dx} (\sinh x) = \dfrac{d}{dx}\left(\dfrac{e^x - e^{-x}}{2}\right) = \tfrac{1}{2}[e^x - (-e^{-x})]$

$$= \tfrac{1}{2}(e^x + e^{-x}) = \cosh x$$

More generally, if $y = \sinh ax$ then $\dfrac{dy}{dx} = a \cosh ax$

(b) $\dfrac{d}{dx} (\cosh x) = \dfrac{d}{dx}\left(\dfrac{e^x + e^{-x}}{2}\right) = \tfrac{1}{2}[e^x + (-e^{-x})]$

$$= \tfrac{1}{2}(e^x - e^{-x}) = \sinh x$$

More generally, if $y = \cosh ax$ then $\dfrac{dy}{dx} = a \sinh ax$

(c) Since $\tanh x = \dfrac{\sinh x}{\cosh x}$ then by the quotient rule:

$$\dfrac{d}{dx} (\tanh x) = \dfrac{d}{dx}\left(\dfrac{\sinh x}{\cosh x}\right) = \dfrac{(\cosh x)(\cosh x) - (\sinh x)(\sinh x)}{\cosh^2 x}$$

$$= \dfrac{\cosh^2 x - \sinh^2 x}{\cosh^2 x} = \dfrac{1}{\cosh^2 x} = \operatorname{sech}^2 x$$

More generally, if $y = \tanh ax$ then $\dfrac{dy}{dx} = a \operatorname{sech}^2 ax$

(d) $\dfrac{d}{dx} (\operatorname{sech} x) = \dfrac{d}{dx}\left(\dfrac{1}{\cosh x}\right) = \dfrac{(\cosh x)(0) - (1)(\sinh x)}{\cosh^2 x} = \dfrac{-\sinh x}{\cosh^2 x}$

$$= -\left(\dfrac{1}{\cosh x}\right)\left(\dfrac{\sinh x}{\cosh x}\right) = -\operatorname{sech} x \tanh x$$

More generally, if $y = \text{sech } ax$ then $\dfrac{dy}{dx} = -a \text{ sech } ax \text{ tanh } ax$

(e) $\dfrac{d}{dx} (\text{cosech } x) = \dfrac{d}{dx}\left(\dfrac{1}{\sinh x}\right) = \dfrac{(\sinh x)(0) - (1)(\cosh x)}{\sinh^2 x} = -\dfrac{\cosh x}{\sinh^2 x}$

$$= -\left(\dfrac{1}{\sinh x}\right)\left(\dfrac{\cosh x}{\sinh x}\right) = -\text{ cosech } x \text{ coth } x$$

More generally, if $y = \text{cosech } ax$ then $\dfrac{dy}{dx} = -a \text{ cosech } ax \text{ coth } ax$

(f) $\dfrac{d}{dx} (\coth x) = \dfrac{d}{dx}\left(\dfrac{\cosh x}{\sinh x}\right) = \dfrac{(\sinh x)(\sinh x) - (\cosh x)(\cosh x)}{\sinh^2 x}$

$$= \dfrac{\sinh^2 x - \cosh^2 x}{\sinh^2 x} = -\dfrac{(\cosh^2 x - \sinh^2 x)}{\sinh^2 x}$$

$$= \dfrac{-1}{\sinh^2 x} = -\text{ cosech}^2 x$$

More generally, if $y = \coth ax$ then $\dfrac{dy}{dx} = -a \text{ cosech}^2 ax$

Summary

y or $f(x)$	$\dfrac{dy}{dx}$ or $f'(x)$
$\sinh ax$	$a \cosh ax$
$\cosh ax$	$a \sinh ax$
$\tanh ax$	$a \text{ sech}^2 ax$
$\text{sech } ax$	$-a \text{ sech } ax \text{ tanh } ax$
$\text{cosech } ax$	$-a \text{ cosech } ax \text{ coth } ax$
$\coth ax$	$-a \text{ cosech}^2 ax$

Worked problems on differentiation of hyperbolic functions

Problem 1. Differentiate the following with respect to x:

(a) $y = 3 \text{ sh } 4x + \tfrac{2}{3} \text{ ch } 2x$ (b) $y = 5 \text{ th } 3x - 4 \coth \dfrac{x}{2}$

(c) $y = \tfrac{1}{2} (\text{sech } 6x - 3 \text{ cosech } 5x)$

(a) $y = 3 \text{ sh } 4x + \tfrac{2}{3} \text{ ch } 2x$

$\dfrac{dy}{dx} = (3)(4) \text{ ch } 4x + (\tfrac{2}{3})(2) \text{ sh } 2x = 12 \text{ ch } 4x + \tfrac{4}{3} \text{ sh } 2x$

(b) $y = 5 \text{ th } 3x - 4 \coth \dfrac{x}{2}$

$$\frac{dy}{dx} = (5)(3)\,\text{sech}^2\,3x - (4)(-\tfrac{1}{2})\,\text{cosech}^2\,\frac{x}{2}$$

$$= 15\,\text{sech}^2\,3x + 2\,\text{cosech}^2\,\frac{x}{2}$$

(c) $y = \tfrac{1}{2}(\text{sech}\,6x - 3\,\text{cosech}\,5x)$

$$\frac{dy}{dx} = \tfrac{1}{2}[-6\,\text{sech}\,6x\,\text{th}\,6x - (3)(-5)\,\text{cosech}\,5x\,\text{coth}\,5x]$$

$$= \tfrac{3}{2}\,[5\,\text{cosech}\,5x\,\text{coth}\,5x - 2\,\text{sech}\,6x\,\text{th}\,6x]$$

Problem 2. Find the differential coefficients of the following with respect to the variable:

(a) $\ln(\text{sh}\,2x) + 3\,\text{ch}^2\,4x$ (b) $3\sin 2t\,\text{ch}\,5t$ (c) $\dfrac{2\,\theta^3}{\text{ch}\,3\theta}$

(a) $f(x) = \ln(\text{sh}\,2x) + 3\,\text{ch}^2\,4x$ (i.e. 'a function of a function')

$$f'(x) = \left(\frac{1}{\text{sh}\,2x}\right)(2\,\text{ch}\,2x) + (3)(2\,\text{ch}\,4x)(4\,\text{sh}\,4x)$$

$$= 2(\coth 2x + 12\,\text{sh}\,4x\,\text{ch}\,4x)$$

(b) $f(t) = 3\sin 2t\,\text{ch}\,5t$ (i.e. a product)
$\quad\;\; f'(t) = (3\sin 2t)(5\,\text{sh}\,5t) + (\text{ch}\,5t)(6\cos 2t)$
$$= 3(5\sin 2t\,\text{sh}\,5t + 2\cos 2t\,\text{ch}\,5t)$$

(c) $f(\theta) = \dfrac{2\,\theta^3}{\text{ch}\,3\theta}$ (i.e. a quotient)

$$f'(\theta) = \frac{(\text{ch}\,3\theta)(6\theta^2) - (2\theta^3)(3\,\text{sh}\,3\theta)}{(\text{ch}\,3\theta)^2} = 6\theta^2\left[\frac{\text{ch}\,3\theta}{\text{ch}^2\,3\theta} - \frac{\theta\,\text{sh}\,3\theta}{\text{ch}^2\,3\theta}\right]$$

$$= 6\theta^2\left[\frac{1}{\text{ch}\,3\theta} - \frac{\theta\,\text{sh}\,3\theta}{\text{ch}\,3\theta}\left(\frac{1}{\text{ch}\,3\theta}\right)\right]$$

$$= 6\theta^2\,\text{sech}\,3\theta\,(1 - \theta\,\text{th}\,3\theta)$$

Further problems on differentiation of hyperbolic functions may be found in Section 9 (Problems 22 to 27), page 100.

7   Solution of equations of the form $a\,\text{ch}\,x + b\,\text{sh}\,x = c$

Equations of the form $a\,\text{ch}\,x + b\,\text{sh}\,x = c$, where a, b and c are constants, may be solved either by plotting graphs of $y = a\,\text{ch}\,x + b\,\text{sh}\,x$ and $y = c$ and noting the points of intersection or, more accurately, as follows:

(i) Change ch x to $\left(\dfrac{e^x + e^{-x}}{2}\right)$ and sh x to $\left(\dfrac{e^x - e^{-x}}{2}\right)$.

(ii) Rearrange the equation into the form $m\,e^x + n\,e^{-x} + p = 0$, where m, n and p are constants.

(iii) Multiply each term by e^x. This produces an equation of the form: $m(e^x)^2 + p\,e^x + n = 0$, since $(e^{-x})(e^x) = 1$.

(iv) Solve the quadratic equation $m(e^x)^2 + p\,e^x + n = 0$ for e^x either by factorising or by using the quadratic formula.

(v) Given $e^x = k$, take logarithms to base e of both sides to give $x = \ln k$.

Worked problems on solving equations of the form $a\,\text{ch}\,x + b\,\text{sh}\,x = c$

Problem 1. Solve the equation sh $x = 1$ correct to four significant figures.

Following the above procedure:

(i) $\text{sh}\,x = \dfrac{e^x - e^{-x}}{2} = 1$

(ii) $e^x - e^{-x} = 2$

i.e. $e^x - e^{-x} - 2 = 0$

(iii) $(e^x)^2 - (e^{-x})e^x - 2e^x = 0$
$(e^x)^2 - 2e^x - 1 = 0$

(iv) $e^x = \dfrac{-(-2) \pm \sqrt{[-2]^2 - 4(1)(-1)]}}{2(1)} = \dfrac{2 \pm \sqrt{(4+4)}}{2}$

$= \dfrac{2 \pm \sqrt{8}}{2} = \dfrac{2 \pm 2.828\,4}{2} = \dfrac{4.828\,4}{2}$ or $\dfrac{-0.828\,4}{2}$

Hence $e^x = 2.414\,2$ or $-0.414\,2$

(v) $x = \ln 2.414\,2$ or $x = \ln(-0.414\,2)$ which has no real solution.

Hence $x = 0.881\,4$ correct to four significant figures.

Problem 2. Solve the equation $3\,\text{ch}\,\theta - 5 = 0$ correct to four decimal places.

(i) $3\,\text{ch}\,\theta - 5 = 0$

$3\left(\dfrac{e^\theta + e^{-\theta}}{2}\right) - 5 = 0$

(ii) $\tfrac{3}{2}e^\theta + \tfrac{3}{2}e^{-\theta} - 5 = 0$
i.e. $3e^\theta + 3e^{-\theta} - 10 = 0$

(iii) $3(e^\theta)^2 + 3 - 10\,e^\theta = 0$
$3(e^\theta)^2 - 10\,e^\theta + 3 = 0$

(iv) $(3\,e^\theta - 1)(e^\theta - 3) = 0$
$e^\theta = \tfrac{1}{3}$ or $e^\theta = 3$

(v) $\theta = \ln\left(\tfrac{1}{3}\right)$ or $\theta = \ln 3$

Hence $\theta = -1.098\,6$ or $1.098\,6$ correct to four decimal places.

Problem 3. Solve the equation 3.0 ch x + 2.0 sh x = 14.31, correct to four decimal places.

(i) 3.0 ch x + 2.0 sh x = 14.31

$$3.0 \; \frac{e^x + e^{-x}}{2} + 2.0 \; \frac{e^x - e^{-x}}{2} \;\; = 14.31$$

(ii) $1.5 \, e^x + 1.5 \, e^{-x} + e^x - e^{-x}$ = 14.31
     $2.5 \, e^x + 0.5 \, e^{-x} - 14.31$ = 0
(iii) $2.5 \, (e^x)^2 - 14.31 \, e^x + 0.5$ = 0

(iv) $e^x \;\; = \dfrac{-(-14.31) \pm \sqrt{[(-14.31)^2 - 4 \,(2.5) \,(0.5)]}}{2 \,(2.5)}$

$$= \frac{14.31 \pm 14.134 \, 22}{5} = \frac{28.444 \, 22}{5} \quad \text{or} \quad \frac{0.175 \, 78}{5}$$

Hence e^x = 5.688 844 or 0.035 156
(v) x = ln 5.688 844 or x = ln 0.035 156

Hence x = **1.738 5 or − 3.348 0 correct to four decimal places.**

Further problems on solving equations of the form a ch x + b sh x = c may be found in Section 9 (Problems 28 to 38), page 100.

8 Series expansions for cosh x and sinh x

e^x is defined by the power series:

$$e^x = 1 + x + \frac{x^2}{2!} + \frac{x^3}{3!} + \frac{x^4}{4!} + \frac{x^5}{5!} + \dots$$

If x is replaced by $-x$ then $e^{-x} = 1 - x + \dfrac{x^2}{2!} - \dfrac{x^3}{3!} + \dfrac{x^4}{4!} - \dfrac{x^5}{5!} + \dots$

$$\cosh x = \tfrac{1}{2}(e^x + e^{-x}) = \tfrac{1}{2}\left[\left(1 + x + \frac{x^2}{2!} + \frac{x^3}{3!} + \frac{x^4}{4!} + \frac{x^5}{5!} + \dots\right)\right.$$

$$\left. + \left(1 - x + \frac{x^2}{2!} - \frac{x^3}{3!} + \frac{x^4}{4!} - \frac{x^5}{5!} + \dots\right)\right]$$

$$= \tfrac{1}{2}\left[2 + \frac{2x^2}{2!} + \frac{2x^4}{4!} + \dots\right]$$

i.e.   **cosh x** $= 1 + \dfrac{x^2}{2!} + \dfrac{x^4}{4!} + \dots$

cosh x is an even function and contains only even powers of x

Similarly, $\sinh x = \tfrac{1}{2}(e^x - e^{-x}) = \tfrac{1}{2}\left[\left(1 + x + \dfrac{x^2}{2!} + \dfrac{x^3}{3!} + \dfrac{x^4}{4!} + \dfrac{x^5}{5!} + \dots\right)\right.$

$$-\left(1 - x + \frac{x^2}{2!} - \frac{x^3}{3!} + \frac{x^4}{4!} - \frac{x^5}{5!} + \ldots\right)\Bigg]$$

$$= \tfrac{1}{2}\left[2x + \frac{2x^3}{3!} + \frac{2x^5}{5!} + \ldots\right]$$

i.e. $\sinh x \qquad = x + \frac{x^3}{3!} + \frac{x^5}{5!} + \ldots$

$\sinh x$ is an odd function and contains only odd powers of x.
The series expansions for $\cosh x$ and $\sinh x$ are true for all values of x

Worked problems on series expansion for $\cosh x$ and $\sinh x$

Problem 1. Using the series expansion for ch x evaluate ch 1 correct to four decimal places.

$$\cosh x = 1 + \frac{x^2}{2!} + \frac{x^4}{4!} + \cdots$$

Let $x = 1$ then $\cosh 1 = 1 + \dfrac{1}{2\times1} + \dfrac{1}{4\times3\times2\times1} + \dfrac{1}{6\times5\times4\times3\times2\times1}$

$$= 1 + 0.5 + 0.041\,67 + 0.001\,389$$

i.e.   **ch 1** $\qquad$ **= 1.543 1 correct to four decimal places,**

which is the same as the value of ch 1 given in Table 1.

Problem 2. Using the series expansion for sh x evaluate sh 2 correct to four decimal places.

$$\sinh x = x + \frac{x^3}{3!} + \frac{x^5}{5!} + \cdots$$

Let $x = 2$ then $\text{sh } 2 = 2 + \dfrac{2^3}{3\times2\times1} + \dfrac{2^5}{5\times4\times3\times2\times1}$

$$+ \dfrac{2^7}{7\times6\times5\times4\times3\times2\times1} + \dfrac{2^9}{9\times8\times7\times6\times5\times4\times3\times2\times1}$$

$$+ \dfrac{2^{11}}{11\times10\times9\times8\times7\times6\times5\times4\times3\times2\times1}$$

$$= 2 + 1.333\,33 + 0.266\,67 + 0.025\,40$$
$$+ 0.001\,41 + 0.000\,05$$

i.e.   **sh 2** $\qquad$ **= 3.626 9 correct to four decimal places,**

which is the same as the value of sh 2 given in Table 1.

98

Problem 3. Find the series expansion for ch 2θ $-$ sh 2θ as far as the term in θ^5.

In the series expansions for sh x and ch x let $x = 2\theta$ then:

$$\text{ch } 2\theta = 1 + \frac{(2\theta)^2}{2!} + \frac{(2\theta)^4}{4!} + \ldots = 1 + 2\theta^2 + \frac{2\theta^4}{3} + \ldots$$

$$\text{sh } 2\theta = 2\theta + \frac{(2\theta)^3}{3!} + \frac{(2\theta)^5}{5!} + \ldots = 2\theta + \tfrac{4}{3}\theta^3 + \tfrac{4}{15}\theta^5 + \ldots$$

Hence ch 2θ $-$ sh 2θ $= (1 + 2\theta^2 + \tfrac{2}{3}\theta^4 + \ldots) - (2\theta + \tfrac{4}{3}\theta^3 + \tfrac{4}{15}\theta^5 + \ldots)$

$$= 1 - 2\theta + 2\theta^2 - \tfrac{4}{3}\theta^3 + \tfrac{2}{3}\theta^4 - \tfrac{4}{15}\theta^5 + \ldots$$

as far as the term in θ^5.

Further problems on series expansions for cosh x and sinh x may be found in the following Section (9) (Problems 39 to 43), page 101.

9 Further problems

Evaluation of hyperbolic functions

In problems 1 to 10, evaluate correct to four significant figures.

1. (a) sh 0.52   (b) sh 2.24   (c) sh 5.36.
   (a)[0.543 8]   (b)[4.643]   (c)[106.4]
2. (a) sh 0.212   (b) sh 7.7   (c) sh 8.6.
   (a)[0.213 6]   (b)[1 104]   (c)[2 716]
3. (a) ch 0.67   (b) ch 1.84   (c) ch 4.78.
   (a)[1.233]   (b)[3.228]   (c)[59.56]
4. (a) ch 0.346   (b) ch 6.5   (c) ch 7.4.
   (a)[1.060]   (b)[332.6]   (c)[818.0]
5. (a) sh 1.527 3   (b) sh 2.689 1   (c) sh 5.326 4.
   (a)[2.194]   (b)[7.325]   (c)[102.8]
6. (a) ch 0.478 8   (b) ch 1.723 1   (c) ch 4.681 5.
   (a)[1.117]   (b)[2.890]   (c)[53.97]
7. (a) th 0.43   (b) th 0.76   (c) th 1.54.
   (a)[0.405 3]   (b)[0.641 1]   (c)[0.912 1]
8. (a) cosech 0.26   (b) cosech 0.624   (c) cosech 2.45.
   (a)[3.804]   (b)[1.503]   (c)[0.173 9]
9. (a) sech 0.15   (b) sech 0.82   (c) sech 2.324.
   (a)[0.988 8]   (b)[0.737 8]   (c)[0.193 9]
10. (a) coth 0.33   (b) coth 0.746   (c) coth 1.168.
    (a)[3.140]   (b)[1.580]   (c)[1.214]

11. A telegraph wire hangs so that its shape is described by

$y = 60 \text{ ch } \dfrac{x}{60}$. Evaluate, correct to four significant figures,

y when x is 30.   [67.66]

12. The length, l of a heavy cable hanging under gravity is given by:

$$l = 2 c \sinh \frac{L}{2c}$$

Find l when $c = 50$ and $L = 40$.   [41.07]

13. The speed, V, of waves over the bottom of shallow water is given by the

formula: $V^2 = 0.55 L \tanh \left(\frac{6.3 d}{L} \right)$ where d is the depth and L is the

wavelength. If $d = 10$ and $L = 100$, calculate the value of V.   [5.540]

14. The increase in resistance of strip conductors due to eddy currents at

power frequencies is given by: $\lambda = \dfrac{\alpha t}{2} \left[\dfrac{\sinh \alpha t + \sin \alpha t}{\cosh \alpha t - \cos \alpha t} \right]$. Calculate λ

correct to five significant figures when $\alpha = 1.08$ and $t = 1$.   [1.007 5]

Hyperbolic identities

In Problems 15 to 20 prove the given identities.

15. (a) ch $(A + B) \equiv$ ch A ch $B +$ sh A sh B.
    (b) ch $2\theta \equiv$ ch$^2\ \theta +$ sh$^2\ \theta$
    (c) $1 + 2$ sh$^2 y \equiv$ ch $2y$.

16. (a) $1 +$ sh$^2\ A \equiv$ ch$^2\ A$.
    (b) sech$^2\ B +$ th$^2 B \equiv 1$
    (c) cosech$^2 C + 1 \equiv$ coth$^2\ C$.

17. (a) th $\alpha\ (1 +$ sh$^2 \alpha) \equiv \frac{1}{2}$ sh 2α.
    (b) $1 +$ ch $2\beta \equiv$ sh 2β coth β
    (c) coth $t - 2$ cosech $2t \equiv$ th t

18. (a) th $2\theta \equiv \dfrac{2\ \text{th}\ \theta}{1 + \text{th}^2\ \theta}$

    (b) ch $2\theta +$ sh $2\theta \equiv \dfrac{1 + \text{th}\ \theta}{1 - \text{th}\ \theta}$

    (c) sh $2\theta\ \equiv\ 2$ sh θ ch θ.

19. (a) sh A ch $B +$ ch A sh $B \equiv$ sh $(A + B)$

    (b) th $(A + B) \equiv \dfrac{\text{th}\ A + \text{th}\ B}{1 + \text{th}\ A\ \text{th}\ B}$

    (c) sh$^2 x \equiv \frac{1}{2}$ (ch $2x - 1$).

20. (a) ch$(A + B) +$ ch$(A - B)\ \equiv\ 2$ ch A ch B
    (b) ch$(A + B) -$ ch$(A - B)\ \equiv\ 2$ sh A sh B
    (c) sh $3A - 3$ sh A $\equiv\ 4$ sh$^3 A$.

21. (a) If $4e^x + 3e^{-x} \equiv P \text{ sh } x + Q \text{ ch } x$, find P and Q. $[P = 1, Q = 7]$
    (b) If $A \text{ ch } x - B \text{ sh } x \equiv 3e^x - 4e^{-x}$, find A and B. $[A = -1, B = -7]$
    (c) If $\alpha e^x - \beta e^{-x} \equiv 5 \text{ ch } x - 2 \text{ sh } x$, find α and β. $[\alpha = 1\frac{1}{2}, \beta = -3\frac{1}{2}]$

Differentiation of hyperbolic functions

In Problems 22 to 27, differentiate the given functions with respect to the variable.

22. (a) $4 \text{ sh } 3x$ (b) $2 \text{ ch } 4t$ (c) $5 \text{ th } 7x$.
    (a)$[12 \text{ ch } 3x]$ (b)$[8 \text{ sh } 4t]$ (c)$[35 \text{ sech}^2 \, 7x]$
23. (a) $\frac{3}{4} \text{ cosech } \frac{3}{2} \theta$ (b) $\frac{5}{8} \text{ sech } 4x$ (c) $3 \text{ coth } 5y$.

    (a)$[-\frac{9}{8} \text{ cosech } \dfrac{3\theta}{2} \text{ coth } \dfrac{3\theta}{2}$ (b)$[-\frac{5}{2} \text{ sech } 4x \tanh 4x]$

    (c)$[-15 \text{ cosech}^2 \, 5y]$

24. (a) $3 \ln (\text{ch } x)$ (b) $2 \ln \text{ th} \dfrac{3x}{2}$ (c) $\text{cosech}^3 \, 2t$.

    (a)$[3 \text{ th } x]$ (b)$[3 \text{ sech } \frac{3}{2} x \text{ cosech } \frac{3}{2} x]$
    (c)$[-6 \text{ cosech}^3 \, 2t \text{ coth } 2t]$

25. (a) $7 \text{ sh}^2 \, x$ (b) $2 \text{ coth}^4 \, 3p$ (c) $\text{ch}^3 \, 4x + 3 \text{ th}^2 \, 2x$.
    (a)$[14 \text{ sh } x \text{ ch } x \text{ or } 7 \text{ sh } 2x]$ (b)$[-24 \text{ coth}^3 \, 3p \text{ cosech}^2 \, 3p]$
    (c)$[12 \text{ ch}^2 \, 4x \text{ sh } 4x + \text{ th } 2x \text{ sech}^2 \, 2x]$

26. (a) $\text{sh } 3x \text{ ch } 3x$ (b) $2 \cos 3t \text{ ch } 3t$ (c) $e^{2x} \text{ ch } 3x$.
    (a)$[3 (\text{sh}^2 \, 3x + \text{ch}^2 \, 3x) \text{ or } 3 \text{ ch } 6x]$ (b) $[6(\cos 3t \text{ sh } 3t - \text{ch } 3t \sin 3t)]$
    (c)$[e^{2x} (3 \text{ sh } 3x + 2 \text{ ch } 3x)]$

27. (a) $\dfrac{\text{sh } 2\theta}{\sin 2\theta}$ (b) $\dfrac{2 \text{ ch } 5t}{3t^4}$ (c) $x^{\text{ch } x}$

    (a) $\left[\dfrac{2 (\sin 2\theta \text{ ch } 2\theta - \text{sh } 2\theta \cos 2\theta)}{\sin^2 2\theta} \right]$

    (b)$\left[\dfrac{2}{3t^5} (5t \text{ sh } 5t - 4 \text{ ch } 5t) \right]$ (c)$\left[x^{\text{ch } x} \left(\dfrac{\text{ch } x}{x} + \text{sh } x \ln x \right) \right]$

Solution of equations of the form $a \text{ ch } x + b \text{ sh } x = c$

In problems 28 to 37 solve the equations correct to four decimal places.

28. $\text{sh } x = 2$. $[1.443\ 6]$
29. $152 \text{ ch } x = 6.70$ $[\pm 2.163\ 4]$
30. $3.0 \text{ sh } x + 2.0 \text{ ch } x = 0$. $[-0.804\ 7]$
31. $\text{ch } x + 2 \text{ sh } x = 4$. $[1.024\ 7]$
32. $6.80 \text{ sh } \theta - 3.72 \text{ ch } \theta = 1.68$. $[0.905\ 2]$
33. $2.97 - 3.16 \text{ sh } x - 4 \text{ ch } x = 0$. $[-0.432\ 7 \text{ or } -1.710\ 2]$
34. $3.94 \text{ ch } x + 10.82 \text{ sh } x = 4.77$. $[0.075\ 6]$
35. $20.2 \text{ sh } y - 16.4 \text{ ch } y = 11.7$ $[2.008\ 3]$

36. th $x = $ sh x [0]

37. 7 th $x - 3 = 0$.   [0.458 1]

38. A chain hangs in the form $y = 50$ ch $\dfrac{x}{50}$. Find, correct to four significant figures, (i) the value of y when $x = 30$, and (ii) the value of x when $y = 64.73$.

     (i) [59.28]   (ii) [37.49]

Series expansions for cosh x and sinh x

39. Use the series expansion for cosh x to evaluate correct to four decimal places: (a) ch 2   (b) ch 0.5   (c) ch 2.5.

    (a) [3.762 2]   (b) [1.127 6]   (c) [6.132 3]

40. Use the series expansion for sinh x to evaluate correct to four decimal places: (a) sh 0.4   (b) sh 1.5   (c) sh 2.4

    (a) [0.410 8]   (b) [2.129 3]   (c) [5.466 2]

41. Expand the following as a power series as far as the term in x^5:

    (a) sh $2x$ (b) ch $4x$ (c) sh $3x$.

$$\text{(a)} \left[2x + \frac{4}{3} x^3 + \frac{4}{15} x^5 \right] \quad \text{(b)} \left[1 + 8x^2 + \frac{32}{3} x^4 \right]$$

$$\text{(c)} \left[3x + \frac{9}{2} x^3 + \frac{81}{40} x^5 \right]$$

42. Prove the following identities, the series being taken as far as the term in x^5 only.

(a) ch $2x -$ ch $x \equiv \dfrac{x^2}{2} \left(3 + \dfrac{5}{4} x^2 \right)$

(b) sh $x +$ sh $2x \equiv 3x \left(1 + \dfrac{x^2}{2} + \dfrac{11}{120} x^4 \right)$

(c) ch $3x +$ ch $2x +$ ch $x \equiv 3 + 7x^2 + \dfrac{49}{12} x^4$

43. The value of C_v for an Einstein crystal is given by

$$C_v = \frac{3No\, \epsilon^2 k}{(kT)^2} \cdot \frac{e^{\epsilon/kT}}{(e^{\epsilon/kT}-1)^2}$$

Express C_v in terms of sinh $\dfrac{\epsilon}{2kT}$, and by using Taylor's series show that when $\dfrac{\epsilon}{2kT}$ is very small $C_v = 3Nok$.

$$\left[C_v = \frac{3Nok}{4} \cdot \frac{(\epsilon/kT)^2}{\sinh^2(\epsilon/2kT)} \right]$$

Chapter 8

The relationship between trigonometric and hyperbolic functions and hyperbolic identities

1 The relationship between trigonometric and hyperbolic functions

In Chapter 6, the relationship between trigonometric and exponential functions are developed, showing that:

$$\cos \theta + j \sin \theta = e^{j\theta}$$
$$\cos \theta - j \sin \theta = e^{-j\theta}$$

Addition of these two equations gives:

$$\cos \theta = \frac{1}{2} \ (e^{j\theta} + e^{-j\theta}) \tag{1}$$

Subtraction gives:

$$\sin \theta = \frac{1}{2j} \ (e^{j\theta} - e^{-j\theta}) \tag{2}$$

In Chapter 7, the hyperbolic functions $\cosh \theta$ and $\sinh \theta$ are defined in terms of exponential functions and are:

$$\cosh \theta = \frac{1}{2} \ (e^{\theta} + e^{-\theta}) \tag{3}$$

$$\sinh \theta = \frac{1}{2} \ (e^{\theta} - e^{-\theta}) \tag{4}$$

Substituting $(j\theta)$ for θ in equation (1) gives:

$$\cos (j\theta) = \frac{1}{2} \ (e^{j(j\theta)} + e^{-j(j\theta)})$$

Since $j^2 = -1$, $\cos(j\theta) = \dfrac{1}{2}(e^{-\theta} + e^{\theta})$

$\quad\quad\quad\quad\quad\quad\quad\quad = \cosh\theta$ from equation (3).

Hence $\quad\quad\mathbf{\cos(j\theta) = \cosh\theta}$ $\quad\quad\quad\quad\quad\quad\quad\quad\quad\quad$ (5)

Substituting $(j\theta)$ for θ in equation (2) gives:

$$\sin(j\theta) = \frac{1}{2j}(e^{j(j\theta)} - e^{-j(j\theta)})$$

$$= \frac{1}{2j}(e^{-\theta} - e^{\theta})$$

$$= -\frac{1}{2j}(e^{\theta} - e^{-\theta})$$

$$= -\frac{1}{j}\sinh\theta$$

Multiplying both numerator and denominator of $\left(-\dfrac{1}{j}\right)$ by j gives:

$$-\frac{1}{j} \times \frac{j}{j} = -\frac{j}{j^2} = -\frac{j}{-1} = j,$$

hence $\mathbf{\sin(j\theta) = j\sinh\theta}$ $\quad\quad\quad\quad\quad\quad\quad\quad\quad$ (6)

Substituting $(j\theta)$ for θ in equation (3) gives:

$$\cosh(j\theta) = \frac{1}{2}(e^{(j\theta)} + e^{-(j\theta)}) = \cos\theta \text{ from equation (1)},$$

i.e. $\mathbf{\cosh(j\theta) = \cos\theta}$ $\quad\quad\quad\quad\quad\quad\quad\quad\quad\quad$ (7)

Substituting $(j\theta)$ for θ in equation (4) gives:

$$\sinh(j\theta) = \frac{1}{2}(e^{(j\theta)} - e^{-(j\theta)})$$

$$= j \cdot \frac{1}{2j}(e^{(j\theta)} - e^{-(j\theta)})$$

$$= j \cdot \sin\theta \text{ from equation (2)},$$

i.e. $\mathbf{\sinh(j\theta) = j\sin\theta}$ $\quad\quad\quad\quad\quad\quad\quad\quad\quad\quad$ (8)

Dividing equation (6) by equation (5) gives:

$$\frac{\sin j\theta}{\cos j\theta} = \tan j\theta = \frac{j\sinh\theta}{\cosh\theta} = j\tanh\theta,$$

i.e. $\mathbf{\tan(j\theta) = j\tanh\theta}$ $\quad\quad\quad\quad\quad\quad\quad\quad\quad\quad$ (9)

Dividing equation (8) by equation (7) gives:

$$\frac{\sinh j\theta}{\cosh j\theta} = \tanh j\theta = \frac{j\sin\theta}{\cos\theta} = j\tan\theta,$$

i.e. $\tanh (j\theta) = j \tan \theta$ (10)

2 Hyperbolic identities

The equations developed in Section 1 may be used to determine hyperbolic identities corresponding to standard trigonometric identities.

(i) Hyperbolic identities corresponding to the type $\cos^2 A + \sin^2 A = 1$

Substituting $j\theta$ for A gives:

$\cos^2 (j\theta) + \sin^2 (j\theta) = 1$

From equations (5) and (6) above, $\cos j\theta = \cosh \theta$ and $\sin (j\theta) = j \sinh \theta$. Hence,

$\cosh^2 \theta + j^2 \sinh^2 \theta = 1$. But $j^2 = -1$, hence

$\cosh^2 \theta - \sinh^2 \theta = 1$

Hyperbolic identities corresponding to $1 + \tan^2 A = \sec^2 A$ and $\cot^2 A + 1 = \operatorname{cosec}^2 A$ may be developed in a similar way and are listed in Section 3.

(ii) Hyperbolic identities corresponding to the compound angle addition and subtraction formulae

One of this group of formulae is:

$\sin (A + B) = \sin A \cos B + \cos A \sin B$.

Substituting $j\theta$ for A and $j\phi$ for B gives:

$\sin j (\theta + \phi) = \sin j\theta \cos j\phi + \cos j\theta \sin j\phi$

From equations (5) and (6), $\cos j\theta = \cosh \theta$ and $\sin j\theta = j \sinh \theta$, hence $j \sinh (\theta + \phi) = j \sinh \theta \cosh \phi + \cosh \theta j \sinh \phi$. Dividing throughout by j gives:

$\sinh (\theta + \phi) = \sinh \theta \cosh \phi + \cosh \theta \sinh \phi$

Identities for $\sinh (\theta - \phi)$, $\cosh (\theta + \phi)$ and $\cosh (\theta - \phi)$ may be developed in a similar way. Identities for $\tanh (\theta + \phi)$ and $\tanh (\theta - \phi)$ may be developed from $\dfrac{\sinh (\theta + \phi)}{\cosh (\theta + \phi)}$ and $\dfrac{\sinh (\theta - \phi)}{\cosh (\theta - \phi)}$. These identities are listed in Section 3.

(iii) Hyperbolic identities corresponding to double angle formulae

One of this group of formulae is:

$\sin 2A = 2 \sin A \cos A$

Substituting $j\theta$ for A gives:

$\sin 2 (j\theta) = 2 \sin (j\theta) \cos (j\theta)$

From equations (5) and (6), $\cos j\theta = \cosh \theta$ and $\sin j\theta = j \sinh \theta$, hence

$$j \sinh 2\theta = 2j \sinh \theta \cosh \theta$$

Dividing throughout by j gives:

$\sinh 2\theta = 2 \sinh \theta \cosh \theta$

An identity for $\cosh 2\theta$ may be developed in a similar way. An identity for $\tanh 2\theta$ may be developed from $\dfrac{\sinh 2\theta}{\cosh 2\theta}$. These identities are listed in Section 3.

(iv) Hyperbolic identities corresponding to changing products to sums or difference formulae.

One of this group of formulae is:

$$\sin A \cos B = \tfrac{1}{2} \left[\sin (A + B) + \sin (A - B)\right]$$

Substituting $j\theta$ for A and $j\phi$ for B gives:

$$\sin j\theta \cos j\phi = \tfrac{1}{2} \left[\sin j(\theta + \phi) + \sin j(\theta - \phi)\right]$$

From equations (5) and (6), $\cos j\theta = \cosh \theta$ and $\sin j\theta = j \sinh \theta$, hence,

$$j \sinh \theta \cosh \phi = \tfrac{1}{2} \left[j \sinh (\theta + \phi) + j \sinh (\theta - \phi)\right]$$

Dividing throughout by j gives:

$\sinh \theta \cosh \phi = \tfrac{1}{2} \left[\sinh (\theta + \phi) + \sinh (\theta - \phi)\right]$

Identities for $\cosh \theta \sinh \phi$, $\cosh \theta \cosh \phi$ and $\sinh \theta \sinh \phi$ may be developed in a similar way, and are listed in Section 3.

(v) Hyperbolic identities corresponding to changing sums or difference to product formulae.

One of this group of formulae is:

$$\sin A + \sin B = 2 \sin\left(\frac{A + B}{2}\right) \cos \left(\frac{A - B}{2}\right)$$

Substituting $j\theta$ for A and $j\phi$ for B gives:

$$\sin j\theta + \sin j\phi = 2 \sin j \left(\frac{\theta + \phi}{2}\right) \cos j \left(\frac{\theta - \phi}{2}\right)$$

From equations (5) and (6), $\cos j\theta = \cosh \theta$ and $\sin j\theta = j \sinh \theta$, hence,

$$j \sinh \theta + j \sinh \phi = 2j \sinh\left(\frac{\theta + \phi}{2}\right) \cosh \left(\frac{\theta - \phi}{2}\right)$$

Dividing throughout by j gives:

$\sinh \theta + \sinh \phi = 2 \sinh \left(\frac{\theta + \phi}{2}\right) \cosh \left(\frac{\theta - \phi}{2}\right)$

Identities for $\sinh\theta - \sinh\phi$, $\cosh\theta + \cosh\phi$ and $\cosh\theta - \cosh\phi$ may be developed in a similar way and are listed in Section 3.

There are three ways in which any hyperbolic identity may be verified. These are:

(a) by substituting $(j\theta)$ and if necessary $(j\phi)$ in the corresponding trigonometric identity, as shown above,
(b) by expressing the identity in terms of exponential functions, or
(c) by applying **Osborne's rule**. This is introduced in Chapter 7, and may be stated as follows:

> *'in formulae connecting trigonometric functions, replace each circular function by the corresponding hyperbolic function and change the sign of every product (or implied product) of two sines'.*

Each of these three ways are demonstrated in the worked problems following. When the corresponding trigonometric identity is known or can be readily proved, then Osborne's rule is the easiest method of determining the corresponding hyperbolic identity.

The relationship between trigonometric and hyperbolic functions and hyperbolic identities will be met in technological subjects dealing with topics such as transmission line theory, wave guide theory and filters.

Worked problems on hyperbolic identities

Problem 1. By substituting $j\theta$ for A, develop the hyperbolic identity corresponding to the trigonometric identity $\dfrac{1 + \sin^4 A - \cos^4 A}{\sin^2 A} = 2$. Check the result by applying Osborne's rule.

Writing $(j\theta)$ for A gives:

$$\frac{1 + \sin^4(j\theta) - \cos^4(j\theta)}{\sin^2 j\theta} = 2$$

But $\sin j\theta = j \sinh\theta$ and $\cos j\theta = \cosh\theta$. Thus,

$$\frac{1 + j^4 \sinh^4\theta - \cosh^4\theta}{j^2 \sinh^2\theta} = 2$$

But $j^4 = 1$ and $j^2 = -1$, hence the required identity is

$$\frac{1 + \sinh^4\theta - \cosh^4\theta}{-\sinh^2\theta} = 2$$

i.e. $\dfrac{1 + \sinh^4\theta - \cosh^4\theta}{\sinh^2\theta} = -2$

Rewriting the trigonometric identity in terms of the products of two sines gives:

$$\frac{1 + (\sin^2 A)(\sin^2 A) - \cos^4 A}{\sin^2 A} = 2$$

Osborne's rule states that each trigonometric function is replaced by its corresponding hyperbolic function, the sign being changed for any terms which are the products of two sines. Thus, the identity becomes:

$$\frac{1 + (-\sinh^2 \theta)(-\sinh^2 \theta) - \cosh^4 \theta}{(-\sinh^2 \theta)} = 2$$

i.e. $\dfrac{1 + \sinh^4 \theta - \cosh^4 \theta}{\sinh^2 \theta}$ $= -2$, as shown above.

Problem 2. Use the exponential form of a hyperbolic function to prove the identity:

$$\frac{1}{\sqrt{(1 + \sinh^2 \theta)}} = \text{sech } \theta$$

From the basic definitions of hyperbolic functions given in equations (3) and (4), $\cosh \theta = \frac{1}{2}(e^\theta + e^{-\theta})$ and $\sinh \theta = \frac{1}{2}(e^\theta - e^{-\theta})$. Substituting for $\sinh \theta$ in the left-hand side of the identity gives:

$$\frac{1}{\sqrt{(1 + \sinh^2 \theta)}} = \frac{1}{\sqrt{\{1 + [\frac{1}{2}(e^\theta - e^{-\theta})]^2\}}}$$

$$= \frac{1}{\sqrt{\{1 + \frac{1}{4}(e^{2\theta} - 2 + e^{-2\theta})\}}}$$

$$= \frac{1}{\sqrt{\{1 + \frac{1}{4}e^{2\theta} - \frac{1}{2} + \frac{1}{4}e^{-2\theta}\}}}$$

$$= \frac{1}{\sqrt{\{\frac{1}{4}e^{2\theta} + \frac{1}{2} + \frac{1}{4}e^{-2\theta}\}}}$$

$$= \frac{1}{\sqrt{\frac{1}{4}(e^{2\theta} + 2 + e^{-2\theta})}}$$

$$= \frac{1}{\sqrt{[\frac{1}{2}(e^\theta + e^{-\theta})]^2}}$$

$$= \frac{1}{\sqrt{(\cosh \theta)^2}}$$

$$= \frac{1}{\cosh \theta} = \text{sech } \theta$$

$$= \text{R.H.S.}$$

Problem 3. Prove that:

$$\sin^2 \theta \cosh^2 \theta + \cos^2 \theta \sinh^2 \theta = \frac{1}{2}(\cosh 2\theta - \cos 2\theta)$$

Writing the left-hand side in terms of $\cos\theta$ and $\cosh\theta$ only, using the identities $\cos^2\theta + \sin^2\theta = 1$ and $\cosh^2\theta - \sinh^2\theta = 1$, gives:

$(1 - \cos^2\theta)\cosh^2\theta + \cos^2\theta\,(\cosh^2\theta - 1)$

i.e. L.H.S. $= \cosh^2\theta - \cos^2\theta\cosh^2\theta + \cos^2\theta\cosh^2\theta - \cos^2\theta$

$\qquad\quad = \cosh^2\theta - \cos^2\theta$

From the double angle formulae listed in the summary in Section 3,

$\cos^2\theta = \frac{1}{2}(1 + \cos 2\theta)$ and $\cosh^2\theta = \frac{1}{2}(1 + \cosh 2\theta)$

Hence, L.H.S. $= \frac{1}{2}(1 + \cosh 2\theta) - \frac{1}{2}(1 + \cos 2\theta)$

$\qquad\qquad = \frac{1}{2} + \frac{1}{2}\cosh 2\theta - \frac{1}{2} - \frac{1}{2}\cos 2\theta$

$\qquad\qquad = \frac{1}{2}(\cosh 2\theta - \cos 2\theta) = $ R.H.S.

Further problems on hyperbolic identities may be found in Section 4 (Problems 1 to 20), page 109.

3 Summary of trigonometric and hyperbolic identities

Trigonometric identity	Corresponding hyperbolic identity
(i) *Standards*	
$\cos^2 A + \sin^2 A = 1$	$\cosh^2\theta - \sinh^2\theta = 1$
$1 + \tan^2 A = \sec^2 A$	$1 - \tanh^2\theta = \operatorname{sech}^2\theta$
$\cot^2 A + 1 = \operatorname{cosec}^2 A$	$\coth^2\theta - 1 = \operatorname{cosech}^2\theta$
(ii) *Compound angle addition and subtraction formulae*	
$\sin(A+B)$	$\sinh(\theta+\phi)$
$\quad = \sin A\cos B + \cos A\sin B$	$\quad = \sinh\theta\cosh\phi + \cosh\theta\sinh\phi$
$\sin(A-B)$	$\sinh(\theta-\phi)$
$\quad = \sin A\cos B - \cos A\sin B$	$\quad = \sinh\theta\cosh\phi - \cosh\theta\sinh\phi$
$\cos(A+B)$	$\cosh(\theta+\phi)$
$\quad = \cos A\cos B - \sin A\sin B$	$\quad = \cosh\theta\cosh\phi + \sinh\theta\sinh\phi$
$\cos(A-B)$	$\cosh(\theta-\phi)$
$\quad = \cos A\cos B + \sin A\sin B$	$\quad = \cosh\theta\cosh\phi - \sinh\theta\sinh\phi$
$\tan(A+B) = \dfrac{\tan A + \tan B}{1 - \tan A\tan B}$	$\tanh(\theta+\phi) = \dfrac{\tanh\theta + \tanh\phi}{1 + \tanh\theta\tanh\phi}$
$\tan(A-B) = \dfrac{\tan A - \tan B}{1 + \tan A\tan B}$	$\tanh(\theta-\phi) = \dfrac{\tanh\theta - \tanh\phi}{1 - \tanh\theta\tanh\phi}$
(iii) *Double angle formulae*	
$\sin 2A = 2\sin A\cos A$	$\sinh 2\theta = 2\sinh\theta\cosh\theta$
$\cos 2A = \cos^2 A - \sin^2 A$	$\cosh 2\theta = \cosh^2\theta + \sinh^2\theta$
$\qquad = 2\cos^2 A - 1$	$\qquad = 2\cosh^2\theta - 1$
$\qquad = 1 - 2\sin^2 A$	$\qquad = 1 + 2\sinh^2\theta$
$\tan 2A = \dfrac{2\tan A}{1 - \tan^2 A}$	$\tanh 2\theta = \dfrac{2\tanh\theta}{1 + \tanh^2\theta}$

Trigonometric identity	Corresponding hyperbolic identity

(iv) *Changing products into sums or differences formulae*

$\sin A \cos B$
$= \frac{1}{2}[\sin(A+B) + \sin(A-B)]$

$\cos A \sin B$
$= \frac{1}{2}[\sin(A+B) - \sin(A-B)]$

$\cos A \cos B$
$= \frac{1}{2}[\cos(A+B) + \cos(A-B)]$

$\sin A \sin B$
$= -\frac{1}{2}[\cos(A+B) - \cos(A-B)]$

$\sinh\theta \cosh\phi$
$= \frac{1}{2}[\sinh(\theta+\phi) + \sinh(\theta-\phi)]$

$\cosh\theta \sinh\phi$
$= \frac{1}{2}[\sinh(\theta+\phi) - \sinh(\theta-\phi)]$

$\cosh\theta \cosh\phi$
$= \frac{1}{2}[\cosh(\theta+\phi) + \cosh(\theta-\phi)]$

$\sinh\theta \sinh\phi$
$= \frac{1}{2}[\cosh(\theta+\phi) - \cosh(\theta-\phi)]$

(v) *Changing sums or differences into products formulae*

$\sin A + \sin B$
$= 2\sin\left(\frac{A+B}{2}\right)\cos\left(\frac{A-B}{2}\right)$

$\sin A - \sin B$
$= 2\cos\left(\frac{A+B}{2}\right)\sin\left(\frac{A-B}{2}\right)$

$\cos A + \cos B$
$= 2\cos\left(\frac{A+B}{2}\right)\cos\left(\frac{A-B}{2}\right)$

$\cos A - \cos B$
$= -2\sin\left(\frac{A+B}{2}\right)\sin\left(\frac{A-B}{2}\right)$

$\sinh\theta + \sinh\phi$
$= 2\sinh\left(\frac{\theta+\phi}{2}\right)\cosh\left(\frac{\theta-\phi}{2}\right)$

$\sinh\theta - \sinh\phi$
$= 2\cosh\left(\frac{\theta+\phi}{2}\right)\sinh\left(\frac{\theta-\phi}{2}\right)$

$\cosh\theta + \cosh\phi$
$= 2\cosh\left(\frac{\theta+\phi}{2}\right)\cosh\left(\frac{\theta-\phi}{2}\right)$

$\cosh\theta - \cosh\phi$
$= 2\sinh\left(\frac{\theta+\phi}{2}\right)\sinh\left(\frac{\theta-\phi}{2}\right)$

4 Further problems

In Problems 1 to 6, state the trigonometric identities corresponding to the hyperbolic identities given in terms of A (and B) and by substituting $j\theta$ for A (and $j\phi$ for B) in these trigonometric identities, prove the hyperbolic identities.

1. (a) $1 - \tanh^2\theta = \text{sech}^2\theta$
   (b) $\coth^2\theta - 1 = \text{cosech}^2\theta$

2. (a) $\sinh(\theta-\phi) = \sinh\theta\cosh\phi - \cosh\theta\sinh\phi$
   (b) $\cosh(\theta+\phi) = \cosh\theta\cosh\phi + \sinh\theta\sinh\phi$
   (c) $\cosh(\theta-\phi) = \cosh\theta\cosh\phi - \sinh\theta\sinh\phi$.

3. (a) $\sinh\theta - \sinh\phi = 2\cosh\left(\frac{\theta+\phi}{2}\right)\sinh\left(\frac{\theta-\phi}{2}\right)$

(b) $\cosh \theta + \cosh \phi = 2 \cosh \left(\dfrac{\theta + \phi}{2} \right) \cosh \left(\dfrac{\theta - \phi}{2} \right)$

(c) $\cosh \theta - \cosh \theta = 2 \sinh \left(\dfrac{\theta + \phi}{2} \right) \sinh \left(\dfrac{\theta - \phi}{2} \right)$

4. (a) $\tanh (\theta + \phi) = \dfrac{\tanh \theta + \tanh \phi}{1 + \tanh \theta \tanh \phi}$

(b) $\tanh (\theta - \phi) = \dfrac{\tanh \theta - \tanh \phi}{1 - \tanh \theta \tanh \phi}$.

5. (a) $\cosh 2\theta = \cosh^2 \theta + \sinh^2 \theta$
(b) $\cosh 2\theta = 2 \cosh^2 \theta - 1$
(c) $\cosh 2\theta = 1 + 2 \sinh^2 \theta$
(d) $\tanh 2\theta = \dfrac{2 \tanh \theta}{1 + \tanh^2 \theta}$.

6. (a) $\cosh \theta \sinh \phi = \frac{1}{2} [\sinh (\theta + \phi) - \sinh (\theta - \phi)]$
(b) $\cosh \theta \cosh \phi = \frac{1}{2} [\cosh (\theta + \phi) + \cosh (\theta - \phi)]$
(c) $\sinh \theta \sinh \phi = \frac{1}{2} [\cosh (\theta + \phi) - \cosh (\theta - \phi)]$.

In Problems 7 to 10, prove the hyperbolic identities given by expressing them in terms of exponential functions.

7. (a) $1 + \cosh x = 2 \cosh^2 \dfrac{x}{2}$

(b) $1 - \cosh x = -2 \sinh^2 \dfrac{x}{2}$

8. $\coth \dfrac{\theta}{2} - \coth \theta = \operatorname{cosech} \theta$.

9. $\cosh^3 \theta = \frac{1}{4} (\cosh 3\theta + 3 \cosh \theta)$.

10. $\dfrac{1 - \tanh^2 \phi}{1 - \coth^2 \phi} = -\tanh^2 \phi$.

Prove the hyperbolic identities given in Problems 11 to 18.

11. $-\left(\dfrac{1 - \cosh A}{1 + \cosh A} \right) = (\operatorname{cosech} A - \coth A)^2$.

12. $\cosh \theta \sqrt{(1 - \operatorname{sech}^2 \theta)} = \sinh \theta$.

13. $\dfrac{\operatorname{cosech} x}{\operatorname{sech} x} - \dfrac{\operatorname{sech} x}{\operatorname{cosech} x} = \dfrac{2}{\sinh 2x}$

14. $\dfrac{1 + \coth \phi}{1 + \tanh \phi} = \coth \phi$.

15. $(\sinh A + \cosh A)^2 + (\sinh A - \cosh A)^2 = 2 \cosh 2A.$

16. $\cosh^2 (A + B) - \cosh^2 (A - B) = \sinh 2A \sinh 2B.$

17. $\cosh^2 A \sin^2 B + \sinh^2 A \cos^2 B = \frac{1}{2} (\cosh 2B - \cos 2A).$

18. $\cosh x + \cosh 2x = \frac{1}{2} \sinh \frac{5x}{2} \operatorname{cosech} \frac{x}{2} - \frac{1}{2}$

(Hint: use changing sums or differences to products formulae.)

19. If $\sin A \cosh B = \cos \theta$ and $\cos A \sinh B = \sin \theta$, show that
$\sinh^2 B = \cos^2 A = \pm \sin \theta.$

(Hint: use the identities $\cos^2 \theta + \sin^2 \theta = 1$ and $\cosh^2 B - \sinh^2 B = 1$.)

20. If $\tan A = \tan \theta \tanh \phi$ and $\tan B = \cot \theta \tanh \phi$. show that
$$\tan (A + B) = \frac{\sinh 2\phi}{\sin 2\theta}.$$

(Hint: use the double angle formulae.)

Chapter 9

Differentiation of implicit functions

1 Implicit functions

In all previous work involving differentiation, the starting point has been
equations of the form $y = f(x)$. When an equation is expressed in this form,
it is said to be an **explicit** function. Thus:

$$y = 3x^3 + 4x - 5$$

$$y = e^x \sin x$$

and $\quad y = 3(x^2 - 1) \ln \dfrac{x}{2}$

are all examples of explicit functions. In many equations it is difficult, or
indeed impossible, to make y the subject of the formula. The equation is
then called an **implicit** function, there being an implied relationship between
x and y. Thus:

$$x^2 - 4xy + 3y^3 = 8$$

$$e^x \sin y + 4xy^2 = 0$$

and $\quad 2 \sin \dfrac{x}{2} \cos y^2 = x^2 - y$

are all examples of implicit functions.

2 Differentiating implicit functions

When differentiating an explicit function, the function of a variable, say x,
is differentiated with respect to x. However, when differentiating an implicit
function, it becomes necessary to differentiate a function of one variable,
say y, with respect to another variable, say x. This is achieved by using the

'chain' rule of differentiation, i.e.

$$\frac{du}{dx} = \frac{du}{dy} \times \frac{dy}{dx} .$$

Thus, to differentiate, say, y^2 with respect to x, let $u = y^2$.

Then $\frac{du}{dy} = 2y$ and $\frac{d(y^2)}{dx} = 2y \frac{dy}{dx}$ by using the 'chain' rule.

The rule when differentiating a function of, say y, with respect to, say x, is:

(i) differentiate the function of y with respect to y, and

(ii) always multiply by $\frac{dy}{dx}$.

This rule can be expressed mathematically as

$$\frac{d}{dx} [f(y)] = \frac{d}{dy} [f(y)] \times \frac{dy}{dx} \tag{1}$$

When a term is a product or a quotient of two functions of, say, x, the product or quotient rules of differentiation must be applied to find its differential coefficient. These same rules apply to products and quotients of terms having two variables (say $f(x, y)$). Thus if $z = x^2 y^3$, the product rule of differentiation must be used to find $\frac{dz}{dx}$, i.e. $\frac{dz}{dx} = u \frac{dv}{dx} + v \frac{du}{dx}$ where $u = x^2$ and $v = y^3$. Since $u = x^2$, $\frac{du}{dx} = 2x$ and since $v = y^3$, $\frac{dv}{dx} = 3y^2 \frac{dy}{dx}$ from equation (1) above.

Hence $\frac{dz}{dx} = x^2 \cdot 3y^2 \frac{dy}{dx} + y^3 \cdot 2x$

$$= xy^2 \left(3x \frac{dy}{dx} + 2y \right)$$

For a quotient, say $z = \frac{3y^2}{\cos x}$, the quotient rule of differentiation is used,

i.e. $\frac{dz}{dx} = \frac{v \frac{du}{dx} - u \frac{dv}{dx}}{v^2}$, where $u = 3y^2$ and $v = \cos x$.

Since $u = 3y^2$, $\frac{du}{dx} = 6y \frac{dy}{dx}$ from equation (1) above and since $v = \cos x$,

$\frac{dv}{dx} = - \sin x$,

thus, $\frac{dz}{dx} = \frac{(\cos x) (6y \frac{dy}{dx}) - (3y^2) (- \sin x)}{\cos^2 x}$

$$= \frac{3y (2 \cos x \frac{dy}{dx} + y \sin x)}{\cos^2 x}$$

When differentiating an implicit function having several terms, each term is differentiated with respect to the variable. Thus, to differentiate $x^2 + 2xy + 3y^2 = 0$ with respect to x, it becomes:

$$\frac{d(x^2)}{dx} + \frac{d(2xy)}{dx} + \frac{d(3y^2)}{dx} = \frac{d(0)}{dx}$$

i.e. $2x + \left(2x \frac{dy}{dx}\right) + y \cdot 2 + 6y \frac{dy}{dx} = 0$

To express $\frac{dy}{dx}$ in terms of x and y, terms containing $\frac{dy}{dx}$ are grouped on the left hand side of the equation, and those not containing $\frac{dy}{dx}$ are grouped on the right-hand side. Thus,

$$2x \frac{dy}{dx} + 6y \frac{dy}{dx} = -2x - 2y$$

$$\frac{dy}{dx}(2x + 6y) = -2x - 2y$$

$$\frac{dy}{dx} = \frac{-2x - 2y}{2x + 6y} = \frac{-(x + y)}{x + 3y}$$

Worked problems on differentiating implicit functions

Problem 1. Differentiate the following functions with respect to x:

(a) y^5 (b) $\cos 2\theta$ (c) $3 \ln \frac{p}{2}$ (d) $\frac{1}{4} e^{2q + 3}$

(a) Let $u = y^5$ then

$$\frac{du}{dx} = \frac{du}{dy} \times \frac{dy}{dx} = \frac{d}{dy}(y^5) \times \frac{dy}{dx} = 5y^4 \frac{dy}{dx}$$

(b) Let $u = \cos 2\theta$.

$$\frac{du}{dx} = \frac{du}{d\theta} \times \frac{d\theta}{dx} = \frac{d}{d\theta}(\cos 2\theta) \times \frac{d\theta}{dx}$$

$$= -2 \sin 2\theta \frac{d\theta}{dx}$$

(c) Let $u = 3 \ln \frac{p}{2}$

$$\frac{du}{dx} = \frac{du}{dp} \times \frac{dp}{dx} = \frac{d}{dp}\left(3 \ln \frac{p}{2}\right) \times \frac{dp}{dx}$$

$$= \frac{3}{p} \frac{dp}{dx}$$

(d) Let $u = \dfrac{1}{4} e^{2q+3}$

$$\frac{du}{dx} = \frac{du}{dq} \times \frac{dq}{dx} = \frac{d}{dq} \left(\frac{1}{4} e^{2q+3} \right) \times \frac{dq}{dx}$$

$$= \frac{1}{2} e^{2q+3} \frac{dq}{dx}$$

Problem 2. If $y = f(m, p)$, differentiate with respect to p:

(a) $y = m^2 \cos 2p$ and (b) $y = \dfrac{(1 - 3p)^{\frac{1}{2}}}{2 \tan 2m}$

(a) Since y is a function of two variables, m and p, the product rule of differentiation is used.

Let $u = m^2$. Now $\dfrac{du}{dp} = \dfrac{du}{dm} \times \dfrac{dm}{dp} = \dfrac{d}{dm}(m^2) \times \dfrac{dm}{dp}$

$$= 2m \frac{dm}{dp}$$

Let $v = \cos 2p$, $\dfrac{dv}{dp} = -2 \sin 2p$.

Applying the product rule, $\dfrac{dy}{dp} = u \dfrac{dv}{dp} + v \dfrac{du}{dp}$ gives

$$\frac{dy}{dp} = m^2 (-2 \sin 2p) + \cos 2p \left(2m \frac{dm}{dp} \right)$$

$$= 2m \left(\cos 2p \frac{dm}{dp} - m \sin 2p \right)$$

(b) Since y is a function of both m and p, the quotient rule of differentiation is used.

Let $u = (1 - 3p)^{\frac{1}{2}}$, $\dfrac{du}{dp} = -\dfrac{3}{2}(1 - 3p)^{-\frac{1}{2}}$

Let $v = 2 \tan 2m$, $\dfrac{dv}{dp} = \dfrac{dv}{dm} \times \dfrac{dm}{dp}$

$$= 4 \sec^2 2m \frac{dm}{dp}$$

Applying the quotient rule, $\dfrac{dy}{dp} = \dfrac{v \dfrac{du}{dp} - u \dfrac{dv}{dp}}{v^2}$, gives

$$\frac{dy}{dp} = \frac{2 \tan 2m \left(-\dfrac{3}{2}(1 - 3p)^{-\frac{1}{2}} \right) - (1 - 3p)^{\frac{1}{2}} \left(4 \sec^2 2m \dfrac{dm}{dp} \right)}{4 \tan^2 2m}$$

$$= - \frac{3}{4(\tan 2m)(1-3p)^{\frac{1}{2}}} - \frac{(1-3p)^{\frac{1}{2}} \quad \sec^2 2m \, \dfrac{dm}{dp}}{\tan^2 2m}$$

Problem 3. Given that $x^2 \sin \theta - 3x^3 = \sec \theta$, determine the value of $\dfrac{dx}{d\theta}$ when $\theta = \pi$.

$$\frac{d}{d\theta} (x^2 \sin \theta) = x^2 \cos \theta + \sin \theta \left(2x \, \frac{dx}{d\theta} \right)$$

$$\frac{d}{d\theta} (3x^3) = 9x^2 \, \frac{dx}{d\theta}$$

$$\frac{d}{d\theta} (\sec \theta) = \sec \theta \tan \theta$$

Hence, $x^2 \cos \theta + 2x \sin \theta \, \dfrac{dx}{d\theta} - 9x^2 \, \dfrac{dx}{d\theta} = \sec \theta \tan \theta$

Grouping only terms containing $\dfrac{dx}{d\theta}$ on the left-hand side gives:

$$2x \sin \theta \, \frac{dx}{d\theta} - 9x^2 \, \frac{dx}{d\theta} = \sec \theta \tan \theta - x^2 \cos \theta$$

$$\frac{dx}{d\theta} (2x \sin \theta - 9x^2) = \sec \theta \tan \theta - x^2 \cos \theta$$

$$\frac{dx}{d\theta} = \frac{\sec \theta \tan \theta - x^2 \cos \theta}{2x \sin \theta - 9x^2}$$

To determine the value of this expression, the value of x when $\theta = \pi$ is needed.

Substituting $\theta = \pi$ in the original equation, gives:

$x^2 \sin \pi - 3x^3 = \sec \pi$.

Now, $\sin \pi = 0$ and $\sec \pi = -1$. Hence, $0 - 3x^3 = -1$, i.e. $x^3 = \dfrac{1}{3}$

i.e. $x = \sqrt[3]{\tfrac{1}{3}}$

Thus $\dfrac{dx}{d\theta} = \dfrac{\sec \pi \tan \pi - \sqrt[3]{(\tfrac{1}{3})^2} \cos \pi}{2\sqrt[3]{(\tfrac{1}{3})} \sin \pi - 9\sqrt[3]{(\tfrac{1}{3})^2}}$

$$= \frac{(-1)(0) - \sqrt[3]{(\tfrac{1}{3})^2} \cdot (-1)}{2\sqrt[3]{(\tfrac{1}{3})}(0) - 9\sqrt[3]{(\tfrac{1}{3})^2}}$$

$$= \frac{\sqrt[3]{(\tfrac{1}{3})^2}}{-9\sqrt[3]{(\tfrac{1}{3})^2}} = -\frac{1}{9}$$

Problem 4. Determine the values of the gradients of the tangents drawn to the circle $x^2 + y^2 - 3x + 4y + 1 = 0$ at $x = 1$, correct to four significant figures.

The equation of the gradient of the tangent is given by $\dfrac{dy}{dx}$.
Differentiating each term with respect to x, gives.

$$2x + 2y\,\frac{dy}{dx} - 3 + 4\,\frac{dy}{dx} + 0 = 0$$

$$\text{i.e. } \frac{dy}{dx}(2y + 4) = 3 - 2x$$

$$\frac{dy}{dx} = \frac{3 - 2x}{2y + 4}$$

The value of y when $x = 1$ is obtained by substituting $x = 1$ in the original equation. Thus:

$$(1)^2 + y^2 - 3(1) + 4y + 1 = 0$$
$$\text{i.e. } y^2 + 4y - 1 = 0$$
$$y = \frac{-4 \pm \sqrt{[4^2 - 4 \times 1 \times (-1)]}}{2}$$
$$y = 0.236\,07 \text{ or } -4.236\,07.$$

Substituting $x = 1, y = 0.236\,07$ in the equation for $\dfrac{dy}{dx}$ gives:

$$\frac{dy}{dx} = \frac{3 - 2(1)}{2 \times 0.236\,07 + 4} = 0.223\,6, \text{ correct to four significant figures.}$$

Substituting $x = 1$ and $y = -4.236\,07$ in the equation for $\dfrac{dy}{dx}$ gives:

$$\frac{dy}{dx} = \frac{3 - 2(1)}{2 \times (-4.236\,07) + 4} = -0.223\,6, \text{ correct to four significant figures.}$$

That is, the gradients to the tangents are $\pm 0.223\,6$, correct to four significant figures.

Further problems on differentiating implicit functions may be found in the following Section (3) (Problems 1 to 20).

3 Further problems

Differentiating implicit functions

1. Differentiate with respect to x:

(a) y^6 (b) $3z^{\frac{1}{2}}$ (c) $\sin\dfrac{p}{3}$.

(a) $\left[6y^5\,\dfrac{dy}{dx}\right]$ (b) $\left[\dfrac{3}{2}z^{-\frac{1}{2}}\dfrac{dz}{dx}\right]$ (c) $\left[\dfrac{1}{3}\cos\dfrac{p}{3}\,\dfrac{dp}{dx}\right]$

2. Differentiate with respect to y:

    (a) $\sin 2p$ (b) $\dfrac{1}{2} \cos 3x$ (c) $2e^{2m+4}$.

    (a) $\left[2 \cos 2p \, \dfrac{dp}{dy} \right]$ (b) $\left[-\dfrac{3}{2} \sin 3x \, \dfrac{dx}{dy} \right]$ (c) $\left[4e^{2m+4} \, \dfrac{dm}{dy} \right]$

3. Differentiate with respect to u:

    (a) $(3 - 4x)^{\frac{1}{2}}$ (b) $2 \ln (3y + 2)$ (c) $\dfrac{1}{4} \tan (4u + \pi)$.

    (a) $\left[-2(3 - 4x)^{-\frac{1}{2}} \, \dfrac{dx}{du} \right]$ (b) $\left[\dfrac{6}{3y + 2} \quad \dfrac{dy}{du} \right]$

    (c) $[\sec^2 (4u + \pi)]$

4. Differentiate with respect to w:

    (a) $-\dfrac{1}{2} \sec 2\theta$ (b) $\dfrac{6}{3y + 2}$ (c) $\dfrac{3}{2} \cot (\dfrac{\pi}{2} - \alpha)$.

    (a) $\left[- \sec 2\theta \tan 2\theta \, \dfrac{d\theta}{dw} \right]$ (b) $\left[\dfrac{-18}{(3y + 2)^2} \quad \dfrac{dy}{dw} \right]$

    (c) $\left[\dfrac{3}{2} \operatorname{cosec}^2 (\dfrac{\pi}{2} - \alpha) \, \dfrac{d\alpha}{dw} \right]$

In Problems 5 to 8, find $\dfrac{dz}{dx}$.

5. $z = x^2 y^3$. $\left[3x^2 y^2 \, \dfrac{dy}{dx} + 2x \, y^3 \right]$

6. $z = 3y^2 \cos x$. $\left[-3y^2 \sin x + 6 y \cos x \, \dfrac{dy}{dx} \right]$

7. $z = x^2 \ln y$. $\left[\dfrac{x^2}{y} \, \dfrac{dy}{dx} + 2x \ln y \right]$

8. $z = \dfrac{2x^{\frac{1}{3}}}{\tan y}$. $\left[\dfrac{\dfrac{2}{3} (\tan y) x^{-\frac{2}{3}} - 2x^{\frac{1}{3}} \sec^2 y \, \dfrac{dy}{dx}}{\tan^2 y} \right]$

In Problems 9 to 12, differentiate with respect to y.

9. $x^2 + \sin x \cos y$. $\left[2x \, \dfrac{dx}{dy} - \sin x \sin y + \cos x \cos y \, \dfrac{dx}{dy} \right]$

10. $e^{2x} \ln y - 4x^2$. $\left[\dfrac{e^{2x}}{y} + 2e^{2x} \ln y \, \dfrac{dx}{dy} - 8x \, \dfrac{dx}{dy} \right]$

11. $x^2 + 3xy - y^2$. $\left[2x \, \dfrac{dx}{dy} + 3x + 3y \, \dfrac{dx}{dy} - 2y \right]$

12. $y^3 - 4x^2 \cos y + \dfrac{\ln x}{\sec y}$.

$$\left[\dfrac{3y^2 + 4x^2 \sin y - 8x \cos y \dfrac{dx}{dy} + \dfrac{1}{x} \dfrac{dx}{dy} - \ln x \tan y}{\sec y} \right]$$

In Problems 13 to 16, find $\dfrac{dy}{dx}$ in terms of x and y.

13. $x^2 + y^2 + 6x + 7y + 3 = 0.$ $\left[\dfrac{dy}{dx} = \dfrac{-2(x+3)}{2y+7} \right]$

14. $2x^3 - 4x^2 y = \cos y.$ $\left[\dfrac{dy}{dx} = \dfrac{2x\,(3x-4y)}{4x^2 - \sin y} \right]$

15. $\sin 2x \cos 3y = x^{\frac{1}{2}} y^{\frac{1}{3}}.$ $\left[\dfrac{dy}{dx} = \dfrac{2 \cos 2x \cos 3y - y^{\frac{1}{3}}/(2x^{\frac{1}{2}})}{3 \sin 2x \sin 3y + x^{\frac{1}{2}}/(3y^{\frac{2}{3}})} \right]$

16. $2y - \dfrac{x^3}{\sec y} = y^3.$ $\left[\dfrac{dy}{dx} = \dfrac{3x^2}{x^3 \tan y + \sec y\,(2 - 3y^2)} \right]$

17. Determine the gradients of the tangents drawn to the circle $x^2 + y^2 = 4$ at the point $x = 1$.

$$\left[\pm \dfrac{1}{\sqrt{3}} = \pm 0.5774 \right]$$

18. Find the gradients of the tangents drawn to the ellipse $2x^2 + y^2 = 9$ at the point $x = 2$ and to the hyperbola $x^2 - y^2 = 8$ at the point $x = 3$.

$$[\pm 4; \pm 3.]$$

19. If the distance moved by a body is given by $x = 3 \tan \theta$, the angular velocity, w, is $\dfrac{d\theta}{dt}$ and the velocity v is $\dfrac{dx}{dt}$, show that $w = \dfrac{v}{3} \cos^2 \theta$.

20. The pressure, p, and volume, V, of a gas are related by the law $p\,V^n = C$, where n and C are constants. Show that the rate of change of pressure,
$$\dfrac{dp}{dt} = -n\,\dfrac{p}{V}\,\dfrac{dV}{dt} .$$

Chapter 10

Differentiation of functions defined parametrically

1 Parametric representation of points

Certain mathematical relationships can be expressed more simply by stating both x and y in terms of a third variable, say, θ. By doing this, subsequent work is frequently simplified. When both x and y are expressed in terms of the same variable, this variable is called a **parameter**. The equation of any point on a circle, centre at $x = 0$ and $y = 0$, and of radius r, is given by $x^2 + y^2 = r^2$.

Such a circle is shown in Fig. 1. Any point on the circumference of the circle may be expressed in terms of the radius and the angle θ and when this is done, θ is the parameter used.

With reference to Fig. 1, from triangle OAB, $x = r \cos \theta$ and $y = r \sin \theta$. These two equations are called the **parametric equations** for a circle. A check may be made when parametric equations have been formed by substituting for x and y in the given equation. Thus, substituting $x = r \cos \theta$ and $y = r \sin \theta$ in the left-hand side of the equation $x^2 + y^2 = r^2$ gives:

$$\text{L.H.S.} = r^2 \cos^2\theta + r^2 \sin^2 \theta \ = r^2 (\cos^2 \theta + \sin^2 \theta)$$
$$= r^2,$$

which is equal to the right-hand side of the original equation. Hence $x = r \cos \theta, y = r \sin \theta$ are suitable parametric equations of circle $x^2 + y^2 = r^2$.

When the parametric equations of a curve are given, the curve may be plotted as shown below. The parametric equations for a parabola are of the form $x = at^2, y = 2 at$. Checking these parametric equations in the equation of a parabola $y^2 = 4 ax$ gives:

L.H.S. $y^2 = (2 at)^2 = 4 a^2 t^2$.
R.H.S. $4 ax = 4a (at^2) = 4 a^2 t^2$.

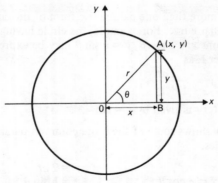

Figure 1

Since L.H.S. = R.H.S., $x = at^2$, $y = 2\,at$ are suitable parametric equations. To plot the graph for values of t between, say, -3 and 3, the values of x and y are calculated from the parametric equations. Thus, when

$t =$	-3	-2	-1	0	1	2	3
$x = at^2 =$	$9a$	$4a$	a	0	a	$4a$	$9a$
$y = 2\,at =$	$-6a$	$-4a$	$-2a$	0	$2a$	$4a$	$6a$

The parabola represented by $x = at^2$, $t = 2\,at$, over a range $-3 \leqslant t \leqslant 3$ is shown in Fig. 2.

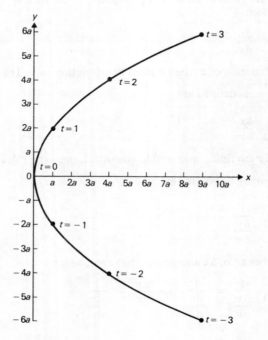

Figure 2

There can be more than one parametric form of the co-ordinates used to represent a particular curve. For example, the circle having parameter θ and parametric equations $x = a \cos \theta$, $y = a \sin \theta$ may be expressed in terms of a different parameter t, as

$$x = \frac{a(1 - t^2)}{1 + t^2} , \quad y = \frac{2at}{1 + t^2} ,$$

where the parameter t is given by $t = \tan \dfrac{\theta}{2}$.

The list below shows some of the more common parametric equations used in mathematics.

Parabola	$x = at^2$,	$y = 2at$
Ellipse	$x = a \cos \theta$,	$y = b \sin \theta$
Hyperbola	$x = a \sec \theta$,	$y = b \tan \theta$
(also	$x = \pm a \cosh u$,	$y = b \sinh u$)
Rectangular $\Big\}$ Hyperbola	$x = ct$,	$y = \dfrac{c}{t}$
Astroid	$x = a \cos^3 \theta$,	$y = a \sin^3 \theta$
Cardioid	$x = a(2 \cos \theta - \cos 2\theta)$,	$y = a(2 \sin \theta - \sin 2\theta)$
Cycloid	$x = a(\theta - \sin \theta)$,	$y = a(1 - \cos \theta)$

Sketches showing the approximate shapes of these curves are shown in Fig. 3.

2 Differentiation in parameters

When x and y are given in terms of a parameter, say, t, then by the chain rule of differentiation:

$$\frac{dy}{dx} = \frac{dy}{dt} \times \frac{dt}{dx}$$

Also, if y is a function of x, then x must be a function of y. The differential coefficient, $\dfrac{dx}{dy}$, is defined as

$$\frac{dx}{dy} = \lim_{\delta y \to 0} \frac{\delta x}{\delta y}$$

Since δx and δy are finite, measurable quantities, the basic rules of fractions apply, and $\dfrac{dx}{dy}$ may be written as

$$\frac{dx}{dy} = \lim_{\delta y \to 0} \frac{1}{\dfrac{\delta y}{\delta x}}$$

As δy approaches zero, δx also approaches zero, hence

$$\frac{dx}{dy} = \lim_{\delta x \to 0} \frac{1}{\dfrac{\delta x}{\delta y}} = \frac{1}{\dfrac{dy}{dx}} , \text{ i.e. } \frac{dy}{dx} = \frac{1}{\dfrac{dx}{dy}}$$

123

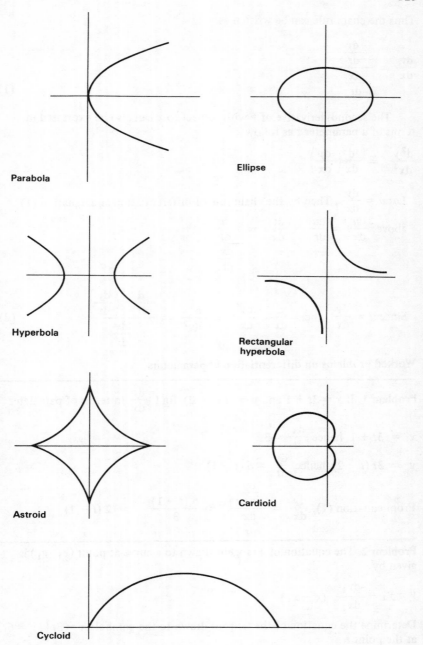

Figure 3 Parabola, ellipse, hyperbola, rectangular hyperbola, astroid, cardioid, cycloid.

Thus the chain rule can be written as:

$$\frac{dy}{dx} = \frac{\dfrac{dy}{dt}}{\dfrac{dx}{dt}}$$

(1)

The second derivative of y with respect to x can also be expressed in terms of a parameter t as follows:

$$\frac{d^2y}{dx^2} = \frac{d}{dx}\left(\frac{dy}{dx}\right)$$

Let $u = \dfrac{dy}{dx}$. Then by the chain rule of differentiation and equation (1) above, $\dfrac{du}{dx} = \dfrac{du}{dt} \times \dfrac{dt}{dx} = \dfrac{\dfrac{du}{dt}}{\dfrac{dx}{dt}}$

Since $u = \dfrac{dy}{dx}$, then $\dfrac{d}{dx}\ \dfrac{dy}{dx} = \dfrac{d^2y}{dx^2} = \dfrac{\dfrac{d}{dt}\left(\dfrac{dy}{dx}\right)}{\dfrac{dx}{dt}}$

(2)

Worked problems on differentiation in parameters

Problem 1. If $x = 3t + 1$ and $y = 3t\,(t - 2)$, find $\dfrac{dy}{dx}$ in terms of parameter t.

$x = 3t + 1$, hence $\dfrac{dx}{dt} = 3$

$y = 3t\,(t - 2)$, hence $\dfrac{dy}{dt} = 6\,(t - 1)$

From equation (1), $\dfrac{dy}{dx} = \dfrac{\dfrac{dy}{dt}}{\dfrac{dx}{dt}} = \dfrac{6\,(t - 1)}{3} = \mathbf{2\,(t - 1)}$

Problem 2. The equation of a tangent drawn to a curve at point (x_1, y_1) is given by:

$$y - y_1 = \frac{dy_1}{dx_1}\ (x - x_1)$$

Determine the equation of the tangent drawn to the parabola $x = 3t^2$, $y = 6t$, at the point t.

At point t, $x_1 = 3t^2$ and $y_1 = 6t$.

$$\frac{dx_1}{dt} = 6t \quad \frac{dy_1}{dt} = 6$$

From equation (1), $\dfrac{dy_1}{dx_1} = \dfrac{\dfrac{dy_1}{dt}}{\dfrac{dx_1}{dt}} = \dfrac{6}{6t} = \dfrac{1}{t}$

Hence, the equation of the tangent is. $y - 6t = \dfrac{1}{t}\ (x - 3t^2)$

Problem 3. The equation of the normal drawn to a curve at point (x_1, y_1) is given by:

$$y - y_1 = \frac{-1}{\dfrac{dy_1}{dx_1}}\ (x - x_1)$$

Determine the equation of the normal drawn to the astroid $x = a\cos^3 \theta$, $y = a\sin^3 \theta$ at the point $\theta = \dfrac{\pi}{4}$.

$x = a\cos^3 \theta, \dfrac{dx}{d\theta} = -3a\cos^2 \theta \sin \theta$

$y = a\sin^3 \theta, \dfrac{dy}{d\theta} = 3a\sin^2 \theta \cos \theta$

From equation (1), $\dfrac{dy}{dx} = \dfrac{\dfrac{dy}{d\theta}}{\dfrac{dx}{d\theta}} = \dfrac{3a\sin^2 \theta \cos \theta}{-3a\cos^2 \theta \sin \theta}$

$$= -\tan \theta$$

When $\theta = \dfrac{\pi}{4}$, $\dfrac{dy_1}{dx_1} = -1$,

$$x_1 = a\cos^3 \frac{\pi}{4} = a\left(\frac{1}{\sqrt{2}}\right)^3 = \frac{a}{2\sqrt{2}}$$

$$y_1 = a\sin^3 \frac{\pi}{4} = a\left(\frac{1}{\sqrt{2}}\right)^3 = \frac{a}{2\sqrt{2}}$$

Hence the equation of the normal is:

$$y - \frac{a}{2\sqrt{2}} = \frac{-1}{-1}\left(x - \frac{a}{2\sqrt{2}}\right)$$

i.e. $\quad x \quad = y$

Problem 4. Determine the values of $\dfrac{dy}{dx}$ and $\dfrac{d^2y}{dx^2}$ for the cycloid

$x = 3(\theta - \sin \theta), y = 3(1 - \cos \theta)$ at the point $\theta = \dfrac{\pi}{3}$.

$$x = 3\,(\theta - \sin\theta), \quad \frac{dx}{d\theta} = 3\,(1 - \cos\theta)$$

$$y = 3\,(1 - \cos\theta), \quad \frac{dy}{d\theta} = 3\sin\theta$$

From equation (1) $\dfrac{dy}{dx} = \dfrac{\dfrac{dy}{d\theta}}{\dfrac{dx}{d\theta}} = \dfrac{3\sin\theta}{3\,(1 - \cos\theta)} = \dfrac{\sin\theta}{1 - \cos\theta}$

When $\theta = \dfrac{\pi}{3}$, $\dfrac{dy}{dx} = \dfrac{\sin\dfrac{\pi}{3}}{1 - \cos\dfrac{\pi}{3}} = \dfrac{\dfrac{\sqrt{3}}{2}}{\left(1 - \dfrac{1}{2}\right)} = \sqrt{3}$

From equation (2) $\dfrac{d^2 y}{dx^2} = \dfrac{\dfrac{d}{d\theta}\left(\dfrac{dy}{dx}\right)}{\dfrac{dx}{d\theta}} = \dfrac{\dfrac{d}{d\theta}\left(\dfrac{\sin\theta}{1 - \cos\theta}\right)}{3\,(1 - \cos\theta)}$

$$= \frac{(1 - \cos\theta)\cos\theta - \sin\theta\,(\sin\theta)}{3\,(1 - \cos\theta)^3}$$

$$= \frac{\cos\theta - (\cos^2\theta + \sin^2\theta)}{3\,(1 - \cos\theta)^3} = -\frac{4}{3}$$

Problem 5. When determining the surface tension of a liquid, the radius of curvature, ρ, of part of the surface is given by

$$\rho = \frac{\left[1 + \left(\dfrac{dy}{dx}\right)^2\right]^{\frac{3}{2}}}{\dfrac{d^2 y}{dx^2}}$$

Determine the radius of curvature of the part of the surface having the parametric equations $x = 2t^2$, $y = 4t$ at the point $t = 1$.

$$x = 2t^2, \quad \frac{dx}{dt} = 4t$$

$$y = 4t, \quad \frac{dy}{dt} = 4$$

From equation (1), $\dfrac{dy}{dx} = \dfrac{\dfrac{dy}{dt}}{\dfrac{dx}{dt}} = \dfrac{4}{4t} = \dfrac{1}{t}$

From equation (2), $\dfrac{d^2 y}{dx^2} = \dfrac{\dfrac{d}{dt}\left(\dfrac{dy}{dx}\right)}{\dfrac{dx}{dt}} = \dfrac{\dfrac{d}{dt}\left(\dfrac{1}{t}\right)}{\dfrac{dx}{dt}}$

$$= \frac{-\dfrac{1}{t^2}}{4t} = \frac{1}{4t^3}$$

$$\rho = \frac{\left[1+\left(\dfrac{dy}{dx}\right)^2\right]^{\frac{3}{2}}}{\dfrac{d^2y}{dx^2}} = \frac{\left[1+\left(\dfrac{1}{t}\right)^2\right]^{\frac{3}{2}}}{-\dfrac{1}{4t^3}}$$

When $t = 1$, $\rho = \dfrac{\sqrt{8}}{-\dfrac{1}{4}} = -4\sqrt{8} = -8\sqrt{2}$

Problem 6. Determine the radius of curvature, ρ, of the cardioid
$x = 3(2\cos\theta - \cos 2\theta), y = 3(2\sin\theta - \sin 2\theta)$ at the point $\theta = \dfrac{\pi}{6}$ radians.
Express the result correct to four decimal places.

$$\left(\rho = \left[1+\left(\frac{dy}{dx}\right)^2\right]^{\frac{3}{2}} \Big/ \frac{d^2y}{dx^2}\right)$$

$x = 3(2\cos\theta - \cos 2\theta)$, $\qquad y = 3(2\sin\theta - \sin 2\theta)$

$\dfrac{dx}{d\theta} = -6\sin\theta + 6\sin 2\theta$, $\qquad \dfrac{dy}{d\theta} = 6\cos\theta - 6\cos 2\theta$

From equation (1), $\dfrac{dy}{dx} = \dfrac{\dfrac{dy}{d\theta}}{\dfrac{dx}{d\theta}} = \dfrac{6\cos\theta - 6\cos 2\theta}{-6\sin\theta + 6\sin 2\theta}$

$$= \frac{\cos\theta - \cos 2\theta}{\sin 2\theta - \sin\theta}$$

From equation (2), $\dfrac{d^2y}{dx^2} = \dfrac{\dfrac{d}{d\theta}\left(\dfrac{dy}{dx}\right)}{\dfrac{dx}{d\theta}}$

$\dfrac{d}{d\theta}\left(\dfrac{dy}{dx}\right) = \dfrac{(\sin 2\theta - \sin\theta)(-\sin\theta + 2\sin 2\theta) - (\cos\theta - \cos 2\theta)(2\cos 2\theta - \cos\theta)}{(\sin 2\theta - \sin\theta)^2}$

$$= \frac{\left[\begin{array}{c}(-\sin 2\theta \sin\theta + 2\sin^2 2\theta + \sin^2\theta - 2\sin\theta \sin 2\theta) \\ -(2\cos\theta\cos 2\theta - \cos^2\theta - 2\cos^2 2\theta + \cos\theta\cos 2\theta)\end{array}\right]}{(\sin 2\theta - \sin\theta)^2}$$

Since $\sin^2\theta + \cos^2\theta = 1$ and $\sin^2 2\theta + \cos^2 2\theta = 1$

$$\frac{d^2y}{dx^2} = \frac{3 - 3\sin\theta\sin 2\theta - 3\cos\theta\cos 2\theta}{(\sin 2\theta - \sin\theta)^2 \, (6) \, (\sin 2\theta - \sin\theta)}$$

$$= \frac{1 - \sin\theta \sin 2\theta - \cos\theta \cos 2\theta}{2(\sin 2\theta - \sin\theta)^3}$$

Radius of curvature, $\rho = \left[1 + \left(\frac{dy}{dx}\right)^2\right]^{3/2} \Big/ \frac{d^2y}{dx^2}$

$$= \frac{\left[1 + \left(\frac{\cos\theta - \cos 2\theta}{\sin 2\theta - \sin\theta}\right)^2\right]^{3/2}}{(1 - \sin\theta \sin 2\theta - \cos\theta \cos 2\theta)/2(\sin 2\theta - \sin\theta)^3}$$

When $\theta = \frac{\pi}{6}$, $\sin\theta = \frac{1}{2}$, $\cos\theta = \frac{\sqrt{3}}{2}$, $\sin 2\theta = \frac{\sqrt{3}}{2}$ and $\cos 2\theta = \frac{1}{2}$

Hence $\rho = \dfrac{\left[1 + \left(\dfrac{\dfrac{\sqrt{3}}{2} - \dfrac{1}{2}}{\dfrac{\sqrt{3}}{2} - \dfrac{1}{2}}\right)^2\right]^{3/2} \left[2\left(\dfrac{\sqrt{3}}{2} - \dfrac{1}{2}\right)^3\right]}{\left(1 - \dfrac{1}{2} \cdot \dfrac{\sqrt{3}}{2} - \dfrac{\sqrt{3}}{2} \cdot \dfrac{1}{2}\right)}$

$$= \frac{(\sqrt{8})(0.098\,08)}{0.133\,97} = \textbf{2.070\,7, correct to four decimal places.}$$

Further problems on differentiating in parameters may be found in the following Section (3) (Problems 1 to 20).

3 Further problems

Differentiation in parameters

In Problems 1 to 5, determine $\frac{dy}{dx}$ in terms of the parameter given and hence find the value of $\frac{dy}{dx}$ at the point stated.

1. The ellipse $x = 2\cos\theta$, $y = 3\sin\theta$ at $\theta = \frac{\pi}{6}$. $\qquad \left[-\frac{3\sqrt{3}}{2}\right]$

2. The parabola $x = \frac{1}{2}t^2$, $y = t$ at $t = \ln 1.3$, correct to four significant figures. [3.811]

3. The rectangular hyperbola $x = 4t$, $y = \frac{4}{t}$ at $t = 0.25$. [−16]

4. The hyperbola $x = 2\sec\theta$, $y = 5\tan\theta$ at $\theta = 1.4$ radians, correct to four decimal places. [2.536 9]

5. The cycloid $x = 3(\theta - \cos\theta)$, $y = 3(1 - \cos\theta)$ at $\theta = 0.57$ radians, correct to four decimal places. [0.350 5]

In problems 6 to 10, given that the equation of the tangent drawn to a curve at point (x_1, y_1) is $y - y_1 = \dfrac{dy_1}{dx_1}(x - x_1)$ and that the equation of the normal drawn to a curve at point (x_1, y_1) is $y - y_1 = \dfrac{-1}{\dfrac{dy_1}{dx_1}}(x - x_1)$, determine the equation stated, expressing it in the form $y = mx + c$.

6. The tangent drawn to the ellipse $x = 2 \cos \theta$, $y = 3 \sin \theta$ at $\theta = \dfrac{\pi}{3}$.
$$\left[y = -\frac{\sqrt{3}}{2} x + 2\sqrt{3} \right]$$

7. The normal to the parabola $x = \dfrac{1}{2} t^2$, $y = t$ at $t = \dfrac{1}{4}$.
$$\left[y = -\frac{1}{4} x + \frac{33}{128} \right]$$

8. The tangent to the rectangular hyperbola $x = 4t$, $y = \dfrac{4}{t}$ at $t = 3$.
$$\left[y = -\frac{1}{9} x + 2\frac{2}{3} \right]$$

9. The normal to the hyperbola $x = 2 \sec \theta$, $y = 5 \tan \theta$ at $\theta = \dfrac{7\pi}{6}$ radians, correct to four decimal places. $\qquad [y = 0.200\,0\,x + 3.348\,6]$

10. The equations of the tangent and normal to the cycloid $x = 3(\theta - \sin\theta)$, $y = 3(1 - \cos \theta)$ at $\theta = 1.3$ radians, correct to four decimal places.
$$\left[\begin{array}{l} \text{Tangent } y = 1.315\,4\,x + 0.869\,8, \\ \text{normal } y = 0.760\,2\,x + 2.964\,8 \end{array} \right]$$

In Problems 11 to 15, find the values of $\dfrac{dy}{dx}$ and $\dfrac{d^2y}{dx^2}$ at the points stated, for the curves represented by the parametric equations given.

11. The hyperbola $x = 2.3 \sec \theta$, $y = 3.4 \tan \theta$ at $\theta = 1$ radian, correct to four decimal places. $\qquad [1.756\,8, -0.391\,3]$

12. The parabola $x = \sqrt{5}t^2$, $y = 2\sqrt{5}t$ at $t = 2.83$, correct to four significant figures. $\qquad [0.353\,4, \pm 0.009\,866]$

13. The ellipse $x = 3.7 \cos \theta$, $y = 4.4 \sin \theta$ at $\theta = 1.43$ radians, correct to four decimal places. $\qquad [-0.168\,5, -0.331\,1]$

14. The rectangular hyperbola $x = \sqrt{3}t$, $y = \dfrac{\sqrt{3}}{t}$ at $t = 0.18$, correct to four significant figures. $\qquad [-30.86, \pm 198.0]$

15. The cardioid $x = 4(2 \cos \theta - \cos 2\theta)$, $y = 4(2 \sin \theta - \sin 2\theta)$ at $\theta = 3.5$ radians, correct to four decimal places. $\qquad [-1.677\,3, 7.228\,2]$

16. If $x = \sqrt{2}(1 - \cos \theta)$ and $y = \sqrt{2} \sin \theta$, show that $\dfrac{dy}{dx} = \cot \theta$ and that
$$\frac{d^2y}{dx^2} = \frac{-\text{cosec}^3\,\theta}{\sqrt{2}}.$$

130

17. A point on a curve is given by $x = 7 \cot t + 3.5 \cos 2t$, $y = 7 \sin t - 3.5 \sin 2t$. Express $\dfrac{d^2y}{dx^2}$ in terms of t.

$$\left[\frac{1}{28} \cos^3\left(\frac{t}{2}\right) \sin \frac{3t}{2} \right]$$

18. For the curve described by $x = 2\sqrt{2}t^2$, $y = 2\sqrt{2}t^3$, find $\dfrac{d^2y}{dx^2}$ in terms of t.

$$\left[\frac{3}{8\sqrt{2}t} \quad \text{or} \quad \frac{3\sqrt{2}}{16t} \right]$$

In Problems 19 and 20, determine the radius of curvature, ρ, for the curves stated in terms of their parameters, given that

$$\rho = \frac{\left[1 + \left(\dfrac{dy}{dx}\right)^2 \right]^{\frac{3}{2}}}{\dfrac{d^2y}{dx^2}}$$

19. The astroid, $x = 5 \cos^3 t$, $y = 5 \sin^3 t$. $\qquad$ [$15 \sin t \cos t$ or $7.5 \sin 2t$]

20. The curve, $x = 3 \sin 2\theta (1 + \cos 2\theta)$, $y = 3 \cos 2\theta (1 - \cos 2\theta)$.
$$[12 \cos 3\theta]$$

Chapter 11

Logarithmic differentiation

1 The laws of logarithms applied to functions

The laws of logarithms may be used to change problems involving products to additions, quotients to subtractions and powers to products. They are stated mathematically as:

$$\log (A \cdot B) = \log A + \log B$$
$$\log \left(\frac{A}{B}\right) = \log A - \log B$$

and $\quad \log A^n = n \log A,$

where 'log' means the logarithm to any base. In calculus, logarithms to the base of 'e' are invariably used. Also, the constants A and B can be changed to functions of x, and the laws stated above are still true. Thus for two functions of x, $f(x)$ and $g(x)$, the laws may be expressed as:

$$\ln (f(x) \cdot g(x)) = \ln f(x) + \ln g(x) \tag{1}$$
$$\ln\left(\frac{f(x)}{g(x)}\right) = \ln f(x) - \ln g(x) \tag{2}$$

and $\quad \ln [f(x)]^n = n \ln f(x) \tag{3}$

If $y = (x^3 + 4) \sin 2x$, let $(x^3 + 4) = f(x)$ and $\sin 2x = g(x)$. Taking logarithms to the base of e of each side of the equation gives:

$\ln y = \ln \{(x^3 + 4) \cdot \sin 2x\}$, and by applying law (1) above:

$\ln y = \ln (x^3 + 4) + \ln (\sin 2x).$

Functions may be made up of a mixture of products and quotients.

When logarithms are applied to numbers,

$$\log\left(\frac{A \cdot B}{C}\right) = \log(A \cdot B) - \log C$$
$$= \log A + \log B - \log C$$

Applying this principle to functions of x, say $f(x)$, $g(x)$ and $F(x)$ gives:

$$y = \frac{f(x) \cdot g(x)}{F(x)}$$

Then, $\ln y = \ln f(x) + \ln g(x) - \ln F(x)$

For a more complicated expression such as

$$y = \frac{f(x) \cdot [g(x)]^n}{F(x) \cdot [G(x)]^m}$$

The three laws of logarithms may be applied as follows:

$$\ln y = \ln \left\{ \frac{f(x) \cdot [g(x)]^n}{F(x) \cdot [G(x)]^m} \right\}$$

Applying law (2), $\ln y = \ln \{f(x) \cdot [g(x)]^n\} - \ln \{F(x) \cdot [G(x)]^m\}$

Applying law (1), $\ln y = \{\ln f(x) + \ln [g(x)]^n\} - \{\ln F(x) + \ln [G(x)]^m\}$

$$= \ln f(x) + \ln [g(x)]^n - \ln F(x) - \ln [G(x)]^m$$

Applying law (3), $\ln y = \ln f(x) + n \ln g(x) - \ln F(x) - m \ln G(x)$.

These principles are used in the worked problems following, and also form the basis of logarithmic differentiation introduced in section 2.

Worked problems on the laws of logarithms applied to functions

In worked problems 1 and 2 below, by taking Naperian logarithms of each side of the equations given, convert the functions from 'product, quotient, power' form to 'addition, subtraction, product' form.

Problem 1. (a) $y = (x^2 + 3) \cos 2x$

(b) $p = \dfrac{3 \tan 2q}{4 e^{\left(\frac{q}{2} - 3\right)}}$

(c) $\alpha = 4 \sin^3 \dfrac{\theta}{2}$

(a) $y = (x^2 + 3) \cos 2x$

Taking logarithms to the base of e of each side of the equation, gives:

$\ln y = \ln [(x^2 + 3) \cdot \cos 2x]$

Applying law (1), $\mathbf{\ln y = \ln (x^2 + 3) + \ln \cos 2x}$

(b) $p = \dfrac{3 \tan 2q}{4\, e^{\left(\frac{q}{2} - 3\right)}}$

Taking Naperian logarithms of each side of the equation gives:

$$\ln p = \ln \left\{ \dfrac{3 \tan 2q}{4\, e^{\left(\frac{q}{2} - 3\right)}} \right\}$$

Applying law (2), $\ln p = \ln (3 \tan 2q) - \ln (4e^{\left(\frac{q}{2} - 3\right)})$

Applying law (1), $\ln p = \ln 3 + \ln \tan 2q - \ln 4 - \ln e^{\left(\frac{q}{2} - 3\right)}$

Applying law (3), $\ln p = \ln 3 + \ln \tan 2q - \ln 4 - \left(\dfrac{q}{2} - 3\right) \ln e$. From the definition of a logarithm to any base, a, if

$$
\begin{aligned}
y &= a^x & (1) \\
\text{then} \quad x &= \log_a y & (2)
\end{aligned}
$$

When $a = y$, giving $\log_a a$ in equation (2), then from equation (1), $y = y^x$, i.e. $x = 1$.

It follows that $\log_a a = 1$, where a is any value, thus $\ln e = \log_e e = 1$.

Hence, **$\ln p = \ln 3 + \ln \tan 2q - \ln 4 - (\frac{q}{2} - 3)$.**

(c) $\qquad \alpha = 4 \sin^3 \dfrac{\theta}{2}$

$$\ln \alpha = \ln \left(4 \sin^3 \dfrac{\theta}{2} \right)$$

$$= \ln \left[4 \left(\sin \dfrac{\theta}{2} \right)^3 \right]$$

Applying law (1), $\ln \alpha = \ln 4 + \ln \left(\sin \dfrac{\theta}{2} \right)^3$

Applying law (3), **$\ln \alpha = \ln 4 + 3 \ln \sin \dfrac{\theta}{2}$**

Problem 2. (a) $(2x^2 - 3x)^{\frac{1}{2}} \sec^3 x$

(b) $\dfrac{x^2 \sin^2 x}{e^{2x}\,(3 - 4x^{\frac{1}{2}})^5}$

(c) $\dfrac{e^{3x} \ln (x^2)}{\ln (x^3) \operatorname{cosec}^3 x}$

(a) Let $y = (2x^2 - 3x)^{\frac{1}{2}} \sec^3 x$

Taking logarithms to the base of e of each side of the equation gives:

$$\ln y = \ln \left[(2x^2 - 3x)^{\frac{1}{2}} \sec^3 x \right]$$

Applying law (1), $\ln y = \ln (2x^2 - 3x)^{\frac{1}{2}} + \ln \sec^3 x$

Applying law (3), $\ln y = \dfrac{1}{2} \ln (2x^2 - 3x) + 3 \ln \sec x$

i.e. $\ln y = \dfrac{1}{2} \ln x (2x - 3) + 3 \ln \sec x$

Applying law (1) to the first term,

$$\ln y = \frac{1}{2} \ln x + \frac{1}{2} \ln (2x - 3) + 3 \ln \sec x$$

(b) Let $y = \dfrac{x^2 \sin^2 x}{e^x (3 - 4x^{\frac{1}{2}})^5}$

Taking logarithms to the base of e of each side of the equation, gives:

$$\ln y = \ln \left[\frac{x^2 \sin^2 x}{e^x (3 - 4x^{\frac{1}{2}})^5} \right]$$

Applying law (2), $\ln y = \ln (x^2 \sin^2 x) - \ln \left[e^x (3 - 4x^{\frac{1}{2}})^5 \right]$

Applying law (1), $\ln y = (\ln x^2 + \ln \sin^2 x) - \left[\ln e^x + \ln (3 - 4x^{\frac{1}{2}})^5 \right]$

Applying law (3), $\ln y = 2 \ln x + 2 \ln \sin x - x \ln e - 5 \ln (3 - 4x^{\frac{1}{2}})$

But $\ln e = 1$, hence $\ln y = 2 (\ln x + \ln \sin x) - x - 5 \ln (3 - 4x^{\frac{1}{2}})$

(c) Let $y = \dfrac{e^{3x} \ln (x^2)}{\ln (x^3) \operatorname{cosec}^3 x}$

Taking logarithms to the base of e of each side of the equation gives:

$$\ln y = \ln \left[\frac{e^{3x} \ln (x^2)}{\ln (x^3) \operatorname{cosec}^3 x} \right]$$

Applying law (3), $\ln y = \ln \left[\dfrac{e^{3x} \cdot 2 \ln x}{3 \ln x \cdot \operatorname{cosec}^3 x} \right]$

$$= \ln \left[\frac{2 e^{3x}}{3 \operatorname{cosec}^3 x} \right]$$

Applying law (2), $\ln y = \ln 2 + \ln e^{3x} - \ln 3 - \ln \operatorname{cosec}^3 x$

Applying law (3), $\ln y = \ln 2 + 3x \ln e - \ln 3 - 3 \ln \operatorname{cosec} x$

But $\ln e = 1$,

Hence, $\qquad \ln y = \ln 2 + 3x - \ln 3 - 3 \ln \operatorname{cosec} x$

Applying law (2), $\ln y = \ln \dfrac{2}{3} + 3x - 3 \ln \operatorname{cosec} x$

Further problems on the laws of logarithms applied to functions may be found in Section 3 (Problems 1 to 5), page 138.

2 Logarithmic differentiation

To differentiate a function which contains products and/or quotients and/or powers, it is often easier to firstly take logarithms to a base of e and to express the function in a different form. For example, a simple function such as, say, $y = x^2 \sin x$ may be differentiated using the product rule of differentiation, giving:

$$\frac{dy}{dx} = x^2 \cos x + 2x \sin x$$

An alternative method of differentiating this function is by logarithmic differentiation as follows:

Taking Naperian logarithms of each side of the equation gives:

$\ln y = \ln (x^2 \sin x)$
i.e. $\ln y = 2 \ln x + \ln \sin x$

from the laws of logarithms introduced in Section 1. In Chapter 9, dealing with implicit functions, it is shown that

$$\frac{d}{dx} (f(y)) = \frac{d}{dy} (f(y)) \times \frac{dy}{dx}$$

Thus the left-hand side, i.e. $\ln y$, becomes $\frac{1}{y} \frac{dy}{dx}$ when differentiated with respect to x. With logarithmic differentiation, the right-hand side of the equation is of the form $\ln f(x) \pm \ln g(x) \pm \ldots$. To differentiate, say, $\ln f(x)$ with respect to x, a substitution method is used.

Thus, let $u = f(x)$, then $\frac{du}{dx} = f'(x)$

Hence, $\frac{d}{dx} (\ln f(x)) = \frac{d}{du} (\ln u) \times \frac{du}{dx}$

$$= \frac{1}{u} \frac{du}{dx}$$

i.e. $\frac{d}{dx} (\ln f(x)) = \frac{f'(x)}{f(x)}$ \hfill (4)

Applying equation (4) to the right-hand side, i.e. $2 \ln x + \ln \sin x$, gives:

$\frac{2}{x} + \frac{\cos x}{\sin x}$ when it is differentiated with respect to x. Thus, applying logarithmic differentiation to the equation $y = x^2 \sin x$, gives:

$$\frac{1}{y}\frac{dy}{dx} = \frac{2}{x} + \frac{\cos x}{\sin x}$$

i.e. $\frac{dy}{dx} = y\left(\frac{2}{x} + \cos x\right)$

But $y = x^2 \sin x$, hence

$$\frac{dy}{dx} = x^2 \sin x \left(\frac{2}{x} + \frac{\cos x}{\sin x}\right)$$

$$= 2x \sin x + x^2 \cos x,$$

as obtained by applying the product rule of differentiation.

In this case, using a comparatively simple function, the process of logarithmic differentiation is longer than that of applying the product rule of differentiation. However, for more complicated functions, logarithmic differentiation is usually the simplest method of differentiating the function.

Procedure for logarithmic differentiation

The procedure for differentiating a function of the form, say,
$y = \frac{f(x)\,[g(x)]^n}{h(x)}$ is shown below.

(i) Take Naperian logarithms of each side of the equation, giving

$$\ln y = \left\{\frac{f(x)\,[g(x)]^n}{h(x)}\right\}$$

(ii) Apply the laws of logarithms to change products to addition, etc. This gives:

$$\ln y = \ln f(x) + n \ln g(x) - \ln h(x)$$

(iii) Differentiate each side with respect to the variable, giving:

$$\frac{1}{y}\frac{dy}{dx} = \frac{f'(x)}{f(x)} + \frac{n g'(x)}{g(x)} - \frac{h'(x)}{h(x)}$$

(iv) Multiply throughout by y, giving

$$\frac{dy}{dx} = y\left\{\frac{f'(x)}{f(x)} + \frac{n g'(x)}{g(x)} - \frac{h'(x)}{h(x)}\right\}$$

(v) Substitute for y in terms of x, giving:

$$\frac{dy}{dx} = \frac{f(x)\,[g(x)]^n}{h(x)}\left\{\frac{f'(x)}{f(x)} + \frac{n g'(x)}{g(x)} - \frac{h'(x)}{h(x)}\right\}$$

(vi) Simplify the expression obtained in (v) where possible.

This procedure is used to determine the differential coefficients in the worked problems following.

Worked problems on logarithmic differentiation

Problem 1. Use logarithmic differentiation to differentiate

$$y = \frac{3\,(4-x^2)}{\tan x}$$

With reference to the procedure for logarithmic differentiation given above:

(i) $\quad \ln y \quad = \quad \ln \left\{ \dfrac{3\,(4-x^2)}{\tan x} \right\}$

(ii) $\quad \ln y \quad = \quad \ln 3 + \ln (4-x^2) - \ln \tan x$

(iii) $\quad \dfrac{1}{y}\dfrac{dy}{dx} \quad = \quad 0 + \dfrac{(-2x)}{4-x^2} - \dfrac{\sec^2 x}{\tan x}$

(iv) $\quad \dfrac{dy}{dx} \quad = \quad y \left\{ \dfrac{-2x}{4-x^2} - \dfrac{\sec^2 x}{\tan x} \right\}$

(v) $\quad \dfrac{dy}{dx} \quad = \quad \dfrac{3\,(4-x^2)}{\tan x} \left\{ -\dfrac{2x}{4-x^2} - \dfrac{\sec^2 x}{\tan x} \right\}$

(vi) $\quad \dfrac{dy}{dx} \quad = -\dfrac{6x}{\tan x} - \dfrac{3\,(4-x^2)\sec^2 x}{\tan^2 x}$

$$\qquad\qquad = \; -\frac{3}{\tan x} \left\{ 2x + \frac{(4-x^2)\sec^2 x}{\tan x} \right\}$$

$$\qquad\qquad = \; -3 \cot x \left\{ 2x + \frac{4-x^2}{\sin x \cos x} \right\}$$

Problem 2. Use logarithmic differentiation to differentiate

(a) $\quad y \;=\; \dfrac{4e^{-2x}\sec x}{\left(x^2 + \dfrac{1}{2}\right)^{3/2}}$

(b) $\quad p \;=\; \dfrac{q^3 \ln 2q}{(2-q)^{1/3}\cosec 2q}$

(a) $\quad y \;=\; \dfrac{4e^{\,2x}\sec x}{\left(x^2 + \dfrac{1}{2}\right)^{3/2}}$

Applying the laws of logarithms:

$$\ln y = \ln 4 + (-2x)\ln e + \ln \sec x - \frac{3}{2}\ln\left(x^2 + \frac{1}{2}\right)$$

Differentiating each term:

$$\frac{1}{y}\frac{dy}{dx} = -2 + \frac{\sec x \tan x}{\sec x} - \frac{3\,(2x)}{2\left(x^2 + \frac{1}{2}\right)}$$

$$\frac{dy}{dx} = \frac{4e^{-2x}\sec x}{\left(x^2 + \frac{1}{2}\right)^{3/2}} \left[-2 + \tan x - \frac{3x}{\left(x^2 + \frac{1}{2}\right)} \right]$$

(b) $\quad p = \dfrac{q^3 \ln 2q}{(2-q)^{\frac{1}{3}} \operatorname{cosec} 2q}$

$$\ln\; p = 3 \ln q + \ln\,(\ln 2q) - \frac{1}{3}\ln\,(2-q) - \ln \operatorname{cosec} 2q$$

$$\frac{1}{p}\frac{dp}{dq} = \frac{3}{q} + \frac{\frac{2}{2q}}{\ln 2q} - \frac{(-1)}{3\,(2-q)} - \frac{-2\operatorname{cosec} 2q \cot 2q}{\operatorname{cosec} 2q}$$

$$\frac{dp}{dq} = \frac{q^3 \ln 2q}{(2-q)^{\frac{1}{3}} \operatorname{cosec} 2q} \left\{ \frac{3}{q} + \frac{1}{q \ln 2q} + \frac{1}{3\,(2-q)} + 2 \cot 2q \right.$$

Further problems on logarithmic differentiation may be found in the following Section (3) (Problems 6 to 17).

3 Further Problems

Laws of logarithms applied to functions

In Problems 1 to 5, by taking Napierian logarithms of each side of the equation, convert the functions from 'product, quotient, power' form to 'addition, subtraction, product' form.

1. $\;y = (3x-4)\tan 2x \qquad [\ln y = \ln\,(3x-4) + \ln\,(\tan 2x)]$

2. $\;p = \dfrac{3 \cot 2q}{e^{3-q}}. \qquad [\ln p = \ln 3 + \ln \cot\, 2q - (3-q)]$

3. $\;m = \dfrac{n^3 e^{2n}}{\operatorname{cosec} \dfrac{n}{2}}. \qquad [\ln m = 3 \ln n + 2n - \ln \operatorname{cosec} \dfrac{n}{2}]$

4. $\;y = \dfrac{2(1+x^{3/2})^{1/2} \cos\,(x-2)}{3\,e^{(3x+4)} \ln 2x}.$

$[\ln y = \ln 2 + \dfrac{1}{2}\ln\,(1+x^{3/2}) + \ln \cos\,(x-2) - \ln 3 - (3x+4) - \ln\,(\ln 2x)]$

5. $\;\alpha = \dfrac{3\theta^{3/2}\sin^3\theta}{(2-\theta^{1/2})^{1/3}\sec^3\theta}.$

$[\ln \alpha = \ln 3 + \dfrac{3}{2}\ln \theta + 3 \ln \sin \theta - \dfrac{1}{3}\ln\,(2-\theta^{1/2}) - 3 \ln \sec \theta]$

Logarithmic Differentiation

In Problems 6 to 11, use logarithmic differentiation to differentiate the functions given.

6. $y = 3\sqrt{x} \sin 2x.$ $[\frac{dy}{dx} = 3\sqrt{x} \ (\frac{\sin 2x}{2x} + 2\cos 2x \ )\ ]$

7. $l = 4\, e^{(2-3m)} \operatorname{cosec} \frac{m}{2}.$

$$[\frac{dl}{dm} = -4e^{(2-3m)} \operatorname{cosec} \frac{m}{2}\,(3 + \frac{1}{2}\cot\frac{m}{2})]$$

8. $p = \dfrac{2(1-q^2)}{3 \sec(\pi-q)}.$

$$\left[\frac{dp}{dq} = \frac{2}{3\sec(\pi-q)} \left\{ (1-q^2)\tan(\pi-q) - 2q \right\} \right]$$

9. $r = \dfrac{2 \ln (4-\frac{s}{3})}{3 \cot (s^2)}.$ $\left[\dfrac{dr}{ds} = \dfrac{2}{3\cot(s^2)} \left\{ -\dfrac{1}{12-s} + \dfrac{2s\ln(4-\frac{s}{3})}{\sin s^2 \cos s^2} \right\} \right]$

10. $u = 1.8\sqrt{\sin^3 v}.$ $[\frac{du}{dv} = 2.7\sqrt{(\sin v)}\cos v \ ]$

11. $\alpha = \dfrac{4}{9}\left[3 - (2\theta)^{1/3} \right]^{1/2}$

$$\left[\frac{d\alpha}{d\theta} = -\frac{4}{27}\left\{ \frac{1}{(2\theta)^{2/3}\,(3-(2\theta)^{1/3})^{1/2}} \right\} \right]$$

12. If $\dfrac{y}{e^{2x}} = \sin^3(4-2x),$ find $\dfrac{dy}{dx}$

$$\left[\frac{dy}{dx} = 2e^{2x}\sin^3(4-2x)\left\{ 1 - 3\cot(4-2x) \right\} \right]$$

13. If $\dfrac{r}{\sec^2\frac{x}{2}} = \ln 3x \tan 3x,$ show that

$$\frac{dr}{dx} = \ln 3x \cdot \sec^2\frac{x}{2}\,\tan 3x \left\{ \frac{1}{x\ln 3x} + \tan\frac{x}{2} + \frac{3\sec^2 3x}{\tan 3x} \right\}$$

14. If $\dfrac{p}{4(3-\frac{q}{4})^{1/4}} = e^{(2q-4)}\sqrt{(\cos(3-2q))},$ find $\dfrac{dp}{dq}$

$$\left[\frac{dp}{dq} = 4(3-\frac{q}{4})^{1/4}\,e^{(2q-4)}\sqrt{(\cos(3-2q))}\left\{ 2 + \tan(3-2q) - \frac{1}{4(12-q)} \right\} \right]$$

15. If $y\left(\ln\frac{x}{4}\right)^2 = (x\sin x)^3,$ show that

$$\frac{dy}{dx} = \frac{(x \sin x)^3}{(\ln \frac{x}{4})^2} \left\{ \frac{3}{x} + 3 \cot x - \frac{2}{x \ln \frac{x}{4}} \right\}.$$

16. If $u(v-1)^{1/2} \cot^3 v = v^3 \sec^2 v$, find $\dfrac{du}{dv}$.

$$\left[\frac{du}{dv} = \frac{v^3 \sec^2 v}{(v-1)^{1/2} \cot^3 v} \left\{ \frac{3}{v} + 2 \tan v - \frac{1}{2(v-1)} + \frac{3}{\sin v \cos v} \right\} \right]$$

17. If $pe^{q/3} \sqrt{(\frac{q}{2})} = 16 \cot^2 q$, show that

$$\frac{dp}{dq} = - \frac{16 \cot^2 q}{\sqrt{(\frac{q}{2})} e^{q/3}} \left\{ \frac{2}{\sin q \cos q} + \frac{1}{2q} + \frac{1}{3} \right\}$$

Chapter 12

Differentiation of inverse trigonometric and inverse hyperbolic functions

1 Inverse functions

If y is a function of x, i.e. $y = f(x)$, it is often possible to transpose the equation to find x in terms of y.

For example, if $y = 2x - 1$ then $x = \dfrac{y + 1}{2}$

The two functions, i.e. $y = 2x - 1$ and $x = \dfrac{y + 1}{2}$, are called **inverse functions.**

Further examples of inverse functions are:

(i) if $y = x^2$ then $x = \sqrt{y}$,
(ii) if $y = 10^x$ then $x = \log_{10} y$,
(iii) if $y = \sin x$ then x is the inverse sine of y, which is written as $x = \arcsin y$,
(iv) if $y = \cosh x$ then x is the inverse hyperbolic cosine of y, which is written as $x = \operatorname{arcosh} y$.

The inverse circular functions are denoted by prefixing the function with 'arc'. For example, $\arcsin x$, $\operatorname{arcsec} x$, etc. The former notation $\sin^{-1} x$, $\tan^{-1} x$, etc., is discouraged because of possible confusion with $(\sin x)^{-1}$, $(\tan x)^{-1}$, etc.

The inverse hyperbolic functions are denoted by prefixing the function with 'ar'. For example, $\operatorname{arsinh} x$, $\operatorname{arcosech} x$, etc. The former notation $\sinh^{-1} x$, $\coth^{-1} x$, etc., is discouraged because of possible confusion with $(\sinh x)^{-1}$, $(\coth x)^{-1}$, etc.

2 Differentiation of inverse trigonometric functions

(a) $y = \arcsin x$

If $y = \arcsin x$ then $x = \sin y$.

Differentiating x with respect to y gives $\dfrac{dx}{dy} = \cos y$.

Now $\cos^2 y + \sin^2 y = 1$. Hence $\cos y = \sqrt{(1 - \sin^2 y)} = \sqrt{(1 - x^2)}$

Hence $\dfrac{dx}{dy} = \sqrt{(1 - x^2)}$

It may be shown that $\dfrac{dy}{dx} = \dfrac{1}{dx/dy}$ (see Chapter 10, page 122).

Thus when $y = \arcsin x, \dfrac{dy}{dx} = \dfrac{1}{\sqrt{(1 - x^2)}}$ and there are two possible values, one positive and one negative.

A sketch of part of the graph $y = \arcsin x$ is shown in Fig. 1(a) and it is seen that there are an infinite number of values for y for a given value of x. When y has more than one value for a given value of x, such a is the relationship $y = \arcsin x$, the numerically least of these values is called the **principal value** of y. If there are two numerically equal least values, the positive one is called the principal value.

The principal value of $\arcsin x$ is defined as the value between $-\dfrac{\pi}{2}$ and $+\dfrac{\pi}{2}$, shown between points P and Q in Fig. 1(a). This range of values covers every possible value of x that can occur and when adopting principal values, if given a particular value of x, only one value of y is possible. The gradient of the curve between P and Q is positive for all values of x, and hence if $\arcsin x$ is understood to mean the principal value of $\arcsin x$ then

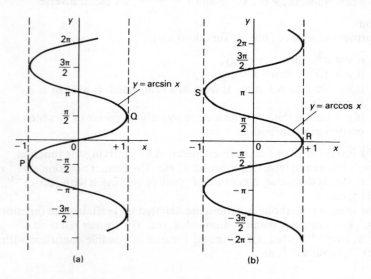

Figure 1

the differential coefficient will be positive and only the positive value is taken when evaluating $\sqrt{(1-x^2)}$.

Hence if $y = \arcsin x$ then $\dfrac{dy}{dx} = \dfrac{1}{\sqrt{(1-x^2)}}$

If $y = \arcsin \dfrac{x}{a}$ then $\dfrac{x}{a} = \sin y$

i.e. $\quad x = a \sin y$

$$\frac{dx}{dy} = a \cos y = a\sqrt{(1 - \sin^2 y)} = a\sqrt{\left[1 - \left(\frac{x}{a}\right)^2\right]}$$

i.e. $\dfrac{dx}{dy} = a\sqrt{\left(\dfrac{a^2 - x^2}{a^2}\right)}$

$$= a\,\frac{\sqrt{(a^2 - x^2)}}{a} = \sqrt{(a^2 - x^2)}$$

Hence $\dfrac{dy}{dx} = \dfrac{1}{dx/dy} = \dfrac{1}{\sqrt{(a^2 - x^2)}}$

This latter form is important when integrating functions of the form

$\dfrac{1}{\sqrt{(a^2 - x^2)}}$, i.e. $\displaystyle\int \dfrac{1}{\sqrt{(a^2 - x^2)}}\ dx = \arcsin \dfrac{x}{a} + c.$

(This is discussed in Chapter 15, page 194.)

The most general form of the differential coefficient of $\arcsin f(x)$ is obtained by using the 'chain rule' of differentiation for 'functions of a function'. That is, given $y = \arcsin f(x)$, let $u = f(x)$

then $\quad \dfrac{du}{dx} = f'(x)$

$$\frac{dy}{du} = \frac{1}{\sqrt{(1-u^2)}}$$

Thus $\dfrac{dy}{dx} = \dfrac{dy}{du}\cdot\dfrac{du}{dx} = \dfrac{1}{\sqrt{(1-u^2)}}\,f'(x)$,

i.e. $\dfrac{dy}{dx} = \dfrac{1}{\sqrt{\{1 - [f(x)]^2\}}}\cdot f'(x)$

(b) $y = \arccos x$

If $y = \arccos x$ then $x = \cos y$.

Differentiating x with respect to y gives:

$\dfrac{dx}{dy} = -\sin y = -\sqrt{(1 - \cos^2 y)}$

$$= -\sqrt{(1 - x^2)}$$

Hence $\quad \dfrac{dy}{dx} = \dfrac{1}{dx/dy} - -\dfrac{1}{\sqrt{(1 - x^2)}}$

A sketch of part of the graph $y = \arccos x$ is shown in Fig. 1(b) and $\arccos x$ is defined as that angle which lies between 0 and π shown between points R and S, i.e. the principal value of $y = \arccos x$ is between 0 and π.

(If the range $-\dfrac{\pi}{2}$ to $+\dfrac{\pi}{2}$ were used, as in the case of arcsin x, then the cosine of an angle would always be positive. Thus $\arccos\left(-\dfrac{1}{3}\right)$, for example, would have no meaning.) The gradient of the curve between R and S is negative for all values of x and hence the differential coefficient is negative as shown above.

It may be shown that if $y = \arccos\dfrac{x}{a}$ then $\dfrac{dy}{dx} = -\dfrac{1}{\sqrt{(a^2 - x^2)}}$

The most general form of the differential coefficient of arccos $f(x)$ is:

$$\dfrac{d}{dx}\ [\arccos f(x)] = -\dfrac{1}{\sqrt{\{1 - [f(x)]^2\}}}\ [f'(x)]$$

(c) $y = \arctan x$

A sketch of part of the graph of $y = \arctan x$ is shown in Fig. 2(a). By definition the principal value of $y = \arctan x$ is defined as that angle which lies between $-\dfrac{\pi}{2}$ and $+\dfrac{\pi}{2}$. Hence the gradient (i.e. $\dfrac{dy}{dx}$) is always positive.

If $y = \arctan x$, then $x = \tan y$

$$\frac{dx}{dy} = \sec^2 y$$

Now $1 + \tan^2 y = \sec^2 y$

Hence $\dfrac{dx}{dy} = 1 + \tan^2 y = 1 + x^2$

$$\frac{dy}{dx} = \frac{1}{dx/dy} = \frac{1}{1 + x^2}$$

If $y = \arctan\dfrac{x}{a}$, then $\dfrac{x}{a} = \tan y$, i.e. $x = a\tan y$

$$\frac{dx}{dy} = a\sec^2 y = a(1 + \tan^2 y) = a\left[1 + \left(\frac{x}{a}\right)^2\right]$$

$$= a\left(\frac{a^2 + x^2}{a^2}\right) = \frac{a^2 + x^2}{a}$$

Hence $\dfrac{dy}{dx} = \dfrac{a}{a^2 + x^2}$

This latter form is important when integrating functions of the form $\dfrac{a}{a^2 + x^2}$,

i.e. $\displaystyle\int \frac{a}{a^2 + x^2}\ dx = \arctan\frac{x}{a} + c$ or $\displaystyle\int \frac{1}{a^2 + x^2}\ dx = \frac{1}{a}\arctan\frac{x}{a} + c.$

(This is discussed in Chapter 15, page 196.)

The most general form of the differential coefficient of arctan $f(x)$ is:

$$\frac{d}{dx}\ [\arctan f(x)] = \frac{1}{\{1 + [f(x)]^2\}}\ [f'(x)]$$

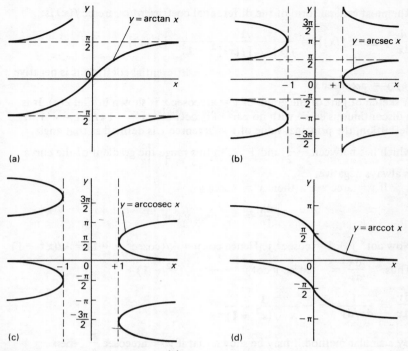

Figure 2 (*a*) $y = \arctan x$ (*b*) $y = \text{arcsec } x$
(*c*) $y = \text{arccosec } x$ (*d*) $y = \text{arccot } x$

(d) $y = \text{arcsec } x$

A sketch of part of the graph of $y = \text{arcsec } x$ is shown in Fig. 2(b). It is a
discontinuous curve with no part of it between $x = -1$ and $x = +1$. By
definition, the principal value of $y = \text{arcsec } x$ is defined as that angle which
lies between 0 and π. In this range the gradient of the curve is always positive.

If $y = \text{arcsec } x$ then $x = \sec y$

$$\frac{dx}{dy} = \sec y \tan y$$

Now $1 + \tan^2 y = \sec^2 y$. Hence $\tan y = \sqrt{(\sec^2 y - 1)} = \sqrt{(x^2 - 1)}$

Thus $\dfrac{dx}{dy} = \sec y \tan y = x \sqrt{(x^2 - 1)}$

$$\frac{dy}{dx} = \frac{1}{dx/dy} = \frac{1}{x \sqrt{(x^2 - 1)}}$$

By a similar method it may be shown that if $y = \text{arcsec } \dfrac{x}{a}$, then

$$\frac{dy}{dx} = \frac{a}{x \sqrt{(x^2 - a^2)}}.$$

The most general form of the differential coefficient of arcsec $f(x)$ is:

$$\frac{d}{dx} [\ \text{arcsec } f(x)] = \frac{1}{f(x)\sqrt{\{[f(x)]^2 - 1\}}} [f'(x)]$$

(e) $y = \text{arccosec } x$

A sketch of part of the graph of $y = \text{arccosec } x$ is shown in Fig. 2(c). It is a discontinuous curve with no part of it between $x = -1$ and $x = +1$. By definition, the principal value of $y = \text{arccosec } x$ is defined as that angle which lies between $-\frac{\pi}{2}$ and $+\frac{\pi}{2}$. In this range the gradient of the curve is always negative.

If $y = \text{arccosec } x$ then $x = \text{cosec } y$

$$\frac{dx}{dy} = -\text{cosec } y \cot y.$$

Now $\cot^2 y + 1 = \text{cosec}^2 y$. Hence $\cot y = \sqrt{(\text{cosec}^2 y - 1)} = \sqrt{(x^2 - 1)}$

Thus $\quad \dfrac{dx}{dy} = -\text{cosec } y \cot y = -x\sqrt{(x^2 - 1)}$

$$\frac{dy}{dx} = \frac{1}{dx/dy} = -\frac{1}{x\sqrt{(x^2 - 1)}}$$

By a similar method it may be shown that if $y = \text{arccosec } \dfrac{x}{a}$, then

$$\frac{dy}{dx} = \frac{-a}{x\sqrt{(x^2 - a^2)}} .$$

The most general form of the differential coefficient of arccosec $f(x)$ is:

$$\frac{d}{dx} [\ \text{arccosec } f(x)] = -\frac{1}{f(x)\sqrt{\{[f(x)]^2 - 1\}}} [f'(x)]$$

(f) $y = \text{arccot } x$

A sketch of part of the graph of $y = \text{arccot } x$ is shown in Fig. 2(d). By definition, the principal value of $y = \text{arccot } x$ is defined as that angle which lies between $-\frac{\pi}{2}$ and $+\frac{\pi}{2}$.

In this range the gradient of the curve is always negative.

If $y = \text{arccot } x$ then $x = \cot y$

$$\frac{dx}{dy} = -\text{cosec}^2 y$$

Now $\cot^2 y + 1 = \text{cosec}^2 y$. Hence $\dfrac{dx}{dy} = -(\cot^2 y + 1) = -(x^2 + 1)$.

Thus $\quad \dfrac{dy}{dx} = \dfrac{1}{dx/dy} = -\dfrac{1}{(1 + x^2)}$

By a similar method it may be shown that if $y = \text{arccot}\,\dfrac{x}{a}$, then

$$\frac{dy}{dx} = \frac{-a}{(a^2 + x^2)}$$

The most general form of the differential coefficient of $\text{arccot}\,f(x)$ is:

$$\frac{d}{dx}\,[\text{arccot}\,f(x)] = -\frac{1}{\{1 + [f(x)]^2\}}\,[f'(x)]$$

Summary of differential coefficients of inverse trigonometrical functions

y or $f(x)$	$\dfrac{dy}{dx}$ or $f'(x)$
(a) $\arcsin \dfrac{x}{a}$	$\dfrac{1}{\sqrt{(a^2 - x^2)}}$
$\arcsin f(x)$	$\dfrac{1}{\sqrt{\{1 - [f(x)]^2\}}}[f'(x)]$
(b) $\arccos \dfrac{x}{a}$	$\dfrac{-1}{\sqrt{(a^2 - x^2)}}$
$\arccos f(x)$	$\dfrac{-1}{\sqrt{\{1 - [f(x)]^2\}}}[f'(x)]$
(c) $\arctan \dfrac{x}{a}$	$\dfrac{a}{a^2 + x^2}$
$\arctan f(x)$	$\dfrac{1}{\{1 + [f(x)]^2\}}\,[f'(x)]$
(d) $\text{arcsec} \dfrac{x}{a}$	$\dfrac{a}{x\sqrt{(x^2 - a^2)}}$
$\text{arcsec}\,f(x)$	$\dfrac{1}{f(x)\sqrt{\{[f(x)]^2 - 1\}}}\,[f'(x)]$
(e) $\text{arccosec} \dfrac{x}{a}$	$\dfrac{-a}{x\sqrt{(x^2 - a^2)}}$
$\text{arccosec}\,f(x)$	$\dfrac{-1}{f(x)\sqrt{\{[f(x)]^2 - 1\}}}\,[f'(x)]$
(f) $\text{arccot} \dfrac{x}{a}$	$\dfrac{-a}{a^2 + x^2}$
$\text{arccot}\,f(x)$	$\dfrac{-1}{\{1 + [f(x)]^2\}}\,[f'(x)]$

Worked problems on differentiating inverse trigonometric functions

Problem 1. Find the differential coefficient of $y = \arcsin 3x^2$

$$\frac{dy}{dx} = \frac{1}{\sqrt{\{1-[f(x)]^2\}}} \, [f'(x)] = \frac{1}{\sqrt{\{1-[3x^2]^2\}}} \, [6x] = \frac{6x}{\sqrt{(1-9x^4)}}$$

Problem 2. Differentiate $y = \ln(\arccos 2t)$.

Let $\quad u = \arccos 2t$.
then $\quad y = \ln u$

$$\frac{dy}{dt} = \frac{dy}{du} \cdot \frac{du}{dt} = \frac{1}{u} \cdot \frac{d}{dt}(\arccos 2t)$$

$$= \left(\frac{1}{\arccos 2t}\right)\left(-\frac{2}{\sqrt{\{1-(2t)^2\}}}\right)$$

Hence $\dfrac{d}{dt}(\ln \arccos 2t) = \dfrac{-2}{\sqrt{(1-4t^2)}\,\arccos 2t}$

Problem 3. Find $\dfrac{d}{d\theta}\left(\arctan\dfrac{2}{\theta^2}\right)$.

$$\frac{d}{d\theta}\left(\arctan\frac{2}{\theta^2}\right) = \left(\frac{1}{1+\left(\frac{2}{\theta^2}\right)^2}\right)\left(\frac{-4}{\theta^3}\right) = \left(\frac{1}{\frac{\theta^4+4}{\theta^4}}\right)\left(\frac{-4}{\theta^3}\right) = \frac{-4\theta}{\theta^4+4}$$

Problem 4. Find the differential coefficient of $f(x) = x \operatorname{arcsec} x$.

$$f'(x) = (x)\left(\frac{1}{x\sqrt{(x^2-1)}}\right) + (\operatorname{arcsec} x)(1), \text{ by the product rule of differentiation}$$

$$= \frac{1}{\sqrt{(x^2-1)}} + \operatorname{arcsec} x.$$

Problem 5. Differentiate $y = \dfrac{\operatorname{arccot} x}{(1+x^2)}$

$$\frac{dy}{dx} = \frac{(1+x^2)\left(\dfrac{-1}{1+x^2}\right) - (\operatorname{arccot} x)(2x)}{(1+x^2)^2}, \text{ by the quotient rule of differentiation}$$

$$= \frac{-(1+2x\operatorname{arccot} x)}{(1+x^2)^2}$$

Problem 6. Show that if $y = \operatorname{arccot}\left(\dfrac{\cos\theta}{1-\sin\theta}\right)$ then $\dfrac{dy}{d\theta} = -\dfrac{1}{2}$

$$\frac{dy}{d\theta} = \frac{-1}{\{1 + [f(x)]^2\}}\ [f'(x)]$$

$$= -\frac{1}{\left\{1 + \dfrac{\cos\theta}{1 - \sin\theta}\right\}^2}\left\{\frac{(1 - \sin\theta)(-\sin\theta) - (\cos\theta)(-\cos\theta)}{[(1 - \sin\theta)^2}\right\}$$

$$= \left\{\frac{-1}{\dfrac{(1 - \sin\theta)^2 + \cos^2\theta}{(1 - \sin\theta)^2}}\right\}\left\{\frac{-\sin\theta + \sin^2\theta + \cos^2\theta}{(1 - \sin\theta)^2}\right\}$$

$$= \left\{\frac{-(1 - \sin\theta)^2}{(1 - \sin\theta)^2 + \cos^2\theta}\right\}\left\{\frac{1 - \sin\theta}{(1 - \sin\theta)^2}\right\}$$

$$= \left\{\frac{-1}{1 - 2\sin\theta + \sin^2\theta + \cos^2\theta}\right\}\ (1 - \sin\theta)$$

$$= \frac{-1}{(2 - 2\sin\theta)}\ (1 - \sin\theta) = \frac{-(1 - \sin\theta)}{2(1 - \sin\theta)} = -\frac{1}{2}$$

Further problems on differentiating inverse trigonometric functions may be found in Section 5, (Problems 1 to 16), page 160.

3 Logarithmic forms of the inverse hyperbolic functions

(i) arsinh $\dfrac{x}{a}$

If $y = \text{arsinh } \dfrac{x}{a}$ then $\dfrac{x}{a} = \sinh y$.

From Chapter 7, $e^y = \cosh y + \sinh y$. Also $\cosh^2 y - \sinh^2 y = 1$, from which $\cosh y = \sqrt{(1 + \sinh^2 y)}$, which is positive since $\cosh y$ is always positive (see Fig. 1, Chapter 7).

Hence $e^y = \sqrt{(1 + \sinh^2 y)} + \sinh y$

i.e. $\qquad e^y = \sqrt{\left[1 + \left(\dfrac{x}{a}\right)^2\right]} + \dfrac{x}{a} = \dfrac{\sqrt{(a^2 + x^2)}}{a} + \dfrac{x}{a}$

Taking Napierian logarithms of both sides gives:

$$y = \ln\left\{\frac{x + \sqrt{(a^2 + x^2)}}{a}\right\}$$

Hence arsinh $\dfrac{x}{a} = \ln\left\{\dfrac{x + \sqrt{(a^2 + x^2)}}{a}\right\}$

(ii) arcosh $\dfrac{x}{a}$

If $y = \text{arcosh } \dfrac{x}{a}$ then $\dfrac{x}{a} = \cosh y$.

$$e^y = \cosh y + \sinh y$$
$$= \cosh y \pm \sqrt{(\cosh^2 y - 1)}, \text{ since } \sinh y \text{ may be positive or negative}$$

(see Fig. 2, Chapter 7).

i.e. $e^y = \dfrac{x}{a} \pm \sqrt{\left[\left(\dfrac{x}{a}\right)^2 - 1\right]} = \dfrac{x}{a} \pm \dfrac{\sqrt{(x^2 - a^2)}}{a}$

Taking Napierian logarithms of both sides gives:

$$y = \ln\left\{\dfrac{x \pm \sqrt{(x^2 - a^2)}}{a}\right\}$$

The two values obtained are $y = \ln\left\{\dfrac{x + \sqrt{(x^2 - a^2)}}{a}\right\}$ and

$$y = \ln\left\{\dfrac{x - \sqrt{(x^2 - a^2)}}{a}\right\}$$

Adding these two values of y gives:

$$\ln\left\{\dfrac{x + \sqrt{(x^2 - a^2)}}{a}\right\} + \ln\left\{\dfrac{x - \sqrt{(x^2 - a^2)}}{a}\right\}$$

$$= \ln\left\{\dfrac{x + \sqrt{(x^2 - a^2)}}{a}\right\}\left\{\dfrac{x - \sqrt{(x^2 - a^2)}}{a}\right\} \text{ from the laws of logarithms}$$

$$= \ln\left\{\dfrac{x^2 - (x^2 - a^2)}{a^2}\right\} = \ln 1 = 0.$$

Hence the two values of y are equal but opposite in sign.

Assuming the principal value, $\mathbf{arccosh}\ \dfrac{x}{a} = \ln\left\{\dfrac{x + \sqrt{(x^2 - a^2)}}{a}\right\}$

(iii) artanh $\dfrac{x}{a}$

If $y = \text{artanh}\ \dfrac{x}{a}$ then $\dfrac{x}{a} = \tanh y$

Now $\tanh y = \dfrac{\sinh y}{\cosh y} = \dfrac{\frac{1}{2}(e^y - e^{-y})}{\frac{1}{2}(e^y + e^{-y})} = \dfrac{e^{2y} - 1}{e^{2y} + 1}$

Then $\dfrac{x}{a} = \dfrac{e^{2y} - 1}{e^{2y} + 1}$ and $x(e^{2y} + 1) = a(e^{2y} - 1)$

Hence $x + a = ae^{2y} - xe^{2y}$

$$e^{2y} = \dfrac{a + x}{a - x}$$

Taking Napierian logarithms of both sides gives:

$$2y = \ln\left(\dfrac{a + x}{a - x}\right)$$

$$y = \tfrac{1}{2}\ln\left(\dfrac{a + x}{a - x}\right)$$

Hence artanh $\dfrac{x}{a} = \tfrac{1}{2}\ln\left(\dfrac{a + x}{a - x}\right)$

Worked problem on evaluating inverse hyperbolic functions

Problem 1. Evaluate, correct to four decimal places:

(a) arsinh $\frac{4}{3}$ (b) arcosh 3   (c) artanh 0.3

(a) In logarithmic form, arsinh $\frac{x}{a} = \ln\left\{\frac{x + \sqrt{(a^2 + x^2)}}{a}\right\}$

If $x = 4$ and $a = 3$ then arsinh $\frac{4}{3} = \ln\left\{\frac{4 + \sqrt{(3^2 + 4^2)}}{3}\right\}$

$$= \ln\left(\frac{4 + 5}{3}\right) = \ln 3 \text{ or } \mathbf{1.098\ 6}$$

(b) In logarithmic form, arcosh $\frac{x}{a} = \ln\left\{\frac{x + \sqrt{(x^2 - a^2)}}{a}\right\}$

If $x = 3$ and $a = 1$ then arcosh $\frac{3}{1} = \ln\left\{\frac{3 + \sqrt{(3^2 - 1^2)}}{1}\right\}$

$$= \ln(3 + \sqrt{8}) = \ln 5.828\ 4 = \mathbf{1.762\ 7}$$

(c) In logarithmic form, artanh $\frac{x}{a} = \frac{1}{2}\ln\left(\frac{a + x}{a - x}\right)$

artanh 0.3 = artanh $\frac{3}{10}$

If $x = 3$ and $a = 10$ then artanh $\frac{3}{10} = \frac{1}{2}\ln\left(\frac{10 + 3}{10 - 3}\right) = \frac{1}{2}\ln\frac{13}{7}$

$$= \mathbf{0.309\ 5}$$

Further problems on evaluating inverse hyperbolic functions may be found in Section 5 (Problems 17 to 19), page 161.

4 Differentiation of inverse hyperbolic functions

(a) $y = \text{arsinh } \frac{x}{a}$

If $y = \text{arsinh } \frac{x}{a}$ then $\frac{x}{a} = \sinh y$ and $x = a \sinh y$.

$\frac{\mathrm{d}x}{\mathrm{d}y} = a \cosh y$. (For differential coefficients of hyperbolic functions, see Chapter 7.)

Now $\cosh^2 y - \sinh^2 y = 1$. Thus $\cosh y = \sqrt{(1 + \sinh^2 y)} = \sqrt{\left\{1 + \left(\frac{x}{a}\right)^2\right\}}$

Hence $\frac{\mathrm{d}x}{\mathrm{d}y} = a \cosh y = (a)\left[\frac{\sqrt{(a^2 + x^2)}}{a}\right] = \sqrt{(a^2 + x^2)}$

$\frac{\mathrm{d}y}{\mathrm{d}x} = \frac{1}{\mathrm{d}x/\mathrm{d}y} = \frac{1}{\sqrt{(a^2 + x^2)}}$, and there are two possible values, one positive and one negative, due to the square root sign. A sketch of part of the graph $y = \text{arsinh } x$ is shown in Fig. 3(a) where it is seen that the gradient (i.e. $\frac{dy}{dx}$) is always positive.

Hence if $y = \text{arsinh } \dfrac{x}{a}$ then $\dfrac{dy}{dx} = \dfrac{1}{\sqrt{(a^2 + x^2)}}$

Alternatively, since $y = \text{arsinh } \dfrac{x}{a} = \ln\left\{\dfrac{x + \sqrt{(a^2 + x^2)}}{a}\right\}$ then

$$\frac{d}{dx}\left(\text{arsinh } \frac{x}{a}\right) = \frac{d}{dx}\left[\ln\left\{\frac{x + \sqrt{(a^2 + x^2)}}{a}\right\}\right]$$

$$= \left(\frac{1}{\frac{x + \sqrt{(a^2 + x^2)}}{a}}\right)\left(\frac{1}{a}\right)[1 + \tfrac{1}{2}(a^2 + x^2)^{-1/2}\,2x]$$

$$= \left(\frac{a}{x + \sqrt{(a^2 + x^2)}}\right)\left(\frac{1}{a}\right)\left(1 + \frac{x}{\sqrt{(a^2 + x^2)}}\right)$$

$$= \left(\frac{1}{x + \sqrt{(a^2 + x^2)}}\right)\left(\frac{\sqrt{(a^2 + x^2)} + x}{\sqrt{(a^2 + x^2)}}\right)$$

$$= \frac{1}{\sqrt{(a^2 + x^2)}} \text{, as above}$$

When $a = 1$, $\dfrac{d}{dx}(\text{arsinh } x) = \dfrac{1}{\sqrt{(1 + x^2)}}$

The most general form of the differential coefficient of $\text{arsinh } f(x)$ is:

$$\frac{d}{dx}[\text{arsinh } f(x)] = \frac{1}{\sqrt{\{1 + [f(x)]^2\}}}[f'(x)]$$

(b) $y = \text{arcosh } \dfrac{x}{a}$

If $y = \text{arcosh } \dfrac{x}{a}$ then $\dfrac{x}{a} = \cosh y$ and $x = a \cosh y$.

$\dfrac{dx}{dy} = a \sinh y = (a)[\sqrt{(\cosh^2 y - 1)}]$, since $\cosh^2 y - \sinh^2 y = 1$.

$\dfrac{dx}{dy} = (a)\sqrt{\left[\left(\dfrac{x}{a}\right)^2 - 1\right]} = (a)\left(\dfrac{\sqrt{(x^2 - a^2)}}{a}\right) = \sqrt{(x^2 - a^2)}$

$\dfrac{dy}{dx} = \dfrac{1}{dx/dy} = \dfrac{1}{\sqrt{(x^2 - a^2)}}$ and there are two possible values, one positive and one negative, due to the square root sign. A sketch of part of the graph $y = \text{arcosh } x$ is shown in Fig. 3(b), where it is seen that when x is greater than $+1$ there are two values of y which correspond to a particular value of x. The positive value of y is defined as the principal value of $\text{arcosh } \dfrac{x}{a}$.

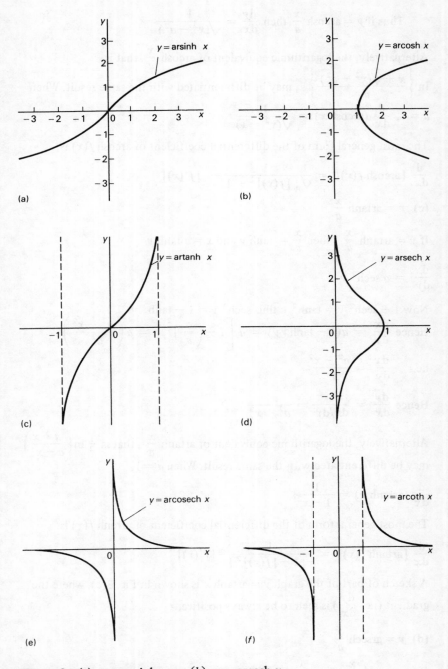

Figure 3 (a) $y = \operatorname{arsinh} x$ (b) $y = \operatorname{arcosh} x$
        (c) $y = \operatorname{artanh} x$ (d) $y = \operatorname{arsech} x$
        (e) $y = \operatorname{arcosech} x$ (f) $y = \operatorname{arcoth} x$

Thus if $y = \text{arcosh } \dfrac{x}{a}$ **then** $\dfrac{dy}{dx} = \dfrac{1}{\sqrt{(x^2 - a^2)}}$

Alternatively, the logarithmic equivalent of $\text{arcosh } \dfrac{x}{a}$, that is

$\ln\left\{\dfrac{x + \sqrt{(x^2 - a^2)}}{a}\right\}$, may be differentiated with the same result. When

$a = 1$, $\dfrac{d}{dx}(\text{arcosh } x) = \dfrac{1}{\sqrt{(x^2 - 1)}}$

The most general form of the differential coefficient of $\text{arcosh } f(x)$ is:

$\dfrac{d}{dx}[\text{arcosh } f(x)] = \dfrac{1}{\sqrt{\{[f(x)]^2 - 1\}}} [f'(x)]$

(c) $y = \text{artanh } \dfrac{x}{a}$

If $y = \text{artanh } \dfrac{x}{a}$, then $\dfrac{x}{a} = \tanh y$ and $x = a \tanh y$

$\dfrac{dx}{dy} = a \text{ sech}^2 y$

Now $1 - \text{sech}^2 y = \tanh^2 y$ thus $\text{sech}^2 y = 1 - \tanh^2 y$

Hence $\dfrac{dx}{dy} = a(1 - \tanh^2 y) = a\left[1 - \left(\dfrac{x}{a}\right)^2\right] = a\left(\dfrac{a^2 - x^2}{a^2}\right)$

i.e. $\dfrac{dx}{dy} = \dfrac{a^2 - x^2}{a}$

Hence $\dfrac{dy}{dx} = \dfrac{1}{dx/dy} = \dfrac{a}{a^2 - x^2}$

Alternatively, the logarithmic equivalent of $\text{artanh } \dfrac{x}{a}$, that is $\frac{1}{2}\ln\left(\dfrac{a + x}{a - x}\right)$, may be differentiated with the same result. When $a = 1$,

$\dfrac{d}{dx}(\text{artanh } x) = \dfrac{1}{1 - x^2}$

The most general form of the differential coefficient of $\text{artanh } f(x)$ is:

$\dfrac{d}{dx}[\text{artanh } f(x)] = \dfrac{1}{\{1 - [f(x)]^2\}} [f'(x)]$

A sketch of part of the graph $y = \text{artanh } x$ is shown in Fig. 3(c), where the gradient (i.e. $\dfrac{dy}{dx}$) is seen to be always positive.

(d) $y = \text{arsech } \dfrac{x}{a}$

If $y = \text{arsech } \dfrac{x}{a}$, then $\dfrac{x}{a} = \text{sech } y$ and $x = a \text{ sech } y$.

$$\frac{dx}{dy} = -a \operatorname{sech} y \tanh y$$

Now $1 - \tanh^2 y = \operatorname{sech}^2 y$, thus $\tanh y = \sqrt{(1 - \operatorname{sech}^2 y)}$

$$= \sqrt{\left[1 - \left(\frac{x}{a}\right)^2\right]} = \frac{\sqrt{(a^2 - x^2)}}{a},$$

the positive value being taken, since the positive value is defined as the principal value.

$$\frac{dx}{dy} = -a \operatorname{sech} y \tanh y = -a\left(\frac{x}{a}\right)\left[\frac{\sqrt{(a^2 - x^2)}}{a}\right]$$

Hence $\dfrac{dy}{dx} = \dfrac{1}{dx/dy} = \dfrac{-a}{x\sqrt{(a^2 - x^2)}}$

When $a = 1$, $\dfrac{d}{dx}(\operatorname{arsech} x) = \dfrac{-1}{x\sqrt{(1 - x^2)}}$

A sketch of part of the graph $y = \operatorname{arsech} x$ is shown in Fig. 3(d), where it is seen that the gradient is negative for principal values of y.

The most general form of the differential coefficient of $\operatorname{arsech} f(x)$ is:

$$\frac{d}{dx}[\operatorname{arsech} f(x)] = \frac{-1}{f(x)\sqrt{\{1 - [f(x)]^2\}}}[f'(x)]$$

(e) $y = \operatorname{arcosech} \dfrac{x}{a}$

If $y = \operatorname{arcosech} \dfrac{x}{a}$, then $\dfrac{x}{a} = \operatorname{cosech} y$ and $x = a \operatorname{cosech} y$.

$$\frac{dx}{dy} = -a \operatorname{cosech} y \coth y.$$

Now $\coth^2 y - 1 = \operatorname{cosech}^2 y$, thus $\coth y = \sqrt{(\operatorname{cosech}^2 y + 1)} = \sqrt{\left[\left(\frac{x}{a}\right)^2 + 1\right]}$

$$= \frac{\sqrt{(x^2 + a^2)}}{a},$$

the positive value being taken since the positive value is defined as the principal value.

Hence $\dfrac{dx}{dy} = -a \operatorname{cosech} y \coth y = -a\left(\dfrac{x}{a}\right)\left[\dfrac{\sqrt{(x^2 + a^2)}}{a}\right] = -\dfrac{x}{a}\sqrt{(x^2 + a^2)}$

$$\frac{dy}{dx} = \frac{1}{dx/dy} = \frac{-a}{x\sqrt{(x^2 + a^2)}}$$

When $a = 1$, $\dfrac{d}{dx}(\operatorname{arcosech} x) = \dfrac{-1}{x\sqrt{(x^2 + 1)}}$

a sketch of part of the graph $y = \operatorname{arcosech} x$ is shown in Fig. 3(e), where it is seen that the gradient is negative for principal values of y.

The most general form of the differential coefficient of arcosech $f(x)$ is:

$$\frac{d}{dx}[\text{arcosech } f(x)] = \frac{-1}{f(x)\sqrt{\{[f(x)]^2 + 1\}}}[f'(x)]$$

(f) $y = \text{arcoth } \dfrac{x}{a}$

If $y = \text{arcoth } \dfrac{x}{a}$ then $\dfrac{x}{a} = \coth y$ and $x = a \coth y$.

$$\frac{dx}{dy} = -a \, \text{cosech}^2 \, y$$

Now $\coth^2 y - 1 = \text{cosech}^2 y$. Hence $\dfrac{dx}{dy} = -a(\coth^2 y - 1)$

$$= -a\left[\left(\frac{x}{a}\right)^2 - 1\right]$$

$$= -\frac{(x^2 - a^2)}{a}$$

$$\frac{dy}{dx} = \frac{1}{dx/dy} = \frac{-a}{x^2 - a^2} = \frac{a}{a^2 - x^2}$$

(i.e. the same result as obtained for artanh $\dfrac{x}{a}$)

When $a = 1$, $\dfrac{d}{dx}(\text{arcoth } x) = \dfrac{1}{1 - x^2}$

A sketch of part of the graph $y = \text{arcoth } x$ is shown in Fig. 3(f), where it is seen that the gradient is negative for positive values of y.

The most general form of the differential coefficent of arcoth $f(x)$ is:

$$\frac{d}{dx}[\text{arcoth } f(x)] = \frac{1}{\{1 - [f(x)]^2\}}[f'(x)]$$

Summary of differential coefficients of inverse hyperbolic functions

y or $f(x)$	$\dfrac{dy}{dx}$ or $f'(x)$
(a) arsinh $\dfrac{x}{a}$	$\dfrac{1}{\sqrt{(x^2 + a^2)}}$
arsinh $f(x)$	$\dfrac{1}{\sqrt{\{[f(x)]^2 + 1\}}}[f'(x)]$
(b) arcosh $\dfrac{x}{a}$	$\dfrac{1}{\sqrt{(x^2 - a^2)}}$
arcosh $f(x)$	$\dfrac{1}{\sqrt{\{[f(x)]^2 - 1\}}}[f'(x)]$

y or $f(x)$	$\dfrac{dy}{dx}$ or $f'(x)$
(c) artanh $\dfrac{x}{a}$	$\dfrac{a}{(a^2 - x^2)}$
artanh $f(x)$	$\dfrac{1}{\{1 - [f(x)]^2\}}\ [f'(x)]$
(d) arsech $\dfrac{x}{a}$	$\dfrac{-a}{x\sqrt{(a^2 - x^2)}}$
arsech $f(x)$	$\dfrac{-1}{f(x)\sqrt{\{1 - [f(x)]^2\}}}\ [f'(x)]$
(e) arcosech $\dfrac{x}{a}$	$\dfrac{-a}{x\sqrt{(x^2 + a^2)}}$
arcosech $f(x)$	$\dfrac{-1}{f(x)\sqrt{\{[f(x)]^2 + 1\}}}\ [f'(x)]$
(f) arcoth $\dfrac{x}{a}$	$\dfrac{a}{a^2 - x^2}$
arcoth $f(x)$	$\dfrac{1}{\{1 - [f(x)]^2\}}\ [f'(x)]$

From the above results it may be seen that:

$$\int \frac{1}{\sqrt{(x^2 + a^2)}}\,dx = \text{arsinh}\ \frac{x}{a} + c \ \text{ or }\ \ln\left\{\frac{x + \sqrt{(x^2 + a^2)}}{a}\right\} + c,$$

$$\int \frac{1}{\sqrt{(x^2 - a^2)}}\,dx = \text{arcosh}\ \frac{x}{a} + c \ \text{ or }\ \ln\left\{\frac{x + \sqrt{(x^2 - a^2)}}{a}\right\} + c,$$

and

$$\int \frac{1}{(a^2 - x^2)}\,dx = \frac{1}{a}\ \text{artanh}\ \frac{x}{a} + c \ \text{ or }\ \frac{1}{2a}\ln\left(\frac{a + x}{a - x}\right) + c.$$

The method of determining such integrals is discussed in Chapter 15, Section 3.

Worked problems on differentiating inverse hyperbolic functions

Problem 1. Find the differential coefficient of (a) arsinh $3x$ and (b) arcosh $\sqrt{(1 + t^2)}$.

(a) $\dfrac{d}{dx}\ [\text{arsinh}\ f(x)] = \dfrac{1}{\sqrt{\{[f(x)]^2 + 1\}}}\ [f'(x)]$

Hence $\dfrac{d}{dx}\ [\text{arsinh}\ 3x] = \dfrac{1}{\sqrt{\{(3x)^2 + 1\}}}\ [3] = \dfrac{3}{\sqrt{(9x^2 + 1)}}$

158

(b) $\dfrac{d}{dt}$ [arcosh $f(t)$] $= \dfrac{1}{\sqrt{\{[f(t)]^2 - 1\}}}$ $[f'(t)]$

Hence $\dfrac{d}{dt}$ [arcosh $\sqrt{(1 + t^2)}$] $= \dfrac{1}{\sqrt{\{[\sqrt{(1+t^2)}]^2 - 1\}}}$ $[\tfrac{1}{2}(1 + t^2)^{-1/2} 2t]$

$$= \dfrac{1}{\sqrt{(1 + t^2 - 1)}} \left[\dfrac{t}{\sqrt{(1 + t^2)}} \right]$$

$$= \dfrac{1}{t} \left[\dfrac{t}{\sqrt{(1 + t^2)}} \right] = \dfrac{1}{\sqrt{(1 + t^2)}}$$

Problem 2. Differentiate (a) artanh $\dfrac{3x}{4}$, (b) arcosech (sinh x).

(a) $\dfrac{d}{dx}$ [artanh $f(x)$] $= \dfrac{1}{\{1 - [f(x)]^2\}}$ $[f'(x)]$

Hence $\dfrac{d}{dx}$ [artanh $\dfrac{3x}{4}$] $= \dfrac{1}{\left\{1 - \left(\dfrac{3x}{4}\right)^2\right\}} \left[\dfrac{3}{4}\right] = \dfrac{1}{\left(1 - \dfrac{9x^2}{16}\right)} \left[\dfrac{3}{4}\right]$

$$= \dfrac{1}{\left(\dfrac{16 - 9x^2}{16}\right)} \left[\dfrac{3}{4}\right] = \dfrac{16}{(16 - 9x^2)} \left[\dfrac{3}{4}\right]$$

$$= \dfrac{12}{16 - 9x^2}$$

(b) $\dfrac{d}{dx}$ [arcosech $f(x)$] $= \dfrac{-1}{f(x)\sqrt{\{[f(x)]^2 + 1\}}} [f'(x)]$

Hence $\dfrac{d}{dx}$ [arcosech (sinh x)] $= \dfrac{-1}{\sinh x \sqrt{\{\sinh^2 x + 1\}}}$ $[\cosh x]$

Now $\cosh^2 x - \sinh^2 x = 1$, hence $\sinh^2 x + 1 = \cosh^2 x$.

Therefore $\dfrac{d}{dx}$ [arcosech (sinh x)] $= \dfrac{-1}{\sinh x \sqrt{(\cosh^2 x)}}$ $[\cosh x]$

$$= \dfrac{-1}{\sinh x} = -\text{cosech } x.$$

Problem 3. Find (a) $\dfrac{d}{dx}$ [arsech $(3x - 1)$] and (b) $\dfrac{d}{d\theta}$ [arcoth $\sqrt{(1 - 2\theta^2)}$].

(a) $\dfrac{d}{dx}$ [arsech $f(x)$] $= \dfrac{-1}{f(x)\sqrt{\{1 - [f(x)]^2\}}}$ $[f'(x)]$

Hence $\dfrac{d}{dx}$ [arsech $(3x - 1)$] $= \dfrac{-1}{(3x - 1)\sqrt{\{1 - (3x - 1)^2\}}}$ [3]

$$= \frac{-3}{(3x-1)\sqrt{\{1-(9x^2-6x+1)\}}}$$

$$= \frac{-3}{(3x-1)\sqrt{(6x-9x^2)}}$$

$$= \frac{-3}{(3x-1)\sqrt{[3x(2-3x)]}}$$

(b) $\frac{d}{d\theta}$ [arcoth $f(\theta)$] $= \frac{1}{\{1-[f(\theta)]^2\}}$ $[f'(\theta)]$

Hence $\frac{d}{d\theta}$ [arcoth $\sqrt{(1-2\theta^2)}$]

$$= \frac{1}{\{1-[\sqrt{(1-2\theta^2)}]^2\}}[\tfrac{1}{2}(1-2\theta^2)^{-1/2}(-4\theta)]$$

$$= \frac{1}{\{1-(1-2\theta^2)\}}\left[\frac{-2\theta}{\sqrt{(1-2\theta^2)}}\right]$$

$$= \frac{1}{2\theta^2}\left[\frac{-2\theta}{\sqrt{(1-2\theta^2)}}\right] = \frac{-1}{\theta\sqrt{(1-2\theta^2)}}$$

Problem 4. Differentiate: (a) $(1-t^2)$ arcoth t, (b) $\frac{\text{arcosh}(\sec x)}{x}$

(a) $f(t) = (1-t^2)$ arcoth t

Hence $f'(t) = (1-t^2)\dfrac{1}{(1-t^2)}$ $+ (\text{arcoth } t)(-2t)$, by the product rule of differentiation

$$= 1 - 2t \text{ arcoth } t.$$

(b) $f(x) = \dfrac{\text{arcosh}(\sec x)}{x}$

Hence $f'(x) = \dfrac{(x)\left[\dfrac{1}{\sqrt{(\sec^2 x - 1)}}(\sec x \tan x)\right] - [\text{arcosh}(\sec x)](1)}{x^2}$,

by the quotient rule of differentiation

$$= \frac{(x)\left(\dfrac{\sec x \tan x}{\tan x}\right) - \text{arcosh}(\sec x)}{x^2}, \text{ since } \sqrt{(\sec^2 x - 1)} = \tan x,$$

$$= \frac{x \sec x - \text{arcosh}(\sec x)}{x^2}$$

Further problems on differentiating inverse hyperbolic functions may be found in the following Section (5) (Problems 20 to 35), page 162.

5 Further problems

Differentiation of inverse trigonometrical functions
In Problems 1 to 8, differentiate with respect to the variable.

1. (a) $\arcsin 3x$ (b) $\arcsin \dfrac{x}{5}$

$$(a)\left[\frac{3}{\sqrt{(1-9x^2)}}\right] \quad (b)\left[\frac{1}{\sqrt{(25-x^2)}}\right]$$

2. (a) $2 \arccos 4x$ (b) $3 \arccos \dfrac{2x}{3}$

$$(a)\left[\frac{-8}{\sqrt{(1-16x^2)}}\right] \quad (b)\left[\frac{-6}{\sqrt{(9-4x^2)}}\right]$$

3. (a) $2 \arctan 2x$ (b) $\arctan \sqrt{(x-1)}$

$$(a)\left[\frac{4}{1+4x^2}\right] \quad (b)\left[\frac{1}{2x\sqrt{(x-1)}}\right]$$

4. (a) $\operatorname{arcsec} 4\theta$ (b) $2 \operatorname{arcsec} \sqrt{x}$.

$$(a)\left[\frac{1}{\theta\sqrt{(16\theta^2-1)}}\right] \quad (b)\left[\frac{1}{x\sqrt{(x-1)}}\right]$$

5. (a) $\operatorname{arccosec} \dfrac{x}{3}$ (b) $\operatorname{arccosec}(x^2+1)$.

$$(a)\left[\frac{-3}{x\sqrt{(x^2-9)}}\right] \quad (b)\left[\frac{-2}{(x^2+1)\sqrt{(x^2+2)}}\right]$$

6. (a) $\operatorname{arccot} x^2$ (b) $3 \operatorname{arccot} \sqrt{(x^2-1)}$.

$$(a)\left[\frac{-2x}{1+x^4}\right] \quad (b)\left[\frac{-3}{x\sqrt{(x^2-1)}}\right]$$

7. (a) $\ln(\arcsin x)$ (b) $3e^{\operatorname{arccot} x}$

$$(a)\left[\frac{1}{\sqrt{(1-x^2)}\arcsin x}\right] \quad (b)\left[\frac{-3e^{\operatorname{arccot} x}}{1+x^2}\right]$$

8. (a) $\arctan\left(\dfrac{2t}{1-t^2}\right)$ (b) $\dfrac{1}{2} \operatorname{arcsec}\left(\dfrac{x^2+1}{x^2-1}\right)$

$$(a)\left[\frac{2}{1+t^2}\right] \quad (b)\left[\frac{-1}{(x^2+1)}\right]$$

9. Show that (a) $\dfrac{\mathrm{d}}{\mathrm{d}x}\left[\arcsin\left(\dfrac{2x^2+3}{5}\right)\right] = \dfrac{2x}{\sqrt{(4-3x^2-x^4)}}$

(b) $\dfrac{\mathrm{d}}{\mathrm{d}t}\left[\operatorname{arcsec}\sqrt{\left(\dfrac{t}{2}\right)}\right] = \dfrac{1}{t\sqrt{[2(t-2)]}}$.

10. Show that (a) if $y = \arctan\left(\dfrac{\sin\theta}{1-\cos\theta}\right)$ then $\dfrac{\mathrm{d}y}{\mathrm{d}\theta} = -\dfrac{1}{2}$,

(b) if $y = 2 \text{ arcsec } \sqrt{(\cos t)}$ then $\dfrac{dy}{dt} = \dfrac{-\tan t}{\sqrt{(\cos t - 1)}}$

In Problems 11 to 16 differentiate with respect to the variable.

11. (a) $x \arcsin \dfrac{x}{2}$ (b) $x^2 \text{ arccosec } 2x$

(a) $\left[\dfrac{x}{\sqrt{(4-x^2)}} + \arcsin \dfrac{x}{2} \right]$

(b) $\left[x \left(2 \text{ arccosec } 2x - \dfrac{1}{\sqrt{(4x^2-1)}} \right) \right]$

12. (a) $t \arccos (2t^2 - 1)$ (b) $3t^3 \text{ arccot } t.$

(a) $\left[\dfrac{-2t}{\sqrt{(1-t^2)}} + \arccos (2t^2 - 1) \right]$

(b) $\left[3t^2 \left(3 \text{ arccot } t - \dfrac{t}{1+t^2} \right) \right]$

13. (a) $(1 + \theta^2) \arctan \theta$ (b) $\sqrt{(1-x^2)} \arcsin x.$

(a) $[1 + 2\theta \arctan \theta]$ (b) $\left[1 - \dfrac{x \arcsin x}{\sqrt{(1-x^2)}} \right]$

14. (a) $\sin \left(\text{arcsec } \dfrac{1}{x} \right)$ (b) $2t^2 \arccos (t - 1).$

(a) $\left[\dfrac{-\cos (\text{arcsec } 1/x)}{\sqrt{(1-x^2)}} \right]$

(b) $\left[4t \arccos (t-1) - \dfrac{2t^2}{\sqrt{[t(2-t)]}} \right]'$

15. (a) $\dfrac{\arcsin 2x}{x}$ (b) $\dfrac{\arccos x}{\sqrt{(1-x^2)}}.$

(a) $\left[\dfrac{1}{x^2} \left(\dfrac{2x}{\sqrt{(1-4x^2)}} - \arcsin 2x \right) \right]$

(b) $\left[\dfrac{-1}{(1-x^2)} + \dfrac{x \arccos x}{\sqrt{(1-x^2)^3}} \right]$

16. (a) $\dfrac{\text{arcsec } \sqrt{(x^2+1)}}{\sqrt{(x^2+1)}}$ (b) $\dfrac{\text{arccot } x}{(1+x^2)}.$

(a) $\left[\dfrac{1 - x \text{ arcsec } \sqrt{(x^2+1)}}{\sqrt{(x^2+1)^3}} \right]$

(b) $\left[\dfrac{-(1 + 2x \text{ arccot } x)}{(1+x^2)^2} \right]$

Evaluation of inverse hyperbolic functions

In Problems 17 to 19 use logarithmic equivalents of inverse hyperbolic functions functions to evaluate correct to four decimal places.

162

17. (a) arsinh $\frac{2}{3}$ (b) arsinh $\frac{3}{4}$ (c) arsinh 0.5

                      (a) [0.625 1]   (b) [0.693 1]   (c) [0.481 2]

18. (a) arcosh $\frac{5}{3}$ (b) arcosh 2   (c) arcosh $\frac{13}{12}$

                 (a) [1.098 6]   (b) [1.317 0]   (c) [0.405 5]

19. (a) artanh $\frac{1}{2}$ (b) artanh $\frac{3}{5}$ (c) artanh 0.3.

                 (a) [0.549 3]   (b) [0.693 1]   (c) [0.309 5]

Differentiation of inverse hyperbolic functions

In Problems 20 to 34 differentiate with respect to the variable.

20. (a) arsinh $\frac{x}{2}$ (b) arsinh $5x$.

$$\text{(a) } \left[\frac{1}{\sqrt{(x^2 + 4)}} \right] \quad \text{(b) } \left[\frac{5}{\sqrt{(25x^2 + 1)}} \right]$$

21. (a) arcosh $\frac{x}{3}$ (b) arcosh $4x$.

$$\text{(a) } \left[\frac{1}{\sqrt{(x^2 - 9)}} \right] \quad \text{(b) } \left[\frac{4}{\sqrt{(16x^2 - 1)}} \right]$$

22. (a) artanh $\frac{2x}{3}$ (b) artanh $3x$.

$$\text{(a) } \left[\frac{6}{9 - 4x^2} \right] \quad \text{(b) } \left[\frac{3}{1 - 9x^2} \right]$$

23. (a) arsech $\frac{3t}{4}$ (b) arsech $3t$.

$$\text{(a) } \left[\frac{-4}{t\sqrt{(16 - 9t^2)}} \right] \quad \text{(b) } \left[\frac{-1}{t\sqrt{(1 - 9t^2)}} \right]$$

24. (a) arcosech $\frac{x}{5}$ (b) arcosech $2x$.

$$\text{(a) } \left[\frac{-5}{x\sqrt{(x^2 + 25)}} \right] \quad \text{(b) } \left[\frac{-1}{x\sqrt{(4x^2 + 1)}} \right]$$

25. (a) arcoth $\frac{2\theta}{5}$ (b) arcoth 4θ.

$$\text{(a) } \left[\frac{10}{25 - 4\theta^2} \right] \quad \text{(b) } \left[\frac{4}{1 - 16\theta^2} \right]$$

26. (a) arsinh $\sqrt{(t^2 - 1)}$ (b) arsech $(2x - 1)$.

$$\text{(a) } \left[\frac{1}{\sqrt{(t^2 - 1)}} \right] \quad \text{(b) } \left[\frac{-1}{(2x - 1)\sqrt{x(1 - x)}} \right]$$

27. (a) arcosh $\sqrt{(1 + x^2)}$ (b) arcosh $(\cosh x)$

(a) $\left[\dfrac{1}{\sqrt{(1+x^2)}}\right]$ (b) [1]

28. (a) $\operatorname{artanh}\sqrt{(2t^2+1)}$ (b) $\operatorname{arcoth}(\sin x)$.

(a) $\left[\dfrac{-1}{t\sqrt{(2t^2+1)}}\right]$ (b) $[\sec x]$

29. (a) $\operatorname{arcosech}\sqrt{(x^2-1)}$ (b) $\operatorname{arcoth}\sqrt{(1-3t^2)}$

(a) $\left[\dfrac{-1}{(x^2-1)}\right]$ (b) $\left[\dfrac{-1}{t\sqrt{(1-3t^2)}}\right]$

30. (a) $x\operatorname{arsinh}x$ (b) $(1-x^2)\operatorname{artanh}x$.

(a) $\left[\dfrac{x}{\sqrt{(x^2+1)}}+\operatorname{arsinh}x\right]$ (b) $[1-2x\operatorname{artanh}x]$

31. (a) $3x\operatorname{arsech}2x$ (b) $x^2\operatorname{arcoth}\sqrt{(1-x^2)}$.

(a) $\left[\dfrac{-3}{\sqrt{(1-4x^2)}}+3\operatorname{arsech}2x\right]$

(b) $\left[2x\operatorname{arcoth}\sqrt{(1-x^2)}\dfrac{-x}{\sqrt{(1-x^2)}}\right]$

32. (a) $\sqrt{(x^2-1)}\operatorname{arcosh}x$ (b) $\sqrt{(x^2-1)}\operatorname{arcosech}\sqrt{(x^2-1)}$.

(a) $\left[1+\dfrac{x\operatorname{arcosh}x}{\sqrt{(x^2-1)}}\right]$

(b) $\left[\dfrac{1}{\sqrt{(x^2-1)}}\,(x\operatorname{arcosech}\sqrt{(x^2-1)}-1)\right]$

33. (a) $\dfrac{\operatorname{arsinh}\sqrt{(x^2-1)}}{x}$ (b) $\dfrac{\operatorname{artanh}(2x-1)}{2x}$

(a) $\left[\dfrac{1}{x^2}\left\{\dfrac{x}{\sqrt{(x^2-1)}}-\operatorname{arsinh}\sqrt{(x^2-1)}\right\}\right]$

(b) $\left[\dfrac{1}{4x^2}\left\{\dfrac{1}{(1-x)}-2\operatorname{artanh}(2x-1)\right\}\right]$

34. (a) $\dfrac{\operatorname{arsech}x}{\sqrt{(1-x^2)}}$ (b) $\dfrac{\operatorname{arcoth}x/2}{4-x^2}$

(a) $\left[\dfrac{1}{(1-x^2)}\left\{-\dfrac{1}{x}+\dfrac{x\operatorname{arsech}x}{\sqrt{(1-x^2)}}\right\}\right]$

(b) $\left[\dfrac{2(1+x\operatorname{arcoth}x/2)}{(4-x^2)^2}\right]$

35. Show that $\dfrac{\mathrm{d}}{\mathrm{d}x}[x\operatorname{arcosh}(\sec x)]=x\sec x+\operatorname{arcosh}(\sec x)$.

Chapter 13

Partial differentiation

1 Differentiating a function having two variables

When differentiating functions previously, problems have always been associated with one variable only. For a given function of x, say $y = f(x)$, the problem has been to find $\dfrac{\mathrm{d}y}{\mathrm{d}x}$, i.e. $f'(x)$. When applying the product or quotient rules, $y = uv$ or $y = \dfrac{u}{v}$, y, u, and v are all functions of x only.

However, there are many formulae and occurrences in mathematics, engineering and science, in which the variation of one function may depend on changes taking place in two or more variables. Some examples include:

  (i) The volume of a cylinder, $V = \pi r^2 h$. The volume of the cylinder will change if either radius r or height h is changed.

 (ii) Resonant frequency, $f_0 = \dfrac{1}{2\pi\sqrt{(LC)}}$. The resonant frequency will change if either inductance L or capacitance C is changed.

(iii) The pressure of an ideal gas, $p = \dfrac{mRT}{V}$. The pressure will change if either the thermodynamic temperature T or the volume V is changed.

(iv) Torque, $T = I\alpha$. The torque on a shaft will change if either the moment of inertia, I, or the angular acceleration, α, is changed, and so on.

    For the formula giving the volume of a right circular cylinder, the value of V depends on the values of r and h and this is expressed mathematically as $V = f(r, h)$, called 'V is some function of r and h'. A statement such as '$V = f(r, h)$ and $V = \pi r^2 h$' indicates that both r and h are variable quantities.

When differentiating a function having two variables, one variable is kept constant and the differential coefficient is found of the other variable, with respect to that variable. The differential coefficient obtained is called a **partial derivative** of the function. Thus, for a right circular cylinder, the increase in volume with respect to radius $\dfrac{dV}{dr}$, when h remains constant, is the partial derivative of V with respect to r, and is written $\dfrac{\partial V}{\partial r}$. The 'curly' dee, ∂, is used to denote a differential coefficient in an expression containing more than one variable. Alternatively,

$$\left[\frac{dV}{dh}\right]_{r \text{ constant}} = \frac{\partial V}{\partial h},$$

the partial derivative of V with respect to h, with r kept constant.

2 First order partial derivatives

By applying the principles given in section 1 to the volume of a right circular cylinder, the partial derivatives $\dfrac{\partial V}{\partial r}$ and $\dfrac{\partial V}{\partial h}$ may be found. These are called **first order partial derivatives** since $n = 1$ when they are written in the form $\dfrac{\partial^n V}{\partial r^n}$. Second order partial derivatives contain at least one term in which $n = 2$, i.e. a term of the form $\dfrac{\partial^2 V}{\partial r^2}$. The partial derivative of V with respect to r is found by keeping h constant, i.e. since $V = \pi r^2 h$, then (πh) becomes the constant term. Thus $\dfrac{\partial V}{\partial r} = (\pi h)\dfrac{d(r^2)}{dr} = 2\pi rh$. The partial derivative of V with respect to h is found by keeping r constant, i.e. since $V = \pi r^2 h$ then (πr^2) becomes the constant term. Thus $\dfrac{\partial V}{\partial h} = (\pi r^2)\dfrac{d(h)}{dh} = \pi r^2$.

Worked problems on first order partial derivatives

Problem 1. The pressure p of a given mass of gas is given by $pV = kT$, where k is a constant, where V is the volume and T is the thermodynamic temperature. Determine equations for $\dfrac{\partial p}{\partial V}$ and $\dfrac{\partial p}{\partial T}$.

Since $pV = kT$, $p = \dfrac{kT}{V}$

To find $\dfrac{\partial p}{\partial V}$, T is kept constant and $p = (kT)\, V^{-1}$.

Hence, $\dfrac{\partial p}{\partial V} = (kT)\dfrac{d(V^{-1})}{dV} = kT(-1)\, V^{-2} = -\dfrac{kT}{V^2}$

To find $\dfrac{\partial V}{\partial T}$, V is kept constant and $p = \left(\dfrac{k}{V}\right) \cdot T$.

Hence, $\dfrac{\partial V}{\partial T} = \left(\dfrac{k}{V}\right) \cdot \dfrac{d\,(T)}{d\,T} = \left(\dfrac{k}{V}\right) \cdot 1 = \dfrac{k}{V}$

Problem 2. If $z = (x + y) \ln \dfrac{x}{y}$, prove that

$$x \frac{\partial z}{\partial x} + y \frac{\partial z}{\partial y} = z.$$

As there is a product of two functions having x in each, $\dfrac{\partial z}{\partial x}$ is found by keeping y constant and applying the product rule of differentiation. The partial differential coefficient of the first function with respect to x is:

$$\frac{\partial (x + y)}{\partial x} = 1,$$

since y is a constant. The partial differential coefficient of the second function with respect to x is:

$$\frac{\partial (\ln x/y)}{\partial x} = \frac{1}{x/y} \times \frac{1}{y} = \frac{1}{x}$$

Alternatively, $\dfrac{\partial (\ln x/y)}{\partial x} = \dfrac{\partial (\ln x - \ln y)}{\partial x} = \dfrac{1}{x}$

since $\ln y$ is a constant.

Hence, by the product rule, $\dfrac{\partial z}{\partial x} = (x + y) \cdot \dfrac{1}{x} + \left(\ln \dfrac{x}{y}\right) \cdot 1$

$\dfrac{\partial z}{\partial y}$ can be found by keeping x constant.

As there are two functions of y in each, the product rule of differentiation is again used. The partial differential coefficient of the first function with respect to y is: $\dfrac{\partial (x + y)}{\partial y} = 1$, since x is a constant. The partial differential coefficient of the second function with respect to y is:

$$\frac{\partial (\ln x/y)}{\partial y} = \frac{\partial (\ln x - \ln y)}{\partial y} = -\frac{1}{y}.$$

Thus, $\dfrac{\partial z}{\partial y} = (x + y) \left(-\dfrac{1}{y}\right) + \left(\ln \dfrac{x}{y}\right) \cdot 1.$

Taking the L.H.S. of the identity and substituting for $\dfrac{\partial z}{\partial x}$ and $\dfrac{\partial z}{\partial y}$ gives:

$$x \left[\frac{1}{x}(x + y) + \ln \frac{x}{y}\right] + y \left[-\frac{1}{y}(x + y) + \ln \frac{x}{y}\right]$$

i.e. L.H.S. $= (x + y) + x \ln \dfrac{x}{y} - (x + y) + y \ln \dfrac{x}{y}$

$$= (x + y) \ln \frac{x}{y} = z = \textbf{R.H.S.}$$

Problem 3. If $z = f(\theta, \phi)$, find $\frac{\partial z}{\partial \theta}$ and $\frac{\partial z}{\partial \phi}$, if $z = 3 \sin 2\theta \cos 3\phi$.

$\frac{\partial z}{\partial \theta}$ may be found by keeping ϕ constant, i.e. $(3 \cos 3\phi)$ is the constant term.

Thus, $\quad \frac{\partial z}{\partial \theta} = 3 \cos 3\phi \cdot \frac{d (\sin 2\theta)}{d\theta}$,

i.e. $\frac{\partial z}{\partial \theta} = 3 \cos 3\phi \cdot 2 \cos 2\theta$.

$\qquad = \textbf{6 cos } 2\theta \textbf{ cos } 3\phi$

$\frac{\partial z}{\partial \phi}$ may be found by keeping θ constant, i.e. $(3 \sin 2\theta)$ is the constant term.

Thus, $\frac{\partial z}{\partial \phi} = 3 \sin 2\theta \cdot \frac{d (\cos 3\phi)}{d\phi}$

$\qquad = 3 \sin 2\theta \cdot (- 3 \sin 3\phi)$

$\qquad = - \textbf{9 sin } 2\theta \textbf{ sin } 3\phi.$

Problem 4. If $z = f(x^2 + y^2)$, show that $x \frac{\partial z}{\partial y} - y \frac{\partial z}{\partial x} = 0$.

The actual function of z is not known and z could be equal to, say, $\sin (x^2 + y^2)$ or $\ln (x^2 + y^2)$ or $(x^2 + y^2)^{3/2}$ or any other function of $(x^2 + y^2)$. However, proofs based on functions are often possible, even though the actual function is not known. Such techniques are sometimes used in solving partial differential equations. Making a substitution such as let $u = (x^2 + y^2)$ usually helps in proving the identity. If such a substitution is made in this case, then $z = f(u)$.

Also, since $u = x^2 + y^2$, $\frac{\partial u}{\partial x} = 2x$ (y is kept constant) and $\frac{\partial u}{\partial y} = 2y$ (x is kept constant). The chain rule of differentiation applies equally to partial derivatives, hence

$\frac{\partial z}{\partial x} = \frac{\partial z}{\partial u} \times \frac{\partial u}{\partial x} = f'(u) \times 2x$

Similarly,

$\frac{\partial z}{\partial y} = \frac{\partial z}{\partial u} \times \frac{\partial u}{\partial y} = f'(u) \times 2y.$

Substituting for $\frac{\partial z}{\partial x}$ and $\frac{\partial z}{\partial y}$ in the L.H.S. of the equation, gives:

$x \cdot 2y f'(u) - y \cdot 2x f'(u) = 0 = \textbf{R.H.S.}$

Further problems on first order partial derivatives may be found in Section 4 (Problems 1 to 7), page. 171.

3 Second order partial derivatives

It is shown in Section 2 that if $z = f(x, y)$, then it is possible to find $\frac{\partial z}{\partial x}$ and $\frac{\partial z}{\partial y}$. The partial derivatives $\frac{\partial z}{\partial x}$ and $\frac{\partial z}{\partial y}$ are themselves functions of x and y and hence can be differentiated again with respect to x or with respect to y. When this is done, four **second order partial derivatives** are possible for a function containing two variables. These are:

(i) differentiating $\frac{\partial z}{\partial x}$ with respect to x, (keeping y constant), gives

$\frac{\partial \left(\frac{\partial z}{\partial x} \right)}{\partial x}$, written as $\frac{\partial^2 z}{\partial x^2}$,

(ii) differentiating $\frac{\partial z}{\partial x}$ with respect to y (keeping x constant) gives

$\frac{\partial \left(\frac{\partial z}{\partial x} \right)}{\partial y}$, written as $\frac{\partial^2 z}{\partial y \partial x}$,

(iii) differentiating $\frac{\partial z}{\partial y}$ with respect to x (keeping y constant) gives

$\frac{\partial \left(\frac{\partial z}{\partial y} \right)}{\partial x}$, written as $\frac{\partial^2 z}{\partial x \partial y}$, and

(iv) differentiating $\frac{\partial z}{\partial y}$ with respect to y (keeping x constant) gives

$\frac{\partial \left(\frac{\partial z}{\partial y} \right)}{\partial y}$, written as $\frac{\partial^2 z}{\partial y^2}$.

To illustrate these four possibilities, consider the function:
$z = \sin xy$.

$\frac{\partial z}{\partial x} = y \cos xy$ and $\frac{\partial z}{\partial y} = x \cos xy$. Then,

(i) $\frac{\partial^2 z}{\partial x^2} = \frac{\partial \left(\frac{\partial z}{\partial x} \right)}{\partial x} = \frac{\partial (y \cos xy)}{\partial x} = -y^2 \sin xy$

(ii) $\frac{\partial^2 z}{\partial y \partial x} = \frac{\partial \left(\frac{\partial z}{\partial x} \right)}{\partial y} = \frac{\partial (y \cos xy)}{\partial y}$

$\qquad = y \left(-x \sin xy \right) + 1 \left(\cos xy \right)$

$\qquad = \cos xy - xy \sin xy$

(iii) $\dfrac{\partial^2 z}{\partial x\,\partial y} = \dfrac{\partial\left(\dfrac{\partial z}{\partial y}\right)}{\partial x} = \dfrac{\partial(x\cos xy)}{\partial x}$

$\qquad\qquad = x\,(-y\sin xy) + 1\,(\cos xy)$

$\qquad\qquad = \cos xy - xy\sin xy$

(iv) $\dfrac{\partial^2 z}{\partial y^2} = \dfrac{\partial\left(\dfrac{\partial z}{\partial y}\right)}{\partial y} = \dfrac{\partial\ (x\cos xy)}{\partial y} = -x^2\sin x\,y.$

It can be seen that the values of $\dfrac{\partial^2 z}{\partial y\,\partial x}$ and $\dfrac{\partial^2 z}{\partial x\partial y}$ are the same in this case. This result is true for all functions which are continuous. (A function which is continuous means that the graph of the function has no sudden jumps or breaks.)

Since $\dfrac{\partial^2 z}{\partial y\,\partial x}$ is always equal to $\dfrac{\partial^2 z}{\partial x\,\partial y}$ for functions being considered at this level, there are in fact only three second order partial derivatives for functions of two variables, namely:

$$\dfrac{\partial^2 z}{\partial x^2},\qquad \dfrac{\partial^2 z}{\partial y^2}\qquad \text{and either}\qquad \dfrac{\partial^2 z}{\partial x\,\partial y}\qquad \text{or}\qquad \dfrac{\partial^2 z}{\partial y\,\partial x}.$$

Second order partial derivatives will be met in such topic areas as entropy and the continuity theorem in thermodynamics, the waveguide theory in electrical engineering and in the solution of partial differential equations.

Worked problems on second order partial derivatives

Problem 1. If $u = f(x, y)$ and $u = \dfrac{x-y}{x+y}$, find

$$\dfrac{\partial^2 u}{\partial x^2},\qquad \dfrac{\partial^2 u}{\partial y^2},\qquad \dfrac{\partial^2 u}{\partial x\,\partial y}\quad\text{and}\quad \dfrac{\partial^2 u}{\partial y\,\partial x}$$

To find $\dfrac{\partial u}{\partial x}$, the quotient rule of differentiation is used, y being treated as a constant. Hence

$$\dfrac{\partial u}{\partial x} = \dfrac{(x+y)\,(1) - (x-y)\,(1)}{(x+y)^2} = \dfrac{2y}{(x+y)^2} = 2y\,(x+y)^{-2}$$

$$\dfrac{\partial^2 u}{\partial x^2} = \dfrac{\partial\left(\dfrac{\partial u}{\partial x}\right)}{\partial x} = \dfrac{\partial\,[2y\,(x+y)^{-2}]}{\partial x}$$

$$= (2y)(-2)(x+y)^{-3}(1) = \frac{-4y}{(x+y)^3}$$

To find $\frac{\partial u}{\partial y}$, the quotient rule of differentiation is used, x being treated as a constant. Hence,

$$\frac{\partial u}{\partial y} = \frac{(x+y)(-1)-(x-y)(1)}{(x+y)^2} = \frac{-2x}{(x+y)^2} = -2x(x+y)^{-2}$$

$$\frac{\partial^2 u}{\partial y^2} = \frac{\partial\left(\frac{\partial u}{\partial y}\right)}{\partial y} = \frac{\partial(-2x(x+y)^{-2})}{\partial y}$$

$$= (-2x)(-2)(x+y)^{-3}(1) = \frac{4x}{(x+y)^3}$$

$$\frac{\partial^2 u}{\partial x \partial y} = \frac{\partial\left(\frac{\partial u}{\partial y}\right)}{\partial x} = \frac{\partial\left(\frac{-2x}{(x+y)^2}\right)}{\partial x}$$

$$= \frac{(x+y)^2(-2)-(-2x)(2)(x+y)(1)}{(x+y)^4} = \frac{2(x-y)}{(x+y)^3}$$

$$\frac{\partial^2 u}{\partial y \partial x} = \frac{\partial\left(\frac{\partial u}{\partial x}\right)}{\partial y} = \frac{\partial\left(\frac{-2y}{(x+y)^2}\right)}{\partial y}$$

$$= \frac{(x+y)^2(2)-2y(2)(x+y)(1)}{(x+y)^4} = \frac{2(x-y)}{(x+y)^3}$$

The results obtained for $\frac{\partial^2 u}{\partial x \partial y}$ and $\frac{\partial^2 u}{\partial y \partial x}$ verify that in this case the relationship stated in the text that $\frac{\partial^2 u}{\partial x \partial y} = \frac{\partial^2 u}{\partial y \partial x}$ is true.

Problem 2. If $z = f(x, y)$ and $z = x\cos(x+y)$, find $\frac{\partial^2 z}{\partial x^2}$ and $\frac{\partial^2 z}{\partial y^2}$. Show that $\frac{\partial^2 z}{\partial x \partial y} = \frac{\partial^2 z}{\partial y \partial x}$.

Applying the product rule, gives:

$$\frac{\partial z}{\partial x} = x(-\sin(x+y)) + 1 \cdot \cos(x+y)$$
$$= -x\sin(x+y) + \cos(x+y).$$

Differentiating $\frac{\partial z}{\partial x}$ with respect to x gives:

$$\frac{\partial^2 x}{\partial x^2} = -x \cos (x + y) + (-1) \sin (x + y) - \sin (x + y)$$

$$= -x \cos (x + y) - 2 \sin (x + y).$$

Differentiating $z = x \cos (x + y)$ with respect to y gives:

$$\frac{\partial z}{\partial y} = -x \sin (x + y).$$

Differentiating $\frac{\partial z}{\partial y}$ with respect to y gives:

$$\frac{\partial^2 z}{\partial y^2} = -x \cos (x + y))$$

$$\frac{\partial^2 z}{\partial x \partial y} = \frac{\partial \left(\frac{\partial z}{\partial y} \right)}{\partial x} = \frac{\partial (-x \sin (x + y))}{\partial x}$$

$$= -x \cos (x + y) + (-1) \sin (x + y)$$

$$= -x \cos (x + y) - \sin (x + y) \qquad (1)$$

$$\frac{\partial^2 z}{\partial y \partial x} = \frac{\partial \left(\frac{\partial z}{\partial x} \right)}{\partial y} = \frac{\partial (-x \sin (x + y) + \cos (x + y))}{\partial y}$$

$$= -x \cos (x + y) - \sin (x + y) \qquad (2)$$

Since equation (1) is equal to equation (2), then

$$\frac{\partial^2 z}{\partial x \partial y} = \frac{\partial^2 z}{\partial y \partial x}$$

Further problems on second order partial derivatives may be found in the following Section (4), (Problems 8 to 19), page 172.

4 Further problems

First order partial derivatives

1. If $z = f(x, y)$, find $\frac{\partial z}{\partial x}$ and $\frac{\partial z}{\partial y}$ for the following equations:

(a) $z = x^2 + 3xy + y^3$ (b) $z = (3x - 2y)^2$ (c) $z = e^y \ln (x + y)$

(a) $\left[\frac{\partial z}{\partial x} = 2x + 3y, \ \frac{\partial z}{\partial y} = 3 (x + y^2) \right]$

(b) $\left[\frac{\partial z}{\partial x} = 6 (3x - 2y), \ \frac{\partial z}{\partial y} = -4 (3x - 2y) \right]$

(c) $\left[\frac{\partial z}{\partial x} = \frac{e^y}{x + y}, \ \frac{\partial z}{\partial y} = e^y \left(\frac{1}{x + y} + \ln (x + y) \right) \right]$

2. If $u = f(p, q)$, find $\dfrac{\partial u}{\partial p}$ and $\dfrac{\partial u}{\partial q}$ for the following equations:

(a) $u = \dfrac{1}{\sqrt{(p^2 + q^2)}}$　　　(b) $u = e^{(4p + q/2)} \tan(3p + 2q)$

(c) $u = pe^q + \dfrac{1}{p} \ln q$.

(a) $\left[\dfrac{\partial u}{\partial p} = - \dfrac{p}{\sqrt{(p^2 + q^2)^3}} \, , \dfrac{\partial u}{\partial q} = - \dfrac{q}{\sqrt{(p^2 + q^2)^3}} \right]$

(b) $\left[\dfrac{\partial u}{\partial p} = e^{(4p + q/2)} \{ 3 \sec^2 (3p + 2q) + 4 \tan(3p + 2q) \} \right]$

$\left[\dfrac{\partial u}{\partial q} = e^{(4p + q/2)} \{ 2 \sec^2 (3p + 2q) + \dfrac{1}{2} \tan(3p + 2q) \} \right]$

(c) $\left[\dfrac{\partial u}{\partial p} = e^q - \dfrac{\ln q}{p^2} \, , \dfrac{\partial u}{\partial q} = p\, e^q + \dfrac{1}{pq} \right]$

3. If $m = f(u, v)$ and $m = \cos uv$, prove that

$\dfrac{1}{v} \dfrac{\partial m}{\partial u} = \dfrac{1}{u} \dfrac{\partial m}{\partial v}.$

4. The volume V of a right circular cone of height h and base radius r is given by:

$V = \dfrac{1}{3} \pi r^2 h.$

Find $\dfrac{\partial V}{\partial r}$ and $\dfrac{\partial V}{\partial h}$.　　$\left[\dfrac{\partial V}{\partial r} = \dfrac{2}{3} \pi rh, \dfrac{\partial V}{\partial h} = \dfrac{1}{3} \pi r^2 \right]$

5. The time of oscillation T of a simple pendulum is given by

$T = 2\pi \sqrt{\left(\dfrac{l}{g} \right)}$, where l is the length of the pendulum and g is the free fall acceleration due to gravity. Find $\dfrac{\partial T}{\partial l}$ and $\dfrac{\partial T}{\partial g}$.

$$\left[\dfrac{\partial T}{\partial l} = \dfrac{\pi}{\sqrt{(lg)}} \, , \dfrac{\partial T}{\partial g} = - \pi \sqrt{\left(\dfrac{l}{g^3} \right)} \right]$$

6. If $z = f(p - q)$, prove that

$\dfrac{\partial z}{\partial p} + \dfrac{\partial z}{\partial q} = 0.$

7. If $z = m^2 f(m - n)$, show that

$\dfrac{\partial z}{\partial m} + \dfrac{\partial z}{\partial n} = \dfrac{2z}{m}.$

Second order partial derivatives

In Problems 8 to 10, find (a) $\dfrac{\partial^2 z}{\partial x^2}$　　(b) $\dfrac{\partial^2 z}{\partial y \partial x}$　　(c) $\dfrac{\partial^2 z}{\partial x \partial y}$ and

(d) $\dfrac{\partial^2 z}{\partial y^2}$, given that $z = f(x, y)$.

8. $z = \ln xy$.   (a) $\left[-\dfrac{1}{x^2}\right]$ (b) $[0]$ (c) $[0]$ (d) $\left[-\dfrac{1}{y^2}\right]$

9. $z = x\cos y - y\cos x$.   (a) $[y\cos x]$ (b) $[\sin x - \sin y]$
    (c) $[\sin x - \sin y]$ (d) $[-x\cos y]$

10. $z = \dfrac{x}{x+y}$.   (a) $\left[-\dfrac{2y}{(x+y)^3}\right]$ (b) $\left[\dfrac{x-y}{(x+y)^3}\right]$

    (c) $\left[\dfrac{x-y}{(x+y)^3}\right]$ (d) $\left[\dfrac{2x}{(x+y)^3}\right]$

11. If $z = f(x, y)$ and $z = e^{-y}\sin x$, show that

$$\dfrac{\partial^2 z}{\partial x^2} + \dfrac{\partial^2 z}{\partial y^2} = 0.$$

12. If $z = f(x, t)$ and $z = \sqrt{\left(\dfrac{x}{4t}\right)}$, find $\dfrac{\partial^2 z}{\partial x^2}$ and show that $\dfrac{\partial^2 z}{\partial x \partial t} = \dfrac{\partial^2 z}{\partial t \partial x}$.

$$\left[-\dfrac{1}{8\sqrt{(tx^3)}}\right]$$

13 If $z = f(p, q)$ and $z = p^3\tan\dfrac{p}{q}$, prove that

$$\dfrac{\partial^2 z}{\partial p \partial q} = \dfrac{\partial^2 z}{\partial q \partial p} = -\dfrac{2p^3}{q^2}\left(\sec^2\dfrac{p}{q}\right)\left\{\dfrac{p}{q}\tan\left(\dfrac{p}{q}\right) + 2\right\}.$$

14. If $u = f(p, q)$ and $u = \dfrac{2p}{q^2}\ln q$, prove that $\dfrac{\partial^2 u}{\partial q \partial p} - \dfrac{\partial^2 u}{\partial p \partial q} = 0$.

15. If $\phi = f(r, \theta)$ and $\phi = (Ar^n + Br^{-n})\sin(n\theta + \alpha)$ where A, B, n and α are constants, show that

$$\dfrac{\partial^2 \phi}{\partial r^2} + \dfrac{1}{r}\dfrac{\partial \phi}{\partial r} + \dfrac{1}{r^2}\dfrac{\partial^2 \phi}{\partial \theta^2} = 0.$$

16. An equation used in thermodynamics is the Benedict–Webb–Rubine equation of state for the expansion of a gas. The equtaion is:

$$p = \dfrac{RT}{V} + \left(B_0 RT - A_0 - \dfrac{C_0}{T^2}\right)\dfrac{1}{V^2} + (bRT - a)\dfrac{1}{V^3} + \dfrac{A\alpha}{V^6}$$

$$+ \dfrac{C(1 + \gamma/V^2)}{T^2}\left(\dfrac{1}{V^3}\right)e^{-\gamma/V^2}$$

Show that $\dfrac{\partial^2 P}{\partial T^2} = \dfrac{6}{V^2 T^4}\left\{\dfrac{C}{V}\left(1 + \dfrac{\gamma}{V^2}\right)e^{-\frac{\gamma}{V^2}} - 0\,C_0\right\}$

17. An equation resulting from plucking a string is

$$y = \sin\dfrac{n\pi x}{l}\left\{k\cos\dfrac{n\pi b}{l}t + c\sin\dfrac{n\pi b}{l}t\right\}$$

174

Determine $\dfrac{\partial y}{\partial t}$ and $\dfrac{\partial y}{\partial x}$.

$$\left[\begin{array}{l} \dfrac{\partial y}{\partial t} = \dfrac{bn\pi}{l} \sin \dfrac{n\pi}{l} x \left\{ c \cos \dfrac{n\pi b}{l} t - k \sin \dfrac{n\pi b}{l} t \right\} \\[3mm] \dfrac{\partial y}{\partial x} = \dfrac{n\pi}{l} \cos \dfrac{n\pi}{l} x \left\{ k \cos \dfrac{n\pi b}{l} t + c \sin \dfrac{n\pi b}{l} t \right\} \end{array}\right]$$

18. The magnetic field vector H due to a steady current I flowing round a circular wire of radius a and at a distance z from its centre is given by:

$$H = \pm \frac{I}{2} \frac{\partial}{\partial z} \left(\frac{z}{\sqrt{(a^2 + z^2)}} \right)$$

Determine H. $\qquad \left[H = \pm \dfrac{a^2 I}{2\sqrt{(a^2 + z^2)^3}} \right]$

19. In a thermodynamic system
$$k = A e^{\left(\frac{T \Delta S - \Delta H}{RT} \right)},$$

where R, k and A are constants.
Find $\dfrac{\partial k}{\partial T}$, $\dfrac{\partial A}{\partial T}$, $\dfrac{\partial (\Delta S)}{\partial T}$ and $\dfrac{\partial (\Delta H)}{\partial T}$.

$$\left[\begin{array}{l} \dfrac{\partial k}{\partial T} = \dfrac{A \Delta H}{RT^2} e^{\left(\frac{T \Delta S - \Delta H}{RT} \right)}; \\[3mm] \dfrac{\partial A}{\partial T} = \dfrac{-k \Delta H}{RT^2} e^{\left(\frac{\Delta H - T \Delta S}{RT} \right)}; \\[3mm] \dfrac{\partial (\Delta S)}{\partial T} = \dfrac{-\Delta H}{T^2}; \quad \dfrac{\partial (\Delta H)}{\partial T} = \Delta S - R \ln \left(\dfrac{k}{A} \right) \end{array}\right]$$

Chapter 14

Total differential, rates of change and small changes

1 Total differential

In Chapter 13 the principles of partial differentiation are introduced for the case where only one variable changes at a time, the other variables being kept constant. In practice, variables may all be changing at the same time. Consider a right cylindrical block of metal being heated for a time δt. Let the initial volume of the block, V, be $\pi r^2 h$, where r is the radius of the base and h is the height of the cylinder. If the metal of the block has a positive temperature coefficient then both r and h will increase as the block is heated. Let the increase in r be δr, the increase in h be δh and the resultant increase in volume be δV. Then, after time δt:

the new volume, $V + \delta V \quad = \pi (r + \delta r)^2 (h + \delta h)$

The change in volume, $\delta V \quad = \pi (r + \delta r)^2 (h + \delta h) - V$

$\qquad \qquad \qquad \qquad \quad = \pi (r^2 + 2r\delta r + \delta r^2)(h + \delta h) - \pi r^2 h$

i.e. $\delta V = \pi r^2 h + 2\pi \, rh\delta r + \pi \, h\delta r^2 + \pi \, r^2 \delta h + 2\pi \, r\delta r\delta h + \pi\delta r^2 \delta h - \pi r^2 h.$

i.e. $\quad \delta V = 2\pi \, rh\delta r + \pi \, h\delta r^2 + \pi \, r^2\delta h + 2\pi \, r\delta r\delta h + \pi \, \delta r^2\delta h \qquad (1)$

Change of volume in unit time, $\dfrac{\delta V}{\delta t}$, is given by:

$$\frac{\delta V}{\delta t} = 2\pi \, rh \frac{\delta r}{\delta t} + \pi \, h\delta r \frac{\delta r}{\delta t} + \pi \, r^2 \frac{\delta h}{\delta t} + 2\pi \, r\delta r \frac{\delta h}{\delta t} + \pi \, \delta r^2 \frac{\delta h}{\delta t} \qquad (2)$$

As $\delta t \to 0$, the ratios $\dfrac{\delta V}{\delta t}, \dfrac{\delta r}{\delta t}$ and $\dfrac{\delta h}{\delta t}$ become $\dfrac{dV}{dt}, \dfrac{dr}{dt}$ and $\dfrac{dh}{dt}$ respectively, and terms containing the factors δr and δr^2 become zero. Hence rate of

volume, $\dfrac{dV}{dt} = 2\pi rh \dfrac{dr}{dt} + \pi r^2 \dfrac{dh}{dt}$. But it is shown in Chapter 13 that if

$V = \pi r^2 h$, then $\dfrac{\partial V}{\partial r} = 2\pi rh$ and $\dfrac{\partial V}{\partial h} = \pi r^2$

Thus, $\dfrac{dV}{dt} = \dfrac{\partial V}{\partial r} \dfrac{dr}{dt} + \dfrac{\partial V}{\partial h} \dfrac{dh}{dt}$ \hfill (3)

Equation (3) may be made independant of time by multiplying throughout by dt giving

$$\dfrac{dV}{dt} \, dt = \dfrac{\partial V}{\partial r} \dfrac{dr}{dt} \, dt + \dfrac{\partial V}{\partial h} \dfrac{dh}{dt} \, dt$$

i.e. $dV = \dfrac{\partial V}{\partial r} \, dr + \dfrac{\partial V}{\partial h} \, dh$ \hfill (4)

In this form, dV is called the **total differential** of V and $\dfrac{\partial V}{\partial r} \, dr$ and $\dfrac{\partial V}{\partial h} \, dh$ are called the **partial differentials** of V. Thus the total differential is the sum of the separate partial differentials. The total differential can be used as a basis for finding the change of volume with respect to some other quantity. For example, to find the change of volume with respect to, say, thermodynamic temperature, T, each term is divided by dT, giving

$$\dfrac{dV}{dT} = \dfrac{\partial V}{\partial r} \dfrac{dr}{dT} + \dfrac{\partial V}{\partial h} \dfrac{dh}{dT}$$

The total differential is also used as a basis for solving partial differential equations.

The result for the total differential of a right circular cylindrical block can be generalised. If $z = f(u, v, w,)$, then from equation (4),

$$dz = \dfrac{\partial z}{\partial u} \, du + \dfrac{\partial z}{\partial v} \, dv + \dfrac{\partial z}{\partial w} \, dw + \qquad (5)$$

Worked problems on total differentials

Problem 1. If $z = f(x, y)$ and $z = x^3 y^2 + \dfrac{y}{x}$, find the total differential, dz.

$\dfrac{\partial z}{\partial x} = y^2 (3x^2) - \dfrac{y}{x^2}$ \quad (y is kept constant)

$\dfrac{\partial z}{\partial y} = x^3 (2y) + \dfrac{1}{x}$ \quad (x is kept constant)

The total differential is the sum of the partial differentials,

i.e. $dz = \dfrac{\partial z}{\partial x} \, dx + \dfrac{\partial z}{\partial y} \, dy$

Thus, $dz = \left(3x^2 y^2 - \dfrac{y}{x^2} \right) dx + \left(2x^3 y + \dfrac{1}{x} \right) dy$

Problem 2. The pressure p, volume V and temperature T of unit mass of a gas are related by the formula $pV = RT$, where R is a constant.

Prove that $dp = \dfrac{p}{T} dT - \dfrac{p}{V} dV$, and also that

$$dT = \frac{T}{V} dV + \frac{T}{p} dp$$

Since $pV = RT$, $p = \dfrac{RT}{V}$, $\dfrac{\partial p}{\partial T} = \dfrac{R}{V}$ and $\dfrac{\partial p}{\partial V} = -\dfrac{RT}{V^2}$

From equation (5), $dp = \dfrac{\partial p}{\partial T} dT + \dfrac{\partial p}{\partial V} dV$.

$$= \frac{R}{V} dT + \left(-\frac{RT}{V^2}\right) dV$$

But from the initial equation, $\dfrac{R}{V} = \dfrac{p}{T}$, hence substituting for $\dfrac{R}{V}$ gives:

$$dp = \frac{p}{T} dT - \left(\frac{p}{T} \cdot \frac{T}{V}\right) dV$$

$$= \frac{p}{T} dT - \frac{p}{V} dV$$

Since $pV = RT$, $T = \dfrac{pV}{R}$, $\dfrac{\partial T}{\partial p} = \dfrac{V}{R}$ and $\dfrac{\partial T}{\partial V} = \dfrac{p}{R}$

From equation (5), $dT = \dfrac{\partial T}{\partial V} dV + \dfrac{\partial T}{\partial p} dp$

$$= \frac{p}{R} dV + \frac{V}{R} dp$$

But from the initial equation, $\dfrac{p}{R} = \dfrac{T}{V}$ and $\dfrac{V}{R} = \dfrac{T}{p}$,

hence, $dT = \dfrac{T}{V} dV + \dfrac{T}{p} dp$.

Further problems on total differentials may be found in Section 4 (Problems 1 to 5), page 182.

2 Rates of change

It is shown in Section 1, equation (3), that for a right circular cylindrical block being heated, the rate of change of volume is given by

$$\frac{dV}{dt} = \frac{\partial V}{\partial r} \frac{dr}{dt} + \frac{\partial V}{\partial h} \frac{dh}{dt}$$

This result may be generalised and if $z = f(u, v, w, \ldots)$ then

$$\frac{dz}{dt} = \frac{\partial z}{\partial u} \frac{du}{dt} + \frac{\partial z}{\partial v} \frac{dv}{dt} + \frac{\partial z}{\partial w} \frac{dw}{dt} + \cdots \qquad (6)$$

Equation (6) may be used to solve problems in which different quantities have different rates of change. The principles used are shown in the worked problems following.

Worked problems on rates of change

Problem 1. If $z = f(u, v)$ and $z = e^u \sin 2v$, find the rate of change of z, correct to four decimal places, when u is increasing at 3 units per second, v is decreasing at 5 units per second, u is 0.1 units and v is 0.5 units.

Since $z = e^u \sin 2v$, $\dfrac{\partial z}{\partial u} = e^u \sin 2v$ and $\dfrac{\partial z}{\partial v} = 2e^u \cos 2v$.

If u is increasing at 3 units/s, then $\dfrac{du}{dt} = 3$. Similarly, $\dfrac{dv}{dt} = -5$

From equation (6), $\dfrac{dz}{dt} = \dfrac{\partial z}{\partial u} \dfrac{du}{dt} + \dfrac{\partial z}{\partial v} \dfrac{dv}{dt}$

$$= (e^u \sin 2v)(3) + (2e^u \cos 2v)(-5)$$

But $u = 0.1$ and $v = 0.5$ units, hence

$$\frac{dz}{dt} = 3e^{0.1} \sin 1 - 10e^{0.1} \cos 1$$

Sin 1 means the sine of 1 radian, i.e. $\sin \left(\dfrac{180}{\pi} \right)^{\circ}$. Also cos 1 means $\cos \left(\dfrac{180}{\pi} \right)^{\circ}$.

Hence, $\dfrac{dz}{dt} = e^{0.1} \left[3 \sin \left(\dfrac{180}{\pi} \right)^{\circ} - 10 \cos \left(\dfrac{180}{\pi} \right)^{\circ} \right]$, i.e. $\dfrac{dz}{dt} = -3.181\,4$, correct to four decimal places.

That is, **z is decreasing at 3.181 4 units per second**, correct to four decimal places.

Problem 2. The radius of a right circular cylinder is increasing at a rate of 2 cm s^{-1} and the height is decreasing at a rate of 3 cm s^{-1}. Find the rate at which the volume is changing when the radius is 8 cm and the height is 5 cm.

The volume of a right cylinder, $V = \pi r^2 h$.

From equation (6), $\dfrac{dV}{dt} = \dfrac{\partial V}{\partial r} \dfrac{dr}{dt} + \dfrac{\partial V}{\partial h} \dfrac{dh}{dt}$

$$\frac{\partial V}{\partial r} = 2 \pi rh \text{ and } \frac{\partial V}{\partial h} = \pi r^2$$

Since the radius is increasing at 2 cm s^{-1}, $\dfrac{dr}{dt} = 2$.

Since the height is decreasing at 3 cm s^{-1}, $\dfrac{dh}{dt} = -3$.

Thus, rate of change of volume, $\dfrac{dV}{dt} = (2\,\pi\,rh)\,(2) + \pi\,r^2\,(-3)$.

When $r = 8$ and $h = 5$, $\dfrac{dV}{dt} = 4\,\pi\,(8)\,(5) - 3\,\pi\,(8)^2$

$$= 160\,\pi - 192\,\pi$$

$$= -32\,\pi$$

i.e. **the volume is decreasing at a rate of $32\,\pi$ cm^3 s^{-1}**.

Problem 3. A rectangular box has sides of length x mm, y mm and z mm. Sides x and y are expanding at rates of 1.5 and 2.5 mm s^{-1} respectively and side z is contracting at a rate of 0.25 mm s^{-1}. Find the rate of change of volume when x is 15 mm, y is 12 mm and z is 10 mm.

   Volume of box, $V = x\,y\,z$.

From equation (6), $\dfrac{dV}{dt} = \dfrac{\partial V}{\partial x}\dfrac{dx}{dt} + \dfrac{\partial V}{\partial y}\dfrac{dy}{dt} + \dfrac{\partial V}{\partial z}\dfrac{dz}{dt}$

$\dfrac{\partial V}{\partial x} = y\,z,\ \dfrac{\partial V}{\partial y} = x\,z$ and $\dfrac{\partial V}{\partial z} = x\,y$

Also, $\dfrac{dx}{dt} = 1.5,\ \dfrac{dy}{dt} = 2.5$ and $\dfrac{dz}{dt} = -0.25$

Hence, $\dfrac{dV}{dt} = y\,z\,(1.5) + x\,z\,(2.5) + x\,y\,(-0.25)$

When $x = 15, y = 12$ and $z = 10$,

$$\dfrac{dV}{dt} = 12 \times 10 \times 1.5 + 15 \times 10 \times 2.5 - 15 \times 12 \times 0.25$$

$$= +510,$$

i.e. **the rate of increase of volume is 510 mm^3 s^{-1}**.

Further problems on rates of change may be found in Section 4, (Problems 6 to 10), page 182.

3 Small changes

It is shown in Section 1, equation (1), that for a right circular cylinder being heated, the change of volume is given by

$$\delta V = 2\,\pi\,rh\delta r + \pi\,h\,\delta r^2 + \pi\,r^2\delta h + 2\,\pi\,r\delta r\delta h + \pi\,\delta r^2\,\delta h$$

If δr and δh are small compared with r and h, then products of these small terms such as δr^2, $\delta r\delta h$ and $\delta r^2\delta h$ will be very small and may be neglected. Thus the change of volume simplifies to:

$$\delta V \simeq 2\,\pi\,rh\delta r + \pi\,r^2\delta h$$

But, $V = \pi r^2 h$, so $\dfrac{\partial V}{\partial r} = 2\pi r h$ and $\dfrac{\partial V}{\partial h} = \pi r^2$.

Hence, $\delta V \simeq \dfrac{\partial V}{\partial r}\,\delta r + \dfrac{\partial V}{\partial h}\,\delta h$

This result may be generalised and if $z = f(u, v, w, \ldots)$, then the approximate change in z, δz, for small changes in u, v, and w, denoted by δu, δv and δw respectively, is given by

$$\delta z \simeq \frac{\partial z}{\partial u}\,\delta u + \frac{\partial z}{\partial v}\,\delta v + \frac{\partial z}{\partial w}\,\delta w + \cdots \tag{7}$$

Worked problems on small changes

Problem 1. The current in an electrical circuit is given by $i = \dfrac{v}{R}$ amperes. Determine the approximate change in current if the voltage falls from 240 V to 238 V and the value of the resistance is increased from 100 Ω to 100.5 Ω.

The relationship is $i = f(v, R)$ and $i = \dfrac{v}{R}$.

From equation (7), the change in current $\delta i \simeq \dfrac{\partial i}{\partial v}\,\delta v + \dfrac{\partial i}{\partial R}\,\delta R$.

Since $i = \dfrac{v}{R}$, $\dfrac{\partial i}{\partial v} = \dfrac{1}{R}$ and $\dfrac{\partial i}{\partial R} = -\dfrac{v}{R^2}$.

The small change in voltage, δv is -2 volts and change in resistance δR is $+0.5$ ohms.

Thus, $\delta i = \dfrac{1}{R}(-2) + \left(-\dfrac{v}{R^2}\right)(0.5)$.

But $R = 100\ \Omega$ and $v = 240$ volts (the initial values),

hence, $\delta i = -\dfrac{2}{100} - \dfrac{240}{2 \times 100^2} = -0.032$.

That is, the approximate change in current is a decrease of 0.032 A.

Problem 2. The pressure p and volume V of a gas are related by the equation $pV^{1.4} = C$.

Find the approximate percentage change in C when the pressure is increased by 2.3 per cent and the volume is decreased by 0.84 per cent.

Let p, V and C refer to the initial values.

From equation (7), $\delta C = \dfrac{\partial C}{\partial p}\,\delta p + \dfrac{\partial C}{\partial V}\,\delta V$.

Since $C = f(p, V)$ and $C = p\,V^{1.4}$, then

$$\frac{\partial C}{\partial p} = V^{1.4} \quad\text{and}\quad \frac{\partial C}{\partial V} = 1.4\,p\,V^{0.4}.$$

The pressure is increased by 2.3 per cent, i.e. the change in pressure
$\delta p = \dfrac{2.3}{100} \cdot p = 0.023\ p$.

The volume is decreased by 0.84 per cent, i.e. the change in volume

$\delta V = -\dfrac{0.84}{100} \cdot V = -0.008\,4\ V$.

Hence, the approximate change in C,

$$\delta C \doteq V^{1.4}\,(0.023\ p) + 1.4\,p\ V^{0.4}(-0.008\,4\ V)$$

$$\doteq p\ V^{1.4}\,(0.023 - 1.4\,(0.008\,4))$$

$$\doteq 0.011\,2\ p\ V^{1.4}$$

i.e. $\delta C \doteq \dfrac{1.12}{100} \cdot C$

That is, the approximate change in C is a 1.12 per cent increase.

Problem 3. The side a of a triangle is calculated from:

$a^2 = b^2 + c^2 - 2\,bc \cos A$. If b, c and A are measured as 2 mm, 4 mm and 60° respectively and the measurement errors which occur are $+0.1$ mm, $+0.15$ mm and $+2°$ respectively, determine the error in the calculated value of a

Since $a^2 = b^2 + c^2 - 2\,bc \cos A$

$a = (b^2 + c^2 - 2\,bc \cos A)^{\frac{1}{2}}$

From equation (7), $\delta a \doteq \dfrac{\partial a}{\partial b}\,\delta b + \dfrac{\partial a}{\partial c}\,\delta c + \dfrac{\partial a}{\partial A}\,\delta A$.

$$\dfrac{\partial a}{\partial b} = \dfrac{1}{2}\,(b^2 + c^2 - 2\,bc \cos A)^{-\frac{1}{2}}\,(2b - 2c \cos A)$$

$$\dfrac{\partial a}{\partial c} = \dfrac{1}{2}\,(b^2 + c^2 - 2bc \cos A)^{-\frac{1}{2}}\,(2c - 2b \cos A)$$

and $\dfrac{\partial a}{\partial A} = \dfrac{1}{2}\,(b^2 + c^2 - 2bc \cos A)^{-\frac{1}{2}}\,(2bc \sin A)$

Since $b = 2$, $c = 4$ and $A = 60° = \dfrac{\pi}{3}$ rad.

$$b^2 + c^2 - 2bc \cos A = 4 + 16 - 2 \times 8 \times \dfrac{1}{2}$$

$$= 12$$

and $\dfrac{1}{2}\,(b^2 + c^2 - 2bc \cos A)^{-\frac{1}{2}} = \dfrac{1}{2\sqrt{12}} \doteq 0.144\,3$

$\therefore \quad \dfrac{\partial a}{\partial b} \; \simeq \; 0.144\,3\,(4-4)=0$

$\dfrac{\partial a}{\partial c} \; \simeq \; 0.144\,3\,(8-2) \simeq 0.866$

$\dfrac{\partial a}{\partial A} \; \simeq \; 0.144\,3\,(16\,\dfrac{\sqrt{3}}{2}\,) \simeq 1.999$

Also $\delta b = 0.1$ $\delta c = 0.15$ and $\delta A = 2° = \dfrac{2 \times \pi}{180} = 0.035$ rad.

Hence, $\delta a \; \simeq \; 0 + 0.866 \times 0.15 + 1.999 \times 0.035$,
$\simeq 0.199$ mm

i.e. **the approximate error in the calculated value of a is $+ 0.20$ mm.**

Further problems on small changes may be found in the following Section (4) (Problems 11 to 22).

4 Further problems

Total differential

1. Determine the total differential, dz, given $x = f(x, y)$ and
$z = 3xy^2 - \dfrac{\sqrt{x}}{y^3}$. $\left[dz = (3y^2 - \dfrac{1}{2\sqrt{x}\,y^3}) \, dx + (6\,xy + \dfrac{3\sqrt{x}}{y^4}) \, dy \right]$

2. Find the total differential, dz, if $z = f(r, \theta)$ and $z = e^r (\cos \theta + j \sin \theta)$.
$[dz = e^r (\cos \theta + j \sin \theta) \, dr + e^r (-\sin \theta + j \cos \theta) \, d\theta]$

3. If $z = f(u, v, w)$ and $z = 2u^3 + 3v^2 - 4w$, find the total differential, dz.
$[dz = 6u^2 \, du + 6v dv - 4dw]$

4. If $z = p\,e^q + \dfrac{1}{p}$ ln q, where both p and q are variables, show that
$dz = (e^q - \dfrac{1}{p^2}$ ln $q) \, dp + (p\,e^q + \dfrac{1}{pq}) \, dq.$

5. If $xyz = k$, where k is a constant, prove that $dz = -z \left(\dfrac{dx}{x} + \dfrac{dy}{y} \right).$

Rates of change

6. If $z = 2p^3$ tan $\dfrac{q}{2}$, p is decreasing at 5 units per second and q is decreasing
at 0.04 units per second, find the rate of change of z when $p = 1$ and
$q = \pi/4$, correct to two decimal places. [z is decreasing at 12.47 units
per second]

7. Determine the rate of change of m, correct to four significant figures,
given the following data:
$m = f(u, v, w)$

$$m = 3 \ln \frac{v}{3} + w^3 \sec u$$

v is decreasing at 0.4 cm s^{-1}
u is increasing at 0.2 cm s^{-1}
w is increasing at 0.7 cm s^{-1}
$u = 3$ cm, $v = 3.7$ cm and $w = 4.4$ cm.
[m is decreasing at 38.94 cm s^{-1}, correct to four significant figures]

8. If $z = f(x, y)$ and $z = e^{\frac{y}{2}} \ln (2x + 3y)$, find the rate of increase of z, correct to four significant figures, when $x = 2$ cm, $y = 3$ cm, x is increasing at 5 cm s^{-1} and y is increasing at 4 cm s^{-1}.   [30.58 cm s^{-1}]

9. The radius of a right circular cone is increasing at a rate of 1 mm s^{-1} and its height is increasing at 2 mm s^{-1}. Determine, correct to three decimal places, the rate at which the volume is increasing in cm^3 s^{-1} when the radius is 1.2 cm and the height is 3.6 cm.   [1.206 cm^3 s^{-1}]

10. The area of a triangle is given by: area $= \frac{1}{2} bc \sin A$, where A is the angle between sides b and c. If b is increasing at 0.4 units s^{-1}, c is decreasing at 0.25 units s^{-1} and A is increasing at 0.1 units s^{-1}, find the rate of change of area of the triangle, correct to four decimal places when b is 2 units, c is 3 units and A is 0.5 units.
   [Increasing at 0.431 1 units^2 s^{-1}]

Small changes

11. An error of 3.5 per cent too large is made when measuring the radius of a sphere. Determine the approximate error in calculating: (a) the volume and (b) the surface area when they are calculated using the correct radius measurement.   (a) [10.5 per cent too large]
                                        (b) [7 per cent too large]

12. The area of a triangle is given by $A = \frac{1}{2} a b \sin C$, where C is the angle between sides a and b of the triangle. Calculate the approximate change in area when: (a) both a and b are increased by 2 per cent and (b) when a is increased by 2 per cent and b is reduced by 2 per cent.
              (a) [4 per cent increase]   (b) [no change.]

13. The moment of inertia of a body about an axis is given by $I = k b d^3$ where k is a constant and b and d are the dimensions of the body. If b and d are measured as 2 m and 0.8 m respectively and the measurement errors are 10 cm in b and -8 mm in d, determine the error in the calculated value of the moment of inertia using the measured values, in terms of k.   [0.02 k m^4]

14. The radius of a cone is reduced from 10 cm to 9.55 cm and its height is increased from 20 cm to 20.3 cm. Determine the approximate percentage change in its volume.   [7.5 per cent reduction]

15. The power developed by an engine is given by $I = b\,PLAN$, where b is a

constant. Find the approximate change in power in terms of b (Pa m^3 min^{-1} units) when:

P is increased from 2×10^5 to 2.1×10^5 Pa
L is reduced from 8.3 to 8.1 cm
A is reduced from 2.7 cm^2 to 2.6 cm^2, and
N is increased from 300 to 302 rev min^{-1}.
[A reduction of $6b$ (Pa m^3 min^{-1}) units]

16. The modulus of rigidity G is given by $G = \dfrac{R^4 \theta}{L}$, where R is the radius, θ the angle of twist and L the length. Find the approximate percentage change in G when R is increased by 1.5 per cent and θ is reduced by 5 per cent. [An increase of 1 per cent]

17. In triangle ABC, AB = 50 mm, AC = 60 mm and angle BAC = 45°. If length can be measured correct to the nearest 0.1 mm, and angles correct to the nearest 0.5°, determine the approximate value of maximum error which can occur in the calculated value of BC based on the measured results. [0.5 mm]

18. Side b of a triangle is calculated using the cosine rule. Determine the approximate error in b using the measured values if the measured values of a, c and B are 120 mm, 180 mm and 32° respectively, and the actual values are 121 mm, 179 mm and 32.25° respectively. [−0.6 mm]

19. The power consumed in an electrical resistor is given by $P = \dfrac{E^2}{R}$ watts, where E is the voltage drop across the resistor in volts and R the resistance of the resistor in ohms. Determine the approximate change in power when E increases by 4 per cent and R decreases by 0.05 per cent, if the original values of E and R are 100 volts and 10 ohms respectively. [An increase of 80.5 watts]

20. The resonant frequency of a series-connected electrical circuit is given by $f = \dfrac{1}{2 \pi \sqrt{(LC)}}$ Hz, where L is the inductance in henrys and C the capacitance in farads. Determine the approximate percentage change in the resonant frequency when L is increased by 1.7 per cent and C is reduced by 3.4 per cent. [An increase of 0.85 per cent.]

21. The rate of flow of gas in a pipe is given by $v = C d^{\frac{1}{2}} T^{-\frac{5}{6}}$, where C is a constant, d is the diameter of the pipe and T is the thermodynamic temperature of the gas. When determining the rate of flow experimentally, d is measured and subsequently found to be in error by + 1.6 per cent, and T has an error of −1.2 per cent. Determine the percentage error in the rate of flow based on the measured values of d and T. [1.8 per cent too large]

22. The volume (V) of a liquid of viscosity coefficient (η) delivered after time, t, when passed through a tube of length l and diameter d by a pressure p is given by:

$$V = \frac{p\,d^4\,t}{128\,\eta\,l}$$

If the error in V, p and l are 1 per cent, 2 per cent and 3 per cent respectively, determine the error in η.   [−2 per cent]

Chapter 15

Integration using substitution and partial fractions

1 Introduction

Revision of standard integrals.

1. $\int ax^n \, dx$ $= a\dfrac{x^{n+1}}{n+1} + c$ (except where $n = -1$)

2. $\int \cos ax \, dx$ $= \dfrac{1}{a} \sin ax + c$

3. $\int \sin ax \, dx$ $= -\dfrac{1}{a} \cos ax + c$

4. $\int \sec^2 ax \, dx$ $= \dfrac{1}{a} \tan ax + c$

5. $\int \operatorname{cosec}^2 ax \, dx$ $= -\dfrac{1}{a} \cot ax + c$

6. $\int \operatorname{cosec} ax \cot ax \, dx$ $= -\dfrac{1}{a} \operatorname{cosec} ax + c$

7. $\int \sec ax \tan ax \, dx$ $= \dfrac{1}{a} \sec ax + c$

8. $\int e^{ax} \, dx$ $= \dfrac{1}{a} e^{ax} + c$

9. $\displaystyle\int \dfrac{1}{x} \, dx$ $= \ln ax + c \ (= \ln x + \ln a + c = \ln x + c')$

Each of the above standard integrals may be readily checked by differentiation. Functions which require integrating are not usually in the standard integral form shown above. In such cases it may be possible by one of four methods to change the function into a form which can be readily integrated.

The methods available are:
- (a) by using algebraic substututions,
- (b) by using trigonometrical and hyperbolic identities and substitutions,
- (c) by using partial fractions,

and (d) integration by parts (see Chapter 16).

The first three methods are discussed in this chapter.

It should be realised that many mathematical functions cannot be integrated by analytical means at all and approximate methods have then to be used.

2 Integration using algebraic substitutions

An integral may often be reduced to a standard integral by making an algebraic substitution. The substitution usually made is to let u be equal to $f(x)$, such that $f(u)\,du$ is a standard integral.

A most important point in the use of substitution is that once it has been made the original variable must be removed completely, because a variable can only be integrated with respect to itself, i.e. we cannot integrate, for example, a function of t with respect to x.

A concept that $\dfrac{du}{dx}$ is a single entity (indicating the change of u with respect to x) has been established in the work done on differentiation. Frequently, in work on integration and on differential equations, $\dfrac{du}{dx}$ is split. Provided that when this is done the original differential coefficient can be reformed by applying the rules of algebra, then it is in order to do so. For example, if $\dfrac{dy}{dx} = x$, then it is in order to write $dy = x\,dx$ since dividing both sides by dx re-forms the original differential coefficient. This principle is shown in the following worked problems.

Worked problems on integration by algebraic substitutions

Type 1

Problem 1. Find (a) $\int \sin (6x + 5)\,dx$, (b) $\int (3t - 2)^8\,dt$

(c) $\int \dfrac{1}{(9x + 4)}\,dx$

(a) $\int \sin (6x + 5)\,dx$

Let $u = 6x + 5$, then $\dfrac{du}{dx} = 6$, i.e. $dx = \dfrac{du}{6}$

Hence $\int \sin (6x + 5)\,dx = \int \sin u\,\dfrac{du}{6} = \dfrac{1}{6} \int \sin u\,du = \dfrac{1}{6}(-\cos u) + c$

Since the original integral is given in terms of x the result should be stated in terms of x.

Since $u = 6x + 5$, $\int \sin (6x + 5) \, dx = -\dfrac{1}{6} \cos (6x + 5) + c$

(b) $\int (3t - 2)^8 \, dt$

Let $u = 3t - 2$, then $\dfrac{du}{dt} = 3$, i.e. $dt = \dfrac{du}{3}$

Hence $\int (3t - 2)^8 \, dt = \int u^8 \, \dfrac{du}{3} = \dfrac{1}{3} \int u^8 \, du = \dfrac{1}{3} \left(\dfrac{u^9}{9} \right) + c = \dfrac{1}{27} u^9 + c$

Since $u = 3t - 2$, $\int (3t - 2)^8 \, dt = \dfrac{1}{27} (3t - 2)^9 + c$

(c) $\displaystyle\int \dfrac{1}{(9x + 4)} \, dx$

Let $u = 9x + 4$, then $\dfrac{du}{dx} = 9$, i.e. $dx = \dfrac{du}{9}$

Hence $\displaystyle\int \dfrac{1}{(9x + 4)} \, dx = \int \dfrac{1}{u} \, \dfrac{du}{9} = \dfrac{1}{9} \ln u + c = \dfrac{1}{9} \ln (9x + 4) + c$

It may be seen from problem 1 that:

If 'x' in a standard integral is replaced by $(ax + b)$, where a and b are constants, then $(ax + b)$ is written for x in the result and the result is multiplied by $\dfrac{1}{a}$.

In general, $\int (ax + b)^n \, dx = \dfrac{1}{a (n + 1)} (ax + b)^{n+1} + c$ [except when $n = -1$ (see (c) above)]

With practice integrals of this type may be determined by inspection. A check can always be made by differentiating the result.

Problem 2. Determine the following by inspection:

(a) $\int 2 \sin (4x - 1) \, dx$ (b) $\int 3 \, e^{7t-2} \, dt$ (c) $\displaystyle\int \dfrac{4}{6\theta + 7} \, d\theta$

(a) $\int 2 \sin (4x - 1) \, dx = \dfrac{2}{4} [-\cos (4x - 1)] + c = -\dfrac{1}{2} \cos (4x - 1) + c$

(b) $\int 3 \, e^{7t-2} \, dt \qquad = 3 (e^{7t-2})(\dfrac{1}{7}) + c = \dfrac{3}{7} e^{7t-2} + c$

(c) $\displaystyle\int \dfrac{4}{6\theta + 7} \, d\theta \qquad = 4 [\ln (6\theta + 7)] (\dfrac{1}{6}) + c = \dfrac{2}{3} \ln (6\theta + 7) + c$

Type 2.

Problem 3. Find (a) $\int 2x (3x^2 + 5)^6 \, dx$ (b) $\int \sin^3 \theta \cos \theta \, d\theta$

(a) $\int 2x (3x^2 + 5)^6 \, dx$

There appears to be two possible choices for the substitution, i.e. either to let $u = 2x$ or to let $u = (3x^2 + 5)^6$. Choosing the latter substitution, if $u = (3x^2 + 5)$, then $\dfrac{du}{dx} = 6x$, i.e. $dx = \dfrac{du}{6x}$

Hence $\int 2x \, (3x^2 + 5)^6 \, dx = \int 2x \, (u)^6 \, \dfrac{du}{6x} = \int \dfrac{2u^6}{6} \, du = \dfrac{1}{3} \int u^6 \, du$

The original variable, x, has been removed completely and the integral is now only in terms of u.

Hence $\dfrac{1}{3} \int u^6 \, du = \dfrac{1}{3} \left(\dfrac{u^7}{7} \right) + c = \dfrac{1}{21} \, (3x^2 + 5)^7 + c$

(b) $\int \sin^3 \theta \, \cos \theta \, d\theta$

Let $u = \sin \theta$, then $\dfrac{du}{d\theta} = \cos \theta$, i.e. $d\theta = \dfrac{du}{\cos \theta}$

Hence $\int \sin^3 \theta \, \cos \theta \, d\theta = \int u^3 \, \cos \theta \, \dfrac{du}{\cos \theta} = \int u^3 \, du = \dfrac{u^4}{4} + c$

$\qquad\qquad\qquad\qquad\qquad\qquad\qquad\qquad = \dfrac{1}{4} \, \sin^4 \theta + c$

Whenever a power of a sine is multiplied by a cosine of power 1, or vice versa, the integral may be determined by inspection.

For example, $\int \cos^5 x \sin x \, dx = -\dfrac{1}{6} \, \cos^6 x + c$,

since $\dfrac{d}{dx} \left\{ -\dfrac{1}{6} \, \cos^6 x + c \right\} = \cos^5 x \sin x$.

It may be seen from problem 3 that:

Integrals of the form $k \int [f(x)]^n \, f'(x) \, dx$ (where k and n are constants) can be integrated by substituting u for $f(x)$.

Type 3
Problem 4. Find (a) $\displaystyle\int \dfrac{x}{1 + x^2} \, dx$ \qquad (b) $\displaystyle\int \dfrac{4x}{\sqrt{(5x^2 - 1)}} \, dx$

(c) $\int \tan \theta \, d\theta$

(a) $\displaystyle\int \dfrac{x}{1 + x^2} \, dx$

Let $u = 1 + x^2$, then $\dfrac{du}{dx} = 2x$, i.e. $\qquad dx = \dfrac{du}{2x}$

Hence $\displaystyle\int \dfrac{x}{1 + x^2} \, dx = \int \dfrac{x}{u} \, \dfrac{du}{2x} = \dfrac{1}{2} \int \dfrac{1}{u} \, du \qquad = \dfrac{1}{2} \, \ln u + c$

$\qquad\qquad\qquad\qquad\qquad\qquad\qquad\qquad\qquad\qquad = \dfrac{1}{2} \, \ln (1 + x^2) + c$

(b) $\displaystyle\int \frac{4x}{\sqrt{(5x^2-1)}} \, dx$

Let $u = 5x^2 - 1$, then $\dfrac{du}{dx} = 10x$, i.e. $dx = \dfrac{du}{10x}$

Hence $\displaystyle\int \frac{4x}{\sqrt{(5x^2-1)}} \, dx = \int \frac{4x}{\sqrt{u}} \, \frac{du}{10x} = \frac{2}{5} \int u^{-1/2} \, du = \frac{2}{5} \frac{u^{1/2}}{1/2} + c$

$$= \frac{4}{5} \sqrt{u} + c$$

$$= \frac{4}{5} \sqrt{(5x^2-1)} + c$$

(c) $\displaystyle\int \tan \theta \, d\theta = \int \frac{\sin \theta}{\cos \theta} \, d\theta$

Let $u = \cos \theta$, then $\dfrac{du}{d\theta} = -\sin \theta$, i.e. $d\theta = -\dfrac{du}{\sin \theta}$

Hence $\displaystyle\int \frac{\sin \theta}{\cos \theta} \, d\theta = \int \frac{\sin \theta}{u} \left(\frac{-du}{\sin \theta} \right) = -\int \frac{1}{u} \, du = -\ln u + c$

$$= -\ln (\cos \theta) + c$$

$$= \ln (\cos \theta)^{-1} + c$$

i.e. $\displaystyle\int \tan \theta \, d\theta \qquad = \ln (\sec \theta) + c$

It may be seen from Problem 4 that:

Integrals of the form $k \displaystyle\int \frac{f'(x)}{[f(x)]^n} \, dx$ **(where k and n are constants) can be intergrated by substituting u for $f(x)$.**

If, in an integral, it is recognised that the numerator is the differential coefficient of the denominator (times a constant) then it may be integrated by inspection.

General problem on definite integrals

Problem 5. Evaluate the following: (a) $\int_1^3 3 \sec^2 (2\theta - 1) \, d\theta$
(b) $\int_0^3 3x \sqrt{(3x^2 + 9)} \, dx$, taking positive values of roots only.
(c) $\displaystyle\int_1^2 \frac{2e^t}{3 + e^t} \, dt$

(a) $\displaystyle\int_1^3 3 \sec^2 (2\theta - 1) \, d\theta = \left[\frac{3}{2} \tan (2\theta - 1) \right]_1^3 = \frac{3}{2} [\tan 5 - \tan 1]$

$$= \frac{3}{2} [\tan 286° \; 29' - \tan 57° \; 18'] \text{, since 'tan 5'}$$

means 'the tangent of 5 radians',

i.e. $\tan 286° \; 29'$.

$$= \frac{3}{2} \left[-3.380\,5 - 1.557\,4\right] = -7.406\,9$$

(b) $\int_0^3 3x \sqrt{(3x^2 + 9)} \, dx$

Let $u = 3x^2 + 9$, then $\dfrac{du}{dx} = 6x$, i.e. $dx = \dfrac{du}{6x}$

Then $\int_0^3 3x \sqrt{(3x^2 + 9)} \, dx = \int 3x \sqrt{u} \dfrac{du}{6x}$

$$= \frac{1}{2} \int \sqrt{u} \, du = \frac{1}{3} \sqrt{u^3}$$

$$= \frac{1}{3} \left[\sqrt{(3x^2 + 9)^3} \ \right]_0^3$$

$$= \frac{1}{3} \left[\sqrt{(36)^3} - \sqrt{(9)^3} \right]$$

$$= \frac{1}{3} \left[216 - 27\right], \text{ taking positive values of roots}$$
$$\text{only.}$$

$$= \mathbf{63}$$

(c) $\displaystyle\int_1^2 \frac{2e^t}{3 + e^t} \, dt = \left[2 \ln (3 + e^t)\right]_1^2$ (by inspection, or by using the substitution $u = 3 + e^t$)

$$= 2 \left[\ln (3 + e^2) - \ln (3 + e^1)\right] = 2 \ln \left(\frac{3 + e^2}{3 + e^1}\right)$$

$$= \mathbf{1.194\,2}$$

Further problems on integration by algebraic substitutions may be found in Section 6, (Problems 1 to 38) page 209.

3 Integration using trigonometrical and hyperbolic identities and substitutions

(A) Integration of $\cos^2 x$, $\sin^2 x$, $\tan^2 x$ and $\cot^2 x$

It may be shown that the compound angle addition formula $\cos (A + B)$ is given by:

$$\cos (A + B) = \cos A \cos B - \sin A \sin B.$$

If $A = B$, then $\cos 2A = \cos^2 A - \sin^2 A$ (1)

Since $\cos^2 A + \sin^2 A = 1$ then $\sin^2 A = 1 - \cos^2 A$
    Hence $\cos 2A = \cos^2 A - (1 - \cos^2 A)$
    i.e. $\cos 2A = 2 \cos^2 A - 1$ (2)

Also $\cos^2 A = 1 - \sin^2 A$
    Hence $\cos 2A = (1 - \sin^2 A) - \sin^2 A$
    i.e. $\cos 2A = 1 - 2 \sin^2 A$ (3)

From equation (2), $\cos^2 A = \frac{1}{2} (1 + \cos 2A)$ (4)

From equation (3), $\sin^2 A = \frac{1}{2} (1 - \cos 2A)$ (5)

192

Thus $\int\cos^2 x \, dx = \int \frac{1}{2}(1+\cos 2x)\,dx = \frac{1}{2}\left(x+\frac{\sin 2x}{2}\right)+c$

Similarly, $\int\sin^2 x \, dx = \int \frac{1}{2}(1-\cos 2x)\,dx = \frac{1}{2}\left(x-\frac{\sin 2x}{2}\right)+c$

From $1+\tan^2 x = \sec^2 x$, $\tan^2 x = \sec^2 x - 1$

Thus $\int\tan^2 x \, dx = \int(\sec^2 x - 1)\,dx = \tan x - x + c$

From $\cot^2 x + 1 = \operatorname{cosec}^2 x$, $\cot^2 x = \operatorname{cosec}^2 x - 1$

Thus $\int \cot^2 x \, dx = \int(\operatorname{cosec}^2 x - 1)\,dx = -\cot x - x + c$

Problem 1. Find (a) $\int\cos^2 5t \, dt$ (b) $\int\tan^2 3\theta \, d\theta$

(a) $\int\cos^2 5t \, dt = \int \frac{1}{2}(1+\cos 10t)\,dt = \frac{1}{2}\left(t+\frac{\sin 10t}{10}\right)+c$

(b) $\int\tan^2 3\theta \, d\theta = \int(\sec^2 3\theta - 1)\,d\theta = \frac{1}{3}\tan 3\theta - \theta + c$

(B) Powers of sines and cosines of the form $\int\cos^m x \sin^n x \, dx$

(i) To evaluate $\int\cos^m x \sin^n x \, dx$ when either m or n is odd (but not both), use is made of the trigonometric identity $\cos^2 x + \sin^2 x = 1$ as shown in Problems 1 and 2.

Problem 1. Find $\int\cos^5 x \, dx$

$\int\cos^5 x \, dx = \int\cos x (\cos^2 x)^2 \, dx$

$= \int\cos x (1-\sin^2 x)^2 \, dx = \int\cos x (1-2\sin^2 x + \sin^4 x) \, dx$

$= \int\cos x - 2\sin^2 x \cos x + \sin^4 x \cos x \, dx$

Hence the integral has been reduced to one which may be determined by inspection (see Section 2, type 2).

Hence $\int\cos^5 x \, dx = \sin x - \frac{2}{3}\sin^3 x + \frac{1}{5}\sin^5 x + c$

Problem 2. Find $\int\sin^3 x \cos^2 x \, dx$

$\int\sin^3 x \cos^2 x \, dx = \int\sin x (\sin^2 x)(\cos^2 x) \, dx$

$= \int\sin x (1-\cos^2 x)(\cos^2 x) \, dx$

$= \int(\sin x \cos^2 x - \sin x \cos^4 x) \, dx$

$= -\frac{1}{3}\cos^3 x + \frac{1}{5}\cos^5 x + c$

(ii) To evaluate $\int\cos^m x \sin^n x \, dx$ when m and n are both even use is made of equations (4) and (5) of Section 3(A), i.e.

$\cos^2 x = \frac{1}{2}(1+\cos 2x)$ *and* $\sin^2 x = \frac{1}{2}(1-\cos 2x)$.

(It follows from these equations that $\cos^2 2x = \frac{1}{2}(1 + \cos 4x)$, $\sin^2 3x = \frac{1}{2}(1 - \cos 6x)$, and so on.)

Problem 3. Find $\int \cos^4 x \, dx$

$$\int \cos^4 x \, dx = \int (\cos^2 x)^2 \, dx = \int [\frac{1}{2}(1 + \cos 2x)]^2 \, dx$$
$$= \frac{1}{4} \int (1 + 2\cos 2x + \cos^2 2x) \, dx$$
$$= \frac{1}{4} \int 1 + 2\cos 2x + \frac{1}{2}(1 + \cos 4x) \, dx$$
$$\text{(since } \cos 4x = 2\cos^2 2x - 1)$$
$$= \frac{1}{4} \int [\frac{3}{2} + 2\cos 2x + \frac{1}{2}\cos 4x] \, dx$$

Hence $\quad \int \cos^4 x \, dx = \frac{1}{4}[\frac{3}{2}x + \sin 2x + \frac{1}{8}\sin 4x] + c$

Problem 4. Find $\int \sin^4 x \cos^2 x \, dx$

$$\int \sin^4 x \cos^2 x \, dx = \int (\sin^2 x)^2 \cos^2 x \, dx = \int \left(\frac{1 - \cos 2x}{2}\right)^2 \left(\frac{1 + \cos 2x}{2}\right) dx$$
$$= \frac{1}{8} \int (1 - 2\cos 2x + \cos^2 2x)(1 + \cos 2x) \, dx$$
$$= \frac{1}{8} \int (1 - 2\cos 2x + \cos^2 2x + \cos 2x - 2\cos^2 2x$$
$$+ \cos^3 2x) \, dx$$
$$= \frac{1}{8} \int (1 - \cos 2x - \cos^2 2x + \cos^3 2x) \, dx$$
$$= \frac{1}{8} \int 1 - \cos 2x - \left(\frac{1 + \cos 4x}{2}\right) + \cos 2x (1 - \sin^2 2x) \, dx$$
$$= \frac{1}{8} \int [\frac{1}{2} - \cos 2x - \frac{1}{2}\cos 4x + \cos 2x - \cos 2x \sin^2 2x] \, dx$$
$$= \frac{1}{8} \int [\frac{1}{2} - \frac{1}{2}\cos 4x - \cos 2x \sin^2 2x] \, dx$$
$$= \frac{1}{8} \left[\frac{x}{2} - \frac{\sin 4x}{8} - \frac{\sin^3 2x}{6} \right] + c$$

Hence $\int \sin^4 x \cos^2 x \, dx = \frac{1}{16}(x - \frac{1}{4}\sin 4x - \frac{1}{3}\sin^3 2x) + c$

(C) Products of sines and cosines

It is shown in *Technician Mathematics level 3* by J.O. Bird and A.J.C. May, Chapter 1, Section 11, that:

$$\sin A \cos B = \frac{1}{2}[\sin (A + B) + \sin (A - B)] \qquad (1)$$
$$\cos A \sin B = \frac{1}{2}[\sin (A + B) - \sin (A - B)] \qquad (2)$$
$$\cos A \cos B = \frac{1}{2}[\cos (A + B) + \cos (A - B)] \qquad (3)$$
$$\sin A \sin B = -\frac{1}{2}[\cos (A + B) - \cos (A - B)] \qquad (4)$$

These formulae are used when integrating products of sines and cosines and are particularly useful for Fourier series calculations (see Chapter 22).

Problem 1. Find $\int \sin 4\theta \cos 3\theta \, d\theta$.

$\int \sin 4\theta \cos 3\theta \, d\theta = \int \frac{1}{2}(\sin 7\theta + \sin \theta) \, d\theta$ from equation (1).

$$= \frac{1}{2} \left(-\frac{\cos 7\theta}{7} - \cos \theta \right) + c$$

$$= -\frac{1}{14}(\cos 7\theta + 7 \cos \theta) + c$$

Problem 2. Find $\int \cos 6x \sin 2x \, dx$

$\int \cos 6x \sin 2x \, dx = \int \frac{1}{2}(\sin 8x - \sin 4x) \, dx$,   from equation (2)

$$= \frac{1}{2} \left(\frac{-\cos 8x}{8} + \frac{\cos 4x}{4} \right) + c$$

$$= \frac{1}{16}(2 \cos 4x - \cos 8x) + c$$

Problem 3. Find $\int \cos 3t \cos t \, dt$

$\int \cos 3t \cos t \, dt = \int \frac{1}{2}(\cos 4t + \cos 2t) \, dt$, from equation (3)

$$= \frac{1}{2} \left(\frac{\sin 4t}{4} + \frac{\sin 2t}{2} \right) + c$$

$$= \frac{1}{8}(\sin 4t + 2 \sin 2t) + c$$

Problem 4. Find $\int \sin 6x \sin 3x \, dx$

$\int \sin 6x \sin 3x \, dx = \int -\frac{1}{2}(\cos 9x - \cos 3x) \, dx$, from equation (4)

$$= -\frac{1}{2} \left(\frac{\sin 9x}{9} - \frac{\sin 3x}{3} \right) + c$$

$$= -\frac{1}{18}(\sin 9x - 3 \sin 3x) + c$$

(D) Integrals containing $\sqrt{(a^2 - x^2)}$ – the 'sine θ' substitution

When an integral contains a term $\sqrt{(a^2 - x^2)}$, the substitution $x = a \sin \theta$ is used. The reasons for this are made obvious by the following worked problems.

Problem 1. Find $\int \frac{1}{\sqrt{(a^2 - x^2)}} \, dx$

Let $x = a \sin \theta$, then $\frac{dx}{d\theta} = a \cos \theta$, i.e. $dx = a \cos \theta \, d\theta$

Hence $\int \dfrac{1}{\sqrt{(a^2-x^2)}}\ \mathrm{d}x = \int \dfrac{1}{\sqrt{(a^2-a^2\sin^2\theta)}}\ a\cos\theta\ \mathrm{d}\theta$

$$= \int \dfrac{a\cos\theta\ \mathrm{d}\theta}{\sqrt{[(a^2\,(1-\sin^2\theta)\}}}$$

$$= \int \dfrac{a\cos\theta}{a\cos\theta}\ \mathrm{d}\theta,\ \text{since}\ 1-\sin^2\theta = \cos^2\theta$$

$$= \int \mathrm{d}\theta = \theta + c$$

Since $x = a\sin\theta$ then $\sin\theta = \dfrac{x}{a}$ and $\theta = \arcsin\dfrac{x}{a}$

Hence $\int \dfrac{1}{\sqrt{(a^2-x^2)}}\ \mathrm{d}x = \arcsin\dfrac{x}{a} + c$

Problem 2. Evaluate $\displaystyle\int_0^2 \dfrac{1}{\sqrt{(16-x^2)}}\ \mathrm{d}x$

Using the result of problem 1, $\displaystyle\int_0^2 \dfrac{1}{\sqrt{(16-x^2)}}\ \mathrm{d}x = \left[\ \arcsin\dfrac{x}{4}\ \right]_0^2$

$$= (\arcsin\tfrac{1}{2} - \arcsin 0)$$

$$= \dfrac{\pi}{6}\ \text{ or }\ 0.523\ 6$$

Problem 3. Find $\int \sqrt{(a^2-x^2)}\ \mathrm{d}x$

Let $x = a\sin\theta$, then $\dfrac{\mathrm{d}x}{\mathrm{d}\theta} = a\cos\theta$, i.e. $\mathrm{d}x = a\cos\theta\ \mathrm{d}\theta$

Thus $\int\sqrt{(a^2-x^2)}\ \mathrm{d}x = \int\sqrt{(a^2-a^2\sin^2\theta)}\,(a\cos\theta\ \mathrm{d}\theta)$

$$= \int\sqrt{[a^2(1-\sin^2\theta)]}\,(a\cos\theta\ \mathrm{d}\theta)$$

$$= \int(a\cos\theta)\,(a\cos\theta)\ \mathrm{d}\theta$$

$$= \int a^2\cos^2\theta\ \mathrm{d}\theta$$

$$= a^2\int\tfrac{1}{2}(1+\cos 2\theta)\ \mathrm{d}\theta$$

$$= \dfrac{a^2}{2}\left(\ \theta + \dfrac{\sin 2\theta}{2}\ \right) + c$$

In the compound angle addition formula $\sin(A+B) = \sin A\cos B + \cos A\sin B$. If $A = B$ then $\sin 2A = 2\sin A\cos A$.

Hence $\int\sqrt{(a^2-x^2)}\ \mathrm{d}x = \dfrac{a^2}{2}\left(\ \theta + \dfrac{2\sin\theta\cos\theta}{2}\ \right) + c$

$$= \dfrac{a^2}{2}(\theta + \sin\theta\cos\theta) + c$$

Since $x = a \sin \theta$ then $\sin \theta = \dfrac{x}{a}$ and $\theta = \arcsin \dfrac{x}{a}$

Also $\cos^2 \theta + \sin^2 \theta = 1$ from which $\cos \theta = \sqrt{(1 - \sin^2 \theta)}$

i.e. $\cos \theta = \sqrt{\left[1 - \left(\dfrac{x}{a} \right)^2 \right]} = \sqrt{\left(\dfrac{a^2 - x^2}{a^2} \right)} = \dfrac{\sqrt{(a^2 - x^2)}}{a}$

Hence $\int \sqrt{(a^2 - x^2)}\, dx = \dfrac{a^2}{2} \left[\arcsin \dfrac{x}{a} + \left(\dfrac{x}{a} \right) \dfrac{\sqrt{(a^2 - x^2)}}{a} \right] + c$

$$= \dfrac{a^2}{2} \arcsin \dfrac{x}{a} + \dfrac{x}{2} \sqrt{(a^2 - x^2)} + c$$

Problem 4. Evaluate $\int_0^3 \sqrt{(9 - x^2)}\, dx$

Using the result of Problem 3:

$\int_0^3 \sqrt{(9 - x^2)}\, dx = \left[\dfrac{9}{2} \arcsin \dfrac{x}{3} + \dfrac{x}{2} \sqrt{(9 - x^2)} \right]_0^3$

$$= \left[\dfrac{9}{2} \arcsin 1 + \dfrac{3}{2} \sqrt{(9 - 9)} \right] - \left[\dfrac{9}{2} \arcsin 0 + 0 \right]$$

$$= \dfrac{9}{2} \arcsin 1$$

'arcsin 1' means 'the angle whose sine is equal to 1', i.e. $\dfrac{\pi}{2}$ radians.

Hence $\int_0^3 \sqrt{(9 - x^2)}\, dx = \dfrac{9}{2} \times \dfrac{\pi}{2} = \dfrac{9\pi}{4}$ **or 7.068 6.**

Problem 5. Find $\displaystyle\int \dfrac{1}{\sqrt{(5 + 4x - x^2)}}\, dx$

$\displaystyle\int \dfrac{1}{\sqrt{(5 + 4x - x^2)}}\, dx = \int \dfrac{1}{\sqrt{[-(x^2 - 4x - 5)]}}\, dx$

$$= \int \dfrac{1}{\sqrt{[-\{(x - 2)^2 - 9\}]}}\, dx$$

$$= \int \dfrac{1}{\sqrt{[(3)^2 - (x - 2)^2]}}\, dx = \arcsin \left(\dfrac{x - 2}{3} \right) + c$$

(E) $\displaystyle\int \dfrac{1}{a^2 + x^2}\, dx$; the '$\tan \theta$' substitution

When an integral is of the form $\dfrac{1}{a^2 + x^2}$ the substitution $x = a \tan \theta$ is used. The reason for this is made obvious by the following worked problem.

Problem 1. Find $\int \dfrac{1}{a^2 + x^2} \, dx$

Let $x = a \tan \theta$, then $\dfrac{dx}{d\theta} = a \sec^2 \theta$, i.e. $dx = a \sec^2 \theta \, d\theta$

$$\int \frac{1}{a^2 + x^2} \, dx = \int \frac{1}{a^2 + a^2 \tan^2 \theta} \; a \sec^2 \theta \, d\theta = \int \frac{a \sec^2 \theta \, d\theta}{a^2 (1 + \tan^2 \theta)}$$

$$= \int \frac{a \sec^2 \theta}{a^2 \sec^2 \theta} \, d\theta, \text{ since } 1 + \tan^2 \theta = \sec^2 \theta$$

$$= \int \frac{1}{a} \, d\theta = \frac{1}{a} (\theta) + c$$

Hence $\int \dfrac{1}{a^2 + x^2} \, dx = \dfrac{1}{a} \arctan \dfrac{x}{a} + c$, since $x = a \tan \theta$.

Problem 2. Evaluate $\int_0^1 \dfrac{1}{9 + 4x^2} \, dx$, correct to four significant figures.

$$\int_0^1 \frac{1}{9 + 4x^2} \, dx = \int_0^1 \frac{1}{4(\frac{9}{4} + x^2)} \, dx = \frac{1}{4} \int_0^1 \frac{1}{(\frac{3}{2})^2 + x^2} \, dx$$

$$= \frac{1}{4} \left[\frac{1}{(\frac{3}{2})} \arctan \frac{x}{(\frac{3}{2})} \right]_0^1$$

from the result of Problem 1

$$= \frac{1}{6} \left[\arctan \frac{2x}{3} \right]_0^1$$

$$= \frac{1}{6} \left[\arctan \frac{2}{3} - \arctan 0 \right]$$

$$= \frac{1}{6} (0.588\,0) = \mathbf{0.098\,0}$$

(F) Integrals containing $\sqrt{(x^2 + a^2)}$ – the 'sinh θ' substitution

When an integral contains a term $\sqrt{(x^2 + a^2)}$ the substitution $x = a \sinh \theta$ is used. The reason for this is made obvious by the following worked problems.

Problem 1. Find (a) $\int \dfrac{1}{\sqrt{(x^2 + a^2)}} \, dx$, and (b) evaluate $\int_0^3 \dfrac{1}{\sqrt{(x^2 + 9)}} \, dx$

(a) Let $x = a \sinh \theta$ then $\dfrac{dx}{d\theta} = a \cosh \theta$, i.e. $dx = a \cosh \theta \, d\theta$

$$\int \frac{1}{\sqrt{(x^2 + a^2)}} \, dx = \int \frac{1}{\sqrt{(a^2 \sinh^2 \theta + a^2)}} \; a \cosh \theta \, d\theta$$

$$= \int \frac{a \cosh \theta \, d\theta}{\sqrt{[a^2 (\sinh^2 \theta + 1)]}}$$

$$= \int \frac{a \cosh \theta}{a \cosh \theta} \, d\theta, \text{ since } \sinh^2 \theta + 1 = \cosh^2 \theta$$

$$= \int d\theta = \theta = \text{arsinh} \frac{x}{a} + c \text{ since } x = a \sinh \theta$$

From Chapter 12, $\text{arsinh} \frac{x}{a} = \ln \left\{ \frac{x + \sqrt{(x^2 + a^2)}}{a} \right\}$ which gives an alternative solution to $\int \frac{1}{\sqrt{(x^2 + a^2)}} \, dx$.

(b) $\int_0^3 \frac{1}{\sqrt{(x^2 + 9)}} \, dx = \left[\text{arsinh} \frac{x}{3} \right]_0^3$ or $\left[\ln \left\{ \frac{x + \sqrt{(x^2 + 9)}}{3} \right\} \right]_0^3$

from part (a)

$$\left[\text{arsinh} \frac{x}{3} \right]_0^3 = (\text{arsinh } 1 - \text{arsinh } 0)$$

To evaluate arsinh 1: If $x = \text{arsinh } 1$

$$\text{then } \sinh x = 1$$

$$\frac{e^x - e^{-x}}{2} = 1$$

$$e^x - e^{-x} = 2$$

Multiplying each term by e^x gives: $(e^x)^2 - 2e^x - 1 = 0$

$$e^x = \frac{-(-2) \pm \sqrt{[(-2)^2 - 4(1)(-1)]}}{2(1)}$$

$$= \frac{2 \pm \sqrt{8}}{2}$$

Taking natural logarithms of both sides gives:

$$x = \ln \left(\frac{2 + \sqrt{8}}{2} \right), \text{ since the logarithms of a negative number cannot be determined in real terms}$$

$$= 0.881\ 4$$

Hence $\text{arsinh } 1 = 0.881\ 4$

Hence $\int_0^3 \frac{1}{\sqrt{(x^2 + 9)}} \, dx = \mathbf{0.881\ 4}$

Using the logarithmic form:

$$\int_0^3 \frac{1}{\sqrt{(x^2 + 9)}} \, dx = \left[\ln \left\{ \frac{x + \sqrt{(x^2 + 9)}}{3} \right\} \right]_0^3$$

$$= \ln \left(\frac{3 + \sqrt{18}}{3} \right) = \mathbf{0.881\ 4}, \text{ as above.}$$

The logarithmic form is thus seen to be easier for evaluating definite integrals.

Problem 2. Find $\int \sqrt{(x^2 + a^2)}\, dx$

Let $x = a \sinh \theta$, then $dx = a \cosh \theta\, d\theta$

$$\int \sqrt{(x^2 + a^2)}\, dx = \int \sqrt{(a^2 \sinh^2 \theta + a^2)}\, a \cosh \theta\, d\theta$$

$$= \int \sqrt{[a^2 (\sinh^2 \theta + 1)]}\, a \cosh \theta\, d\theta$$

$$= \int (a \cosh \theta)(a \cosh \theta)\, d\theta, \text{ since } \sinh^2 \theta + 1 = \cosh^2 \theta$$

$$= \int a^2 \cosh^2 \theta\, d\theta = a^2 \int \left(\frac{1 + \cosh 2\theta}{2} \right) d\theta$$

since $\cosh 2\theta = 2 \cosh^2 \theta - 1$ (see Chapter 7)

$$= \frac{a^2}{2} \left[\theta + \frac{\sinh 2\theta}{2} \right] + c$$

Hence $\int \sqrt{(x^2 + a^2)}\, dx = \frac{a^2}{2}\, [\theta + \sinh \theta \cosh \theta] + c$, since

$$\sinh 2\theta = 2 \sinh \theta \cosh \theta$$

Since $x = a \sinh \theta$ then $\sinh \theta = \dfrac{x}{a}$ and $\theta = \operatorname{arsinh} \dfrac{x}{a}$.

Since $\cosh^2 \theta - \sinh^2 \theta = 1$, then $\cosh \theta = \sqrt{(1 + \sinh^2 \theta)}$

$$= \sqrt{\left[1 + \left(\frac{x}{a} \right)^2 \right]} = \frac{1}{a} \sqrt{(a^2 + x^2)}$$

Hence $\int \sqrt{(x^2 + a^2)}\, dx = \dfrac{a^2}{2} \left[\operatorname{arsinh} \dfrac{x}{a} + \left(\dfrac{x}{a} \right) \left(\dfrac{1}{a}\, \sqrt{(a^2 + x^2)} \right) \right] + c$

$$= \frac{a^2}{2}\, \operatorname{arsinh} \frac{x}{a} + \frac{x}{2} \sqrt{(x^2 + a^2)} + c$$

Problem 3. Evaluate $\displaystyle\int_{-1}^{1} \frac{1}{x^2 \sqrt{(1 + x^2)}}\, dx$

Since the integral contains a term in $\sqrt{(a^2 + x^2)}$, let $x = \sinh \theta$, then $dx = \cosh \theta\, d\theta$.

Hence $\displaystyle\int \frac{1}{x^2 \sqrt{(1 + x^2)}}\, dx = \int \frac{\cosh \theta\, d\theta}{(\sinh^2 \theta) \sqrt{(1 + \sinh^2 \theta)}}$

$$= \int \operatorname{cosech}^2 \theta\, d\theta = -\coth \theta + c$$

Since $\cosh^2 \theta - \sinh^2 \theta = 1$ then $\cosh \theta = \sqrt{(1 + \sinh^2 \theta)} = \sqrt{(1 + x^2)}$

Hence $-\coth \theta = \dfrac{-\cosh \theta}{\sinh \theta} = \dfrac{-\sqrt{(1 + x^2)}}{x}$

Therefore $\displaystyle\int_{-1}^{1} \frac{1}{x^2 \sqrt{(1 + x^2)}}\, dx = -\left[\frac{\sqrt{(1 + x^2)}}{x} \right]_{-1}^{1} = -\left[\frac{\sqrt{2}}{1} - \frac{\sqrt{2}}{-1} \right]$

$$= -2\sqrt{2} = -2.828\,4$$

(G) Intergrals containing $\sqrt{(x^2 - a^2)}$; the 'cosh θ' substitution

When an integral contains a term $\sqrt{(x^2 - a^2)}$ the substitution $x = a \cosh \theta$ is used. The reason for this is made obvious by the following worked problems.

Problem 1. Find $\displaystyle\int \frac{1}{\sqrt{(x^2 - a^2)}} \, dx$

Let $x = a \cosh \theta$, then $dx = a \sinh \theta \, d\theta$

$$\int \frac{1}{\sqrt{(x^2 - a^2)}} \, dx = \int \frac{1}{\sqrt{(a^2 \cosh^2 \theta - a^2)}} \, a \sinh \theta \, d\theta$$

$$= \int \frac{a \sinh \theta \, d\theta}{\sqrt{[a^2 (\cosh^2 \theta - 1)]}} = \int \frac{a \sinh \theta \, d\theta}{a \sinh \theta}$$

$$\text{since } \cosh^2 \theta - 1 = \sinh^2 \theta$$

$$= \int d\theta = \theta + c$$

Hence $\displaystyle\int \frac{1}{\sqrt{(x^2 - a^2)}} \, dx = \textbf{arcosh } \frac{x}{a} + c$, since $x = a \cosh \theta$

From Chapter 12, $\text{arcosh } \dfrac{x}{a} = \ln \left\{ \dfrac{x + \sqrt{(x^2 - a^2)}}{a} \right\}$ which gives an alternative solution to $\displaystyle\int \frac{1}{\sqrt{(x^2 - a^2)}} \, dx$.

Problem 2. Find $\int \sqrt{(x^2 - a^2)} \, dx$

Let $x = a \cosh \theta$, then $dx = a \sinh \theta \, d\theta$.

$$\int \sqrt{(x^2 - a^2)} \, dx = \int \sqrt{(a^2 \cosh^2 \theta - a^2)} \, a \sinh \theta \, d\theta$$

$$= \int (a \sinh \theta)(a \sinh \theta) \, d\theta = \int a^2 \sinh^2 \theta \, d\theta$$

$$= a^2 \int \left(\frac{\cosh 2\theta - 1}{2} \right) d\theta, \text{ since } \cosh 2\theta = 1 + 2 \sinh^2 \theta$$

$$= \frac{a^2}{2} \left[\frac{\sinh 2\theta}{2} - \theta \right] + c$$

$$= \frac{a^2}{2} \left[\sinh \theta \cosh \theta - \theta \right] + c, \text{ since } \sinh 2\theta =$$
$$2 \sinh \theta \cosh \theta$$

Since $x = a \cosh \theta$ then $\cosh \theta = \dfrac{x}{a}$ and $\theta = \text{arcosh } \dfrac{x}{a}$.

Since $\cosh^2 \theta - \sinh^2 \theta = 1$ then $\sinh \theta = \sqrt{(\cosh^2 \theta - 1)} = \sqrt{\left[\left(\frac{x}{a} \right)^2 - 1 \right]}$

$$= \frac{1}{a} \sqrt{(x^2 - a^2)}$$

Hence $\int \sqrt{(x^2 - a^2)}\,dx = \dfrac{a^2}{2}\left[\dfrac{1}{a}\sqrt{(x^2 - a^2)}\left(\dfrac{x}{a}\right) - \text{arcosh}\,\dfrac{x}{a}\right] + c$

$$= \dfrac{x}{2}\sqrt{(x^2 - a^2)} - \dfrac{a^2}{2}\,\text{arcosh}\,\dfrac{x}{a} + c$$

Problem 3. Find $\displaystyle\int \dfrac{4x - 1}{\sqrt{(x^2 + 4x - 12)}}\,dx$

With this type, two fractions are created, one of the form $k\displaystyle\int \dfrac{f'(x)}{f(x)}\,dx$, and the other to keep the ratio the same.

Thus $\displaystyle\int \dfrac{4x - 1}{\sqrt{(x^2 + 4x - 12)}}\,dx$

$$= \int \left\{\dfrac{2(2x + 4)}{\sqrt{(x^2 + 4x - 12)}} - \dfrac{7}{\sqrt{(x^2 + 4x - 12)}}\right\}dx$$

$$= 4\sqrt{(x^2 + 4x - 12)} - 7\int \dfrac{1}{\sqrt{[(x + 2)^2 - (4)^2]}}\,dx$$

$$= 4\sqrt{(x^2 + 4x - 12)} - 7\,\text{arcosh}\left(\dfrac{x + 2}{4}\right) + c$$

Further problems on integration using trigonometric and hyperbolic identities and substitutions may be found in Section 6 (Problems 39 to 58) page 211.

Summary

$f(x)$	$\int f(x)\,dx$	Method
$\cos^2 x$	$\dfrac{1}{2}\left(x + \dfrac{\sin 2x}{2}\right) + c$	Use $\cos 2x = 2\cos^2 x - 1$
$\sin^2 x$	$\dfrac{1}{2}\left(x - \dfrac{\sin 2x}{2}\right) + c$	Use $\cos 2x = 1 - 2\sin^2 x$
$\tan^2 x$	$\tan x - x + c$	Use $1 + \tan^2 x = \sec^2 x$
$\cot^2 x$	$-\cot x - x + c$	Use $\cot^2 x + 1 = \text{cosec}^2 x$
$\cos^m x \sin^n x$	If m or n is odd, use $\cos^2 x + \sin^2 x = 1$ If both m and n are even, use either $\cos 2x = 2\cos^2 x - 1$ or $\cos 2x = 1 - 2\sin^2 x$	
$\sin A \cos B$		Use $\frac{1}{2}\,[\sin (A + B) + \sin (A - B)]$
$\cos A \sin B$		Use $\frac{1}{2}\,[\sin (A + B) + \sin (A - B)]$

$f(x)$	$\int f(x)\,\mathrm{d}x$	Method
$\cos A \cos B$		Use $\frac{1}{2}\left[\cos\left(A+B\right)+\cos\left(A-B\right)\right]$
$\sin A \sin B$		Use $-\frac{1}{2}\left[\cos\left(A+B\right)-\cos\left(A-B\right)\right]$
$\dfrac{1}{\sqrt{(a^2-x^2)}}$	$\arcsin\dfrac{x}{a}+c$	Use $x=a\sin\theta$ substitution
$\sqrt{(a^2-x^2)}$	$\dfrac{a^2}{2}\arcsin\dfrac{x}{a}+\dfrac{x}{2}\sqrt{(a^2-x^2)}+c$	Use $x=a\sin\theta$ substitution
$\dfrac{1}{a^2+x^2}$	$\dfrac{1}{a}\arctan\dfrac{x}{a}+c$	Use $x=\tan\theta$ substitution
$\dfrac{1}{\sqrt{(x^2+a^2)}}$	$\arsinh\dfrac{x}{a}+c$ or $\ln\left\{\dfrac{x+\sqrt{(x^2+a^2)}}{a}\right\}+c$	Use $x=a\sinh\theta$ substitution
$\sqrt{(x^2+a^2)}$	$\dfrac{a^2}{2}\arsinh\dfrac{x}{a}+\dfrac{x}{2}\sqrt{(x^2+a^2)}+c$	Use $x=a\sinh\theta$ substitution
$\dfrac{1}{\sqrt{(x^2-a^2)}}$	$\arcosh\dfrac{x}{a}+c$ or $\ln\left\{\dfrac{x+\sqrt{(x^2-a^2)}}{a}\right\}+c$	Use $x=a\cosh\theta$ substitution
$\sqrt{(x^2-a^2)}$	$\dfrac{x}{2}\sqrt{(x^2-a^2)}-\dfrac{a^2}{2}\arcosh\dfrac{x}{a}+c$	Use $x=a\cosh\theta$ substitution

4. Change of limits of integration by a substitution

When evaluating definite integrals involving substitutions it is often easier to change the limits of the integral as shown in the following worked problems.

Problem 1. Evaluate $\displaystyle\int_0^2 \frac{t}{\sqrt{(2t^2+1)}}\ \mathrm{d}t$, taking positive values of square roots only.

Let $u = 2t^2+1$, then $\dfrac{\mathrm{d}u}{\mathrm{d}t}=4t$, i.e. $\mathrm{d}t=\dfrac{\mathrm{d}u}{4t}$

Hence $\displaystyle\int_0^2 \frac{t}{\sqrt{(2t^2+1)}}\ \mathrm{d}t = \int_{t=0}^{t=2}\frac{t}{\sqrt{u}}\frac{\mathrm{d}u}{4t}=\frac{1}{4}\int_{t=0}^{t=2}u^{-1/2}\ \mathrm{d}u$

However if $u = 2t^2 + 1$ then: when $t = 2, u = 9$ and when $t = 0, u = 1$.

Hence $\dfrac{1}{4} \displaystyle\int_{t=0}^{t=2} u^{-1/2} \, du = \dfrac{1}{4} \int_{u=1}^{u=9} u^{-1/2} \, du$, i.e. the limits have been change

$$= \frac{1}{4} \left[\frac{u^{-1/2}}{\frac{1}{2}} \right]_1^9 = \frac{1}{2} (\sqrt{9} - \sqrt{1}) = 1$$

taking positive values of square roots only.

When the limits are changed it makes it unnecessary to change back to the original variable (in this case, t) after integration.

Problem 2. Evaluate $\displaystyle\int_0^5 \frac{1}{\sqrt{(25 - x^2)}} \, dx$

Let $x = 5 \sin \theta$, then $dx = 5 \cos \theta \, d\theta$

When $x = 5$, $\sin \theta = 1$ and $\theta = \dfrac{\pi}{2}$

When $x = 0$, $\sin \theta = 0$ and $\theta = 0$.

Hence $\displaystyle\int_{x=0}^{x=5} \frac{1}{\sqrt{(25 - x^2)}} \, dx = \int_{\theta=0}^{\theta=\pi/2} \frac{5 \cos \theta \, d\theta}{\sqrt{[25 \, (1 - \sin^2 \theta)]}}$

$$= [\theta]_0^{\pi/2} = \frac{\pi}{2} \text{ or } 1.570\,8$$

Problem 3. Evaluate $\displaystyle\int_0^\infty \frac{1}{(x^2 + 4)} \, dx$

Let $x = 2 \tan \theta$, then $dx = 2 \sec^2 \theta \, d\theta$

When $x = +\infty$, $\theta = \dfrac{\pi}{2}$ and when $x = 0, \theta = 0$.

Hence $\displaystyle\int_{x=0}^{x=\infty} \frac{1}{(x^2 + 4)} \, dx = \int_{\theta=0}^{\theta=\pi/2} \frac{2 \sec^2 \theta \, d\theta}{4(\tan^2 \theta + 1)} = \int_0^{\pi/2} \frac{1}{2} \, d\theta$

$$= \left[\frac{\theta}{2} \right]_0^{\pi/2} = \frac{1}{2} \left(\frac{\pi}{2} - 0 \right) = \frac{\pi}{4} \text{ or } 0.785\,4$$

Problem 4. Evaluate $\displaystyle\int_0^1 \sqrt{(4 - x^2)} \, dx$

Let $x = 2 \sin \theta$, then $dx = 2 \cos \theta \, d\theta$

When $x = 1$, $\sin \theta = \dfrac{1}{2}$ and $\theta = \dfrac{\pi}{6}$

When $x = 0$, $\sin \theta = 0$ and $\theta = 0$

Hence $\displaystyle\int_{x=0}^{x=1} \sqrt{(4 - x^2)} \, dx = \int_{\theta=0}^{\theta=\pi/6} \sqrt{[4 \, (1 - \sin^2 \theta)]} \, 2 \cos \theta \, d\theta$

$$= \int_0^{\pi/6} 4 \cos^2 \theta \, d\theta$$

$$= \int_0^{\pi/6} 4 \left(\frac{1 + \cos 2\theta}{2} \right) d\theta$$

$$= 2 \left[\theta + \frac{\sin 2\theta}{2} \right]_0^{\pi/6}$$

$$= 2 \left[\left(\frac{\pi}{6} + \frac{\sin \pi/3}{2} \right) - (0) \right]$$

$$= \frac{\pi}{3} + \frac{\sqrt{3}}{2} \quad \text{or} \quad 1.913\,2$$

Problem 5. Evaluate $\int_1^5 \frac{dx}{2x^2 + 4x + 26}$

$$\int_1^5 \frac{dx}{2x^2 + 4x + 26} = \frac{1}{2} \int_1^5 \frac{dx}{x^2 + 2x + 13} = \frac{1}{2} \int_1^5 \frac{dx}{(x+1)^2 + 12}$$

$$= \frac{1}{2} \int_1^5 \frac{dx}{(x+1)^2 + (2\sqrt{3})^2}$$

Let $x + 1 = 2\sqrt{3} \tan \theta$, then $dx = 2\sqrt{3} \sec^2 \theta \, d\theta$

When $x = 5$, $\theta = \frac{\pi}{3}$ and when $x = 1$, $\theta = \frac{\pi}{6}$

Hence $\dfrac{1}{2} \displaystyle\int_1^5 \frac{dx}{(x+1)^2 + (2\sqrt{3})^2}$

$$= \frac{1}{2} \int_{\pi/6}^{\pi/3} \frac{2\sqrt{3} \sec^2 \theta \, d\theta}{(2\sqrt{3} \tan \theta)^2 + (2\sqrt{3})^2}$$

$$= \frac{1}{2} \int_{\pi/6}^{\pi/3} \frac{2\sqrt{3} \sec^2 \theta \, d\theta}{(2\sqrt{3})^2 (\tan^2 \theta + 1)}$$

$$= \frac{1}{2} \int_{\pi/6}^{\pi/3} \frac{1}{2\sqrt{3}} \, d\theta = \frac{1}{4\sqrt{3}} \left[\theta \right]_{\pi/6}^{\pi/3}$$

$$= \frac{\pi}{24\sqrt{3}} \quad \text{or} \quad 0.075\,57$$

Further problems on evaluating definite integrals may be found in Section 6 (Problems 59 to 69) page 213.

5 Integration using partial fractions

The process of expressing a fraction in terms of simple fractions — called partial fractions — was explained in Chapter 2. Certain functions can only be integrated when they have been resolved into partial fractions.

Worked problems on integration using partial fractions

(a) Type 1. Denominator containing linear factors

Problem 1. Find $\int \dfrac{x-8}{x^2-x-2}\ dx$

It was shown on page 15 that $\dfrac{x-8}{x^2-x-2} \equiv \dfrac{3}{x+1} - \dfrac{2}{x-2}$

Hence $\int \dfrac{x-8}{x^2-x-2}\ dx = \int \left(\dfrac{3}{x+1} - \dfrac{2}{x-2} \right) dx$

$$= 3 \ln (x+1) - 2 \ln (x-2) + c$$

$$\text{or } \ln \left\{ \dfrac{(x+1)^3}{(x-2)^2} \right\} + c$$

Problem 2. Determine $\int \dfrac{6x^2+7x-25}{(x-1)(x+2)(x-3)}\ dx$

It was shown on page 16 that $\dfrac{6x^2+7x-25}{(x-1)(x+2)(x-3)}$

$$\equiv \dfrac{2}{(x-1)} - \dfrac{1}{(x+2)} + \dfrac{5}{(x-3)}$$

Hence $\int \dfrac{6x^2+7x-25}{(x-1)(x+2)(x-3)}\ dx = \int \left(\dfrac{2}{(x-1)} - \dfrac{1}{(x+2)} + \dfrac{5}{(x-3)} \right) dx$

$$= 2 \ln (x-1) - \ln (x+2) + 5 \ln (x-3) + c$$

$$\text{or } \ln \left\{ \dfrac{(x-1)^2 (x-3)^5}{(x+2)} \right\} + c$$

Problem 3. Evaluate $\int_3^4 \dfrac{x^3-x^2-5x}{x^2-3x+2}\ dx$

It was shown on page 17 that $\dfrac{x^3-x^2-5x}{x^2-3x+2} \equiv x+2+\dfrac{5}{(x-1)} - \dfrac{6}{(x-2)}$

Hence $\int_3^4 \dfrac{x^3-x^2-5x}{x^2-3x+2}\ dx = \int_3^4 \left(x+2+\dfrac{5}{(x-1)} - \dfrac{6}{(x-2)} \right) dx$

$$= \left[\dfrac{x^2}{2} + 2x + 5 \ln (x-1) - 6 \ln (x-2) \right]_3^4$$

$$= (8 + 8 + 5 \ln 3 - 6 \ln 2) - (\tfrac{9}{2} + 6 + 5 \ln 2 - 6 \ln 1)$$

$$= 3.368\ 4$$

206

(b) Type 2. Denominator containing repeated linear factors

Problem 4. Find $\int \dfrac{x+5}{(x+3)^2} \, dx$

It was shown on page 17 that $\dfrac{x+5}{(x+3)^2} \equiv \dfrac{1}{(x+3)} + \dfrac{2}{(x+3)^2}$

Hence $\int \dfrac{x+5}{(x+3)^2} \, dx = \int \left(\dfrac{1}{(x+3)} + \dfrac{2}{(x+3)^2} \right) dx$

$$= \ln(x+3) - \dfrac{2}{(x+3)} + c$$

Problem 5. Find $\int \dfrac{5x^2 - 19x + 3}{(x-2)^2 (x+1)} \, dx$

It was shown on page 18 that $\dfrac{5x^2 - 19x + 3}{(x-2)^2 (x+1)}$

$$\equiv \dfrac{2}{(x-2)} - \dfrac{5}{(x-2)^2} + \dfrac{3}{(x+1)}$$

Hence $\int \dfrac{5x^2 - 19x + 3}{(x-2)^2 (x+1)} \, dx = \int \left(\dfrac{2}{(x-2)} - \dfrac{5}{(x-2)^2} + \dfrac{3}{(x+1)} \right) dx$

$$= 2 \ln(x-2) + \dfrac{5}{(x-2)} + 3 \ln(x+1) + c$$

$$= \ln(x-2)^2 (x+1)^3 + \dfrac{5}{(x-2)} + c$$

Problem 6. Evaluate $\int_5^6 \dfrac{2x^2 - 13x + 13}{(x-4)^3} \, dx$

It was shown on page 19 that $\dfrac{2x^2 - 13x + 13}{(x-4)^3}$

$$\equiv \dfrac{2}{(x-4)} + \dfrac{3}{(x-4)^2} - \dfrac{7}{(x-4)^3}$$

Hence $\int_5^6 \dfrac{2x^2 - 13x + 13}{(x-4)^3} = \int_5^6 \left(\dfrac{2}{(x-4)} + \dfrac{3}{(x-4)^2} - \dfrac{7}{(x-4)^3} \right) dx$

$$= \left[2 \ln(x-4) - \dfrac{3}{(x-4)} + \dfrac{7}{2(x-4)^2} \right]_5^6$$

$$= (2 \ln 2 - \tfrac{3}{2} + \tfrac{7}{8}) - (2 \ln 1 - \tfrac{3}{1} + \tfrac{7}{2})$$

$$= \mathbf{0.261\ 3}$$

(c) Type 3. Denominator containing a quadratic factor

Problem 7. Find $\int \dfrac{8x^2 - 3x + 19}{(x^2 + 3)(x - 1)} \, dx$

It was shown on page 20 that $\dfrac{8x^2 - 3x + 19}{(x^2 + 3)(x - 1)} \equiv \dfrac{2x - 1}{(x^2 + 3)} + \dfrac{6}{(x - 1)}$

Hence $\int \dfrac{8x^2 - 3x + 19}{(x^2 + 3)(x - 1)} \, dx = \int \left(\dfrac{2x - 1}{(x^2 + 3)} + \dfrac{6}{(x - 1)} \right) \, dx$

$= \int \left(\dfrac{2x}{(x^2 + 3)} - \dfrac{1}{(x^2 + 3)} + \dfrac{6}{(x - 1)} \right) \, dx$

$= \ln (x^2 + 3) - \dfrac{1}{\sqrt{3}} \arctan \dfrac{x}{\sqrt{3}} + 6 \ln (x - 1) + c$

$= \ln (x^2 + 3)(x - 1)^6 - \dfrac{1}{\sqrt{3}} \arctan \dfrac{x}{\sqrt{3}} + c$

Problem 8. Find $\int \dfrac{2 + x + 6x^2 - 2x^3}{x^2 (x^2 + 1)} \, dx$

It was shown on page 20 that $\dfrac{2 + x + 6x^2 - 2x^3}{x^2 (x^2 + 1)} \equiv \dfrac{1}{x} + \dfrac{2}{x^2} + \dfrac{4 - 3x}{x^2 + 1}$

Hence $\int \dfrac{2 + x + 6x^2 - 2x^3}{x^2 (x^2 + 1)} \, dx = \int \left(\dfrac{1}{x} + \dfrac{2}{x^2} + \dfrac{4 - 3x}{x^2 + 1} \right) \, dx$

$= \int \left(\dfrac{1}{x} + \dfrac{2}{x^2} + \dfrac{4}{x^2 + 1} - \dfrac{3x}{x^2 + 1} \right) \, dx$

$= \ln x - \dfrac{2}{x} + 4 \arctan x - \dfrac{3}{2} \ln (x^2 + 1) + c$

$= \ln \left\{ \dfrac{x}{(x^2 + 1)^{3/2}} \right\} - \dfrac{2}{x} + 4 \arctan x + c$

Problem 9. Find $\int \dfrac{5 (x^2 + x + 3)}{x (x^2 + 2x + 5)} \, dx$

Let $\dfrac{5 (x^2 + x + 3)}{x (x^2 + 2x + 5)} \equiv \dfrac{A}{x} + \dfrac{Bx + C}{x^2 + 2x + 5}$

$\equiv \dfrac{A (x^2 + 2x + 5) + (Bx + C)(x)}{x(x^2 + 2x + 5)}$

Hence $5x^2 + 5x + 15 \equiv A (x^2 + 2x + 5) + (Bx + C)(x)$ by equating numerators.

Let $x = 0$, then $A = 3$

Equating the coefficients of x^2 gives $5 = A + B$, i.e. $B = 2$

Equating the coefficients of x gives: $5 = 2A + C$, i.e. $C = -1$.

208

Hence $\int \dfrac{5\,(x^2 + x + 3)}{x\,(x^2 + 2x + 5)}\ dx = \int \left(\dfrac{3}{x} + \dfrac{2x - 1}{x^2 + 2x + 5} \right)\ dx$

Now $\int \dfrac{2x - 1}{x^2 + 2x + 5}\ dx = \int \dfrac{2x + 2}{(x^2 + 2x + 5)}\ dx - \int \dfrac{3}{(x^2 + 2x + 5)}\ dx$

i.e. the numerator of the first integral on the right-hand side has deliberately been made equal to the differential coefficient of the denominator so that the integral will integrate as $\ln (x^2 + 2x + 5)$.

$\int \dfrac{3}{(x^2 + 2x + 5)}\ dx = \int \dfrac{3}{(x + 1)^2 + 4}\ dx = \int \dfrac{3}{(x + 1)^2 + (2)^2}\ dx$

$$= \dfrac{3}{2}\ \text{arctan}\ \dfrac{(x + 1)}{2}$$

Hence $\int \dfrac{5\,(x^2 + x + 3)}{x\,(x^2 + 2x + 5)}\ dx = 3 \ln x + \ln (x^2 + 2x + 5)$

$$- \dfrac{3}{2}\ \text{arctan}\ \dfrac{(x + 1)}{2} + c$$

$$= \ln x^3\,(x^2 + 2x + 5) - \dfrac{3}{2}\ \text{arctan}\ \dfrac{(x + 1)}{2} + c$$

(d) Integrals of the form $\int \dfrac{1}{x^2 - a^2}\ dx$ and $\int \dfrac{1}{a^2 - x^2}\ dx$

Problem 10. Determine (a) $\int \dfrac{1}{x^2 - a^2}\ dx$, (b) $\int_{1}^{2} \dfrac{1}{(9 - x^2)}\ dx$.

(a) Let $\dfrac{1}{x^2 - a^2} \equiv \dfrac{A}{(x - a)} + \dfrac{B}{(x + a)} \equiv \dfrac{A\,(x + a) + B\,(x - a)}{(x - a)\,(x + a)}$

Hence $1 = A\,(x + a) + B\,(x - a)$ by equating numerators

Let $x = a$, then $A = \dfrac{1}{2a}$. Let $x = -a$, then $B = -\dfrac{1}{2a}$

Hence $\int \dfrac{1}{x^2 - a^2}\ dx = \dfrac{1}{2a} \int \left(\dfrac{1}{(x - a)} - \dfrac{1}{(x + a)} \right) dx$

$$= \dfrac{1}{2a}\ [\ln (x - a) - \ln (x + a)] + c$$

$$= \dfrac{1}{2a}\ \ln \left(\dfrac{x - a}{x + a} \right) + c$$

Similarly, it may be shown that $\int \dfrac{1}{a^2 - x^2}\ dx = \dfrac{1}{2a}\ \ln \left(\dfrac{a + x}{a - x} \right) + c$

or $\dfrac{1}{a}\ \text{artanh}\ \dfrac{x}{a} + c$ (by using the substitution $x = a \tanh \theta$)

(b) $\int_{1}^{2} \dfrac{1}{9-x^2} \, dx = \int_{1}^{2} \dfrac{1}{3^2-x^2} \, dx = \dfrac{1}{2(3)} \left[\ln \left(\dfrac{3+x}{3-x} \right) \right]_{1}^{2}$ from (a)

$$= \dfrac{1}{6} \ln \dfrac{5}{2} \text{ or } 0.152\,7$$

Further problems on integration using partial fractions may be found in the following Section (6) (Problems 70 to 100) page 215.

6. Further problems

Integrating using algebraic substitution

In Problems 1 to 24 integrate with respect to the variable.

1. $\sin(5x - 3)$. $[-\frac{1}{5} \cos(5x - 3) + c]$
2. $4 \cos(3t - 1)$. $[\frac{4}{3} \sin(3t - 1) + c]$
3. $5 \sec^2(7\theta - 2)$. $[\frac{5}{7} \tan(7\theta - 2) + c]$
4. $(3x + 2)^9$. $[\frac{1}{30}(3x + 2)^{10} + c]$
5. $14(7t + 4)^6$. $[\frac{2}{7}(7t + 4)^7 + c]$
6. $\dfrac{1}{3d + 1}$. $[\frac{1}{3} \ln(3d + 1) + c]$
7. $\dfrac{-2}{8x + 3}$. $[-\frac{1}{4} \ln(8x + 3) + c]$
8. $3e^{4x-5}$. $[\frac{3}{4} e^{4x-5} + c]$
9. $6e^{3-2t}$. $[-3e^{3-2t} + c]$
10. $4x(2x^2 - 4)^6$. $[\frac{1}{7}(2x^2 - 4)^7 + c]$
11. $18t(1 - 3t^2)^5$. $[-\frac{1}{2}(1 - 3t^2)^6 + c]$
12. $\frac{2}{3} \sin^3 \theta \cos \theta$. $[\frac{1}{6} \sin^4 \theta + c]$
13. $3 \cos^4 t \sin t$. $[-\frac{3}{5} \cos^5 t + c]$
14. $\sec^2 2x \tan 2x$. $[\frac{1}{4} \tan^2 2x + c]$
15. $4x\sqrt{(4x^2 + 1)}$. $[\frac{1}{3}\sqrt{(4x^2 + 1)^3} + c]$
16. $(9x^2 - 4x)\sqrt{(3x^3 - 2x^2 + 1)}$. $[\frac{2}{3}\sqrt{(3x^3 - 2x^2 + 1)^3} + c]$
17. $\dfrac{2 \ln t}{t}$. $[(\ln t)^2 + c]$
18. $\dfrac{4x + 3}{(2x^2 + 3x - 1)^4}$. $\left[-\dfrac{1}{3(2x^2 + 3x - 1)^3} + c\right]$
19. $\dfrac{t}{\sqrt{(t^2 - 3)}}$. $[\sqrt{(t^2 - 3)} + c]$
20. $\dfrac{x^2 - 1}{\sqrt{(x^3 - 3x + 4)}}$. $[\frac{2}{3}\sqrt{(x^3 - 3x + 4)} + c]$

21. $\dfrac{4e^t}{\sqrt{(2 + e^t)}}$. $[8\sqrt{(2 + e^t)} + c]$

22. $5 \tan 2\alpha$. $[\frac{5}{2} \ln (\sec 2\alpha) + c]$

23. $(5\theta + 2) \sec^2 (5\theta^2 + 4\theta)$. $[\frac{1}{2} \tan (5\theta^2 + 4\theta) + c]$

24. $2t\, e^{3t^2 - 2}$. $[\frac{1}{3} e^{3t^2 - 2} + c]$

In Problems 25 to 35 evaluate the definite integrals.

25. $\int_0^1 (2x - 1)^6 \, dx$. $[\frac{1}{7}]$

26. $\int_0^2 t\sqrt{(2t^2 + 1)} \, dt$. $[4\frac{1}{3}]$

27. $\displaystyle\int_0^{\pi/2} \sin (2\theta + \dfrac{\pi}{3}) \, d\theta$. $[\frac{1}{2}]$

28. $\int_1^3 2 \cos (2t - 1) \, dt$. $[-1.800\ 4]$

29. $\int_0^2 (2x^2 - 1)\sqrt{(2x^3 - 3x)} \, dx$. $[7.027\ 3]$

30. $\displaystyle\int_1^2 \dfrac{5 \ln x}{x} \, dx$. $[1.201\ 1]$

31. $\displaystyle\int_1^3 \dfrac{(x - 2)}{(x^2 - 4x + 1)^4} \, dx$. $[0]$

32. $\int_0^1 2t\, e^{4t^2 - 3} \, dt$. $[0.667\ 1]$

33. $\int_0^{\pi/3} 2 \sin^5 \theta \cos \theta \, d\theta$. $[0.140\ 6]$

34. $\int_0^1 \theta \sec^2 (5\theta^2) \, d\theta$. $[-0.338\ 1]$

35. $\displaystyle\int_1^2 \left[\dfrac{e^{(2x-1)} - e^{(-2x+1)}}{2} \right] \, dx$. $[4.262\ 3]$

36. The entropy change for an ideal gas is given by the equation

$$\Delta S = \int_{T_1}^{T_2} C_v \dfrac{dT}{T} - R \int_{V_1}^{V_2} \dfrac{dV}{V}$$

where T is the thermodynamic temperature, V is the volume and $R = 8.314$.

   Determine the entropy change when a gas expands from 2 litres to 3 litres for a temperature rise from 200 K to 400 K given that

$C_V = 45 + 6 \times 10^{-3}\, T + 8 \times 10^{-6}\, T^2$. $[29.5]$

37. The electrostatic potential on all parts of a conducting circular disc of radius r is given by the equation

$$V = 2\pi\sigma \int_0^9 \frac{R\,dR}{\sqrt{(R^2 + r^2)}}$$

Solve the equation by determining the integral.

$[V = 2\pi\sigma\,[\sqrt{(9^2 + r^2)} - r]]$

38. In the study of a rigid rotor the following integration occurs:

$$Z_r = \int_0^\infty (2J + 1)\,e^{-J(J+1)h^2/8\pi^2 Ikt}\,dJ$$

Determine Z_r for constant temperature T assuming h, I, and k are constants.

$$\left[\frac{8\pi^2 IkT}{h^2}\right]$$

Integration using trigonometric and hyperbolic identities and substitutions
In Problems 39 to 58 integrate with respect to the variable.

39. (a) $\cos^2 2x$ (b) $\sin^2 3x$ (c) $\tan^2 4x$ (d) $\cot^2 5x$

(a) $\left[\frac{1}{2}\left(x + \frac{\sin 4x}{4}\right) + c\right]$ (b) $\left[\frac{1}{2}\left(x - \frac{\sin 6x}{6}\right) + c\right]$

(c) $\left[\frac{1}{4}\tan 4x - x + c\right]$ (d) $\left[-\frac{1}{5}\cot 5x - x + c\right]$

Powers of sines and cosines
40. (a) $\cos^3 \theta$ (b) $\sin^3 2\theta$.

(a) $\left[\sin\theta - \frac{\sin^3\theta}{3} + c\right]$ (b) $\left[\frac{1}{6}\cos^3 2\theta - \frac{1}{2}\cos 2\theta + c\right]$

41. (a) $\cos^3 t \sin^2 t$ (b) $\sin^4 x \cos^3 x$.

(a) $[\frac{1}{3}\sin^3 t - \frac{1}{5}\sin^5 t + c]$ (b) $[\frac{1}{5}\sin^5 x - \frac{1}{7}\sin^7 x + c]$

42. (a) $\sin^4 2x$ (b) $\cos^2 x \sin^2 x$.

(a) $\left[\frac{3x}{8} - \frac{1}{8}\sin 4x + \frac{1}{64}\sin 8x + c\right]$ (b) $\left[\frac{x}{8} - \frac{1}{32}\sin 4x + c\right]$

43. (a) $3\cos^4 3t \sin^2 3t$ (b) $\frac{1}{2}\sin^4 2\theta \cos^2 2\theta$.

(a) $[\frac{3}{16}(t - \frac{1}{12}\sin 12t + \frac{1}{9}\sin^3 6t) + c]$

(b) $[\frac{1}{32}(\theta - \frac{1}{8}\sin 8\theta - \frac{1}{6}\sin^3 4\theta) + c]$

Products of sines and cosines
44. (a) $\sin 3t \cos t$ (b) $2\cos 4t \cos 2t$.

(a) $[-\frac{1}{8}(\cos 4t + 2\cos 2t) + c]$ (b) $[\frac{1}{6}\sin 6t + \frac{1}{2}\sin 2t + c]$

45. (a) $3 \cos 6x \cos 2x$ (b) $4 \sin 5x \sin 2x$.

(a) $\left[\frac{3}{16}\left(\sin 8x + 2 \sin 4x\right) + c\right]$ (b) $\left[2\left(\frac{1}{3}\sin 3x - \frac{1}{7}\sin 7x\right) + c\right]$

46. (a) $\frac{1}{2}\sin 9t \sin 3t$ (b) $2 \sin 4\theta \cos 3\theta$.

(a) $\left[\frac{1}{48}\left(2 \sin 6t - \sin 12t\right) + c\right]$ (b) $\left[-\frac{1}{7}\cos 7\theta - \cos \theta + c\right]$

47. (a) $9 \cos 5t \cos 4t$ (b) $\frac{3}{2}\cos 2x \sin x$

(a) $\left[\frac{9}{2}\left(\frac{1}{9}\sin 9t + \sin t\right) + c\right]$ (b) $\left[\frac{3}{4}\left(\cos x - \frac{1}{3}\cos 3x\right) + c\right]$

'Sine θ' substitution

48. (a) $\dfrac{2}{\sqrt{(4-x^2)}}$ (b) $\dfrac{1}{\sqrt{(9-4x^2)}}$.

(a) $\left[2 \arcsin \dfrac{x}{2} + c\right]$ (b) $\left[\dfrac{1}{2}\arcsin \dfrac{2x}{3} + c\right]$

49. (a) $\sqrt{(16-x^2)}$ (b) $\sqrt{(16-9x^2)}$.

(a) $\left[8 \arcsin \dfrac{x}{4} + \dfrac{x}{2}\sqrt{(16-x^2)} + c\right]$

(b) $\left[\dfrac{8}{3}\arcsin \dfrac{3x}{4} + \dfrac{x}{2}\sqrt{(16-9x^2)} + c\right]$

50. (a) $\dfrac{1}{\sqrt{(4+2x-x^2)}}$ (b) $\sqrt{(4+2x-x^2)}$

(a) $\left[\arcsin \dfrac{(x-1)}{\sqrt{5}} + c\right]$

(b) $\left[\dfrac{5}{2}\arcsin \dfrac{(x-1)}{\sqrt{5}} + \left(\dfrac{x-1}{2}\right)\sqrt{(4+2x-x^2)} + c\right]$

'Tan θ' substitution

51. (a) $\dfrac{2}{1+x^2}$ (b) $\dfrac{3}{16+x^2}$.

(a) $[2 \arctan x + c]$ (b) $\left[\dfrac{3}{4}\arctan \dfrac{x}{4} + c\right]$

52. (a) $\dfrac{1}{9+16x^2}$ (b) $\dfrac{1}{2x^2+4x+18}$.

(a) $\left[\dfrac{1}{12}\arctan \dfrac{4x}{3} + c\right]$ (b) $\left[\dfrac{1}{4\sqrt{2}}\arctan \dfrac{x+1}{2\sqrt{2}} + c\right]$

'Sinh θ' substitution

53. (a) $\dfrac{1}{\sqrt{(x^2+4)}}$ (b) $\dfrac{1}{\sqrt{(9+2x^2)}}$.

(a) $\left[\text{arsinh}\, \dfrac{x}{2} + c \text{ or } \ln\left\{ \dfrac{x + \sqrt{(x^2 + 4)}}{2} \right\} + c \right]$

(b) $\left[\dfrac{1}{\sqrt{2}} \text{arsinh}\, \dfrac{\sqrt{2}x}{3} + c \text{ or } \dfrac{1}{\sqrt{2}} \ln\left\{ \dfrac{x + \sqrt{(\frac{9}{2} + x^2)}}{3/\sqrt{2}} \right\} + c \right]$

54. (a) $\sqrt{(x^2 + 16)}$ (b) $\sqrt{(4x^2 + 25)}$.

(a) $\left[8\,\text{arsinh}\, \dfrac{x}{4} + \dfrac{x}{2}\, \sqrt{(x^2 + 16)} + c \right]$

(b) $\left[\dfrac{25}{4} \text{arsinh}\, \dfrac{2x}{5} + x\sqrt{(x^2 + \dfrac{25}{4})} + c \right]$

55. (a) $\sqrt{(x^2 + 2x + 5)}$ (b) $\dfrac{2}{3x^2\,\sqrt{(4 + x^2)}}$ (by letting $x = 2\sinh\theta$).

(a) $\left[2\,\text{arsinh}\, \dfrac{x+1}{2} + \left(\dfrac{x+1}{2} \right) \sqrt{(x^2 + 2x + 5)} + c \right]$

(b) $\left[\dfrac{-\sqrt{(4 + x^2)}}{6x} + c \right]$

'Cosh θ' substitution

56. (a) $\dfrac{1}{\sqrt{(x^2 - 9)}}$ (b) $\dfrac{1}{\sqrt{(4x^2 - 9)}}$.

(a) $\left[\text{arcosh}\, \dfrac{x}{3} + c \text{ or } \ln\left\{ \dfrac{x + \sqrt{(x^2 - 9)}}{3} \right\} + c \right]$

(b) $\left[\dfrac{1}{2} \text{arcosh}\, \dfrac{2x}{3} + c \text{ or } \dfrac{1}{2} \ln\left\{ \dfrac{x + \sqrt{(x^2 - \frac{9}{4})}}{\frac{3}{2}} \right\} + c \right]$

57. (a) $\sqrt{(x^2 - 4)}$ (b) $\sqrt{(9x^2 - 16)}$.

(a) $\left[\dfrac{x}{2}\, \sqrt{(x^2 - 4)} - 2\,\text{arcosh}\, \dfrac{x}{2} + c \right]$

(b) $\left[3\left\{ \dfrac{x}{2}\, \sqrt{(x^2 - \frac{16}{9})} - \dfrac{8}{9}\, \text{arcosh}\, \dfrac{3x}{4} \right\} + c \right]$

58. (a) $\dfrac{1}{\sqrt{(x^2 + 4x - 5)}}$ (b) $\sqrt{(x^2 + 6x + 5)}$.

(a) $\left[\text{arcosh}\, \dfrac{x+2}{3} + c \text{ or } \ln\left\{ \dfrac{(x+2) + \sqrt{(x^2 + 4x - 5)}}{3} \right\} + c \right]$

(b) $\left[\left(\dfrac{x+3}{2} \right) \sqrt{(x^2 + 6x + 5)} - 2\,\text{arcosh}\left(\dfrac{x+3}{2} \right) + c \right]$

Definite integrals

In Problems 59 to 69 evaluate the definite integrals.

59. (a) $\int_0^{\pi/2} \sin^2 \theta \, d\theta$ (b) $\int_0^1 2 \cos^2 \theta \, d\theta$

(a) $[\frac{\pi}{4}$ or $0.785\ 4\ ]$ (b) $[1.454\ 6\ ]$

60. (a) $\int_0^{\pi/4} \sin^3 t \cos t \, dt$ (b) $\int_0^{\pi/3} \cos^3 2t \, dt.$

(a) $[\ 0.062\ 5\ ]$ (b) $[\ 0.324\ 8]$

61. (a) $\int_{\pi/4}^{\pi/2} 4 \sin^2 \theta \cos^2 \theta \, d\theta$ (b) $\int_0^{\pi/2} 3 \sin^2 2t \cos^4 2t \, dt.$

(a) $[\frac{\pi}{8}$ or $0.392\ 7\ ]$ (b) $[\frac{3\pi}{32}$ or $0.294\ 5\ ]$

62. (a) $\int_0^{\pi/4} \sin 5\theta \cos 3\theta \, d\theta$ (b) $\int_0^{\pi/3} 4 \cos 6t \sin 3t \, dt$

(a) $[\frac{1}{4}\ ]$ (b) $[-\frac{8}{9}\ ]$

63. (a) $\int_0^1 2 \cos 7x \cos 2x \, dx$ (b) $\int_0^{\pi/2} \sin 8\alpha \sin 5\alpha \, d\alpha.$

(a) $[-0.146\ 0]$ (b) $[-0.205\ 1]$

64. (a) $\int_0^3 \frac{1}{\sqrt{(9-x^2)}} \, dx$ (b) $\int_0^3 \sqrt{(9-x^2)} \, dx.$

(a) $[\frac{\pi}{2}$ or $1.570\ 8]$ (b) $[\frac{9\pi}{4}$ or $7.068\ 6]$

65. (a) $\int_0^{1/2} \frac{2}{\sqrt{(1-x^2)}} \, dx$ (b) $\int_0^{3/4} \sqrt{(9-16x^2)} \, dx.$

(a) $[\frac{\pi}{3}$ or $1.047\ 2]$ (b) $[\frac{9\pi}{16}$ or $1.767\ 1]$

66. (a) $\int_0^1 \frac{1}{1+x^2} \, dx$ (b) $\int_0^2 \frac{3}{4+x^2} \, dx.$

(a) $[\frac{\pi}{4}$ or $0.785\ 4]$ (b) $[\frac{3\pi}{8}$ or $1.178\ 1]$

67. (a) $\int_{-1}^2 \frac{1}{2x^2+4x+20} \, dx$ (b) $\int_{-3}^0 \frac{5x-2}{x^2+6x+13} \, dx.$

(a) $[\frac{\pi}{24}$ or $0.130\ 9]$ (b) $[-5.407\ 1]$

68. (a) $\int_0^2 \frac{1}{\sqrt{(x^2+4)}} \, dx$ (b) $\int_0^1 \sqrt{(x^2+9)} \, dx.$

(a) $[0.881\ 4]$ (b) $[3.054\ 7]$

69. (a) $\displaystyle\int_{1}^{2} \dfrac{1}{\sqrt{(x^2-1)}}\ dx$ (b) $\displaystyle\int_{2}^{3} \sqrt{(x^2-4)}\ dx.$

(a) [1.317 0]   (b) [1.429 3]

Integration using partial fractions

In Problems 70 to 89 integrate after resolving into partial fractions.

70. $\dfrac{8}{x^2-4}$. $\left[\ 2\ln(x-2)-2\ln(x+2)+c\ \text{or}\ 2\ln\left(\dfrac{x-2}{x+2}\right)+c\ \right]$

71. $\dfrac{3x+5}{x^2+2x-3}$.

$[\ 2\ln(x-1)+\ln(x+3)+c\ \text{or}\ \ln\{(x-1)^2\,(x+3)\}\ +\ c]$

72. $\dfrac{y-13}{y^2-y-6}$.

$\left[\ 3\ln(y+2)-2\ln(y-3)+c\ \text{or}\ \ln\left\{\dfrac{(y+2)^3}{(y-3)^2}\right\}\ +c\ \right]$

73. $\dfrac{17x^2-21x-6}{x\,(x+1)\,(x-3)}$.

$[2\ln x+8\ln(x+1)+7\ln(x-3)+c\ \text{or}\ \ln\{x^2\,(x+1)^8\,(x-3)^7\}+c]$

74. $\dfrac{6x^2+7x-49}{(x-4)\,(x+1)\,(2x-3)}$.

$[3\ln(x-4)-2\ln(x+1)+2\ln(2x-3)+c\ \text{or}$

$\ln\left\{\dfrac{(x-4)^3(2x-3)^2}{(x+1)^2}\right\}+c\ ]$

75. $\dfrac{x^2+2}{(x+4)\,(x-2)}$.

$[x-3\ln(x+4)+\ln(x-2)+c\ \text{or}\ x+\ln\left\{\dfrac{(x-2)}{(x+4)^3}\right\}+c\ ]$

76. $\dfrac{2x^2+4x+19}{2\,(x-3)\,(x+4)}$.

$[x+\tfrac{7}{2}\ln(x-3)-\tfrac{5}{2}\ln(x+4)+c\ \text{or}\ x+\ln\left\{\dfrac{(x-3)^{7/2}}{(x+4)^{5/2}}\right\}+c\ ]$

77. $\dfrac{2x^3+7x^2-2x-27}{(x-1)\,(x+4)}$.

$[x^2+x-4\ln(x-1)+7\ln(x+4)+c\ \text{or}\ x^2+x+\ln\left\{\dfrac{(x+4)^7}{(x-1)^4}\right\}+c]$

78. $\dfrac{2t-1}{(t+1)^2}$. $[2\ln(t+1)+\dfrac{3}{(t+1)}+c\,]$

79. $\dfrac{8x^2+12x-3}{(x+2)^3}$. $[8\ln(x+2)+\dfrac{20}{(x+2)}-\dfrac{5}{2(x+2)^2}+c\,]$

80. $\dfrac{6x+1}{(2x+1)^2}$. $[\frac{3}{2}\ln(2x+1)+\dfrac{1}{(2x+1)}+c]$

81. $\dfrac{1}{x^2(x+2)}$.

$$\left[-\frac{1}{2x}-\frac{1}{4}\ln x+\frac{1}{4}\ln(x+2)+c \;\text{ or }\; \frac{1}{4}\ln\left(\frac{x+2}{x}\right)-\frac{1}{2x}+c\right]$$

82. $\dfrac{9x^2-73x+150}{(x-7)(x-3)^2}$.

$$[5\ln(x-7)+4\ln(x-3)+\frac{3}{(x-3)}+c \text{ or}$$

$$\ln\{(x-7)^5(x-3)^4\}+\frac{3}{x-3}+c\,]$$

83. $\dfrac{-(9x^2+4x+4)}{x^2(x^2-4)}$.

$$[\ln x-\frac{1}{x}+2\ln(x+2)-3\ln(x-2)+c \;\text{ or }\; \ln\left\{\frac{x(x+2)^2}{(x-2)^3}\right\}-\frac{1}{x}+c]$$

84. $\dfrac{-(a^2+5a+13)}{(a^2+5)(a-2)}$.

$$[\ln(a^2+5)-\frac{1}{\sqrt{5}}\arctan\frac{a}{\sqrt{5}}-3\ln(a-2)+c \text{ or}$$

$$\ln\left\{\frac{(a^2+5)}{(a-2)^3}\right\}-\frac{1}{\sqrt{5}}\arctan\frac{a}{\sqrt{5}}+c\,]$$

85. $\dfrac{3-x}{(x^2+3)(x+3)}$.

$$\left[\frac{1}{2\sqrt{3}}\arctan\frac{x}{\sqrt{3}}-\frac{1}{4}\ln(x^2+3)+\frac{1}{2}\ln(x+3)+c \text{ or}\right.$$

$$\left.\ln\left\{\frac{(x+3)^{1/2}}{(x^2+3)^{1/4}}\right\}+\frac{1}{2\sqrt{3}}\arctan\frac{x}{\sqrt{3}}+c\right]$$

86. $\dfrac{12-2x-5x^2}{(x^2+x+1)(3-x)}$.

$$[\ln(x^2+x+1)+\frac{8}{\sqrt{3}}\arctan\frac{2(x+\frac{1}{2})}{\sqrt{3}}+3\ln(3-x)+c]$$

87. $\dfrac{x^3 + 7x^2 + 8x + 10}{x(x^2 + 2x + 5)}$.

$[\, x + 2 \ln x + \frac{3}{2} \ln (x^2 + 2x + 5) - 2 \arctan \left(\dfrac{x+1}{2} \right) + c \,]$

88. $\dfrac{5x^3 - 3x^2 + 41x - 64}{(x^2 + 6)(x - 1)^2}$.

$\left[\dfrac{2}{\sqrt{6}} \arctan \dfrac{x}{\sqrt{6}} - \dfrac{3}{2} \ln (x^2 + 6) + 8 \ln (x - 1) + \dfrac{3}{(x - 1)} + c \right]$

89. $\dfrac{6x^3 + 5x^2 + 4x + 3}{(x^2 + x + 1)(x^2 - 1)}$.

$[\ln (x^2 + x + 1) - \dfrac{4}{\sqrt{3}} \arctan \dfrac{2(x + \frac{1}{2})}{\sqrt{3}} + 3 \ln (x - 1) + \ln (x + 1) + c]$

In Problems 90 to 99 evaluate the definite integrals correct to four decimal places.

90. $\displaystyle\int_4^6 \dfrac{x - 7}{x^2 - 2x - 3} \, dx.$ $[-0.425\ 7]$

91. $\displaystyle\int_2^3 \dfrac{x^3 - 2x^2 - 3x - 2}{(x + 2)(x - 1)} \, dx.$ $[-0.993\ 7]$

92. $\displaystyle\int_3^4 \dfrac{4x^2 + 15x - 1}{(x + 1)(x - 2)(x + 3)} \, dx.$ $[2.371\ 6]$

93. $\displaystyle\int_3^5 \dfrac{x^2 + 1}{x^2 + x - 6} \, dx.$ $[2.523\ 2]$

94. $\displaystyle\int_6^8 \dfrac{1}{(x^2 - 25)} \, dx.$ $[0.093\ 2]$

95. $\displaystyle\int_2^3 \dfrac{1}{(16 - x^2)} \, dx.$ $[0.105\ 9]$

96. $\displaystyle\int_1^2 \dfrac{2 + x + 6x^2 - 2x^3}{x^2(x^2 + 1)} \, dx.$ $[1.605\ 7]$

97. $\displaystyle\int_2^3 \dfrac{2x^2 - x - 2}{(x - 1)^3} \, dx.$ $[2.511\ 3]$

98. $\displaystyle\int_3^4 \dfrac{2x^3 + 4x^2 - 8x - 4}{x^2(x^2 - 4)} \, dx.$ $[1.041\ 8]$

99. $\displaystyle\int_0^1 \dfrac{-(4x^2 + 9x + 8)}{(x + 1)^2(x + 2)} \, dx.$ $[-2.546\ 5]$

100. The velocity constant k of a given chemical reaction is given by

218

$$kt = \int \frac{dx}{(3 - 0.4x)(2 - 0.6x)}$$

where $x = 0$ when $t = 0$.

Determine kt.

$$\left[\ln \left\{ \frac{2(3 - 0.4x)}{3(2 - 0.6x)} \right\} \right]$$

Chapter 16

Integration by parts

1 Introduction

When differentiating the product uv, where u and v are both functions of x, then:

$$\frac{\mathrm{d}}{\mathrm{d}x}(uv) = v\frac{\mathrm{d}u}{\mathrm{d}x} + u\frac{\mathrm{d}v}{\mathrm{d}x}$$

This is known as the **product rule for differentiation.**
Rearranging this formula gives:

$$u\frac{\mathrm{d}v}{\mathrm{d}x} = \frac{\mathrm{d}}{\mathrm{d}x}(uv) - v\frac{\mathrm{d}u}{\mathrm{d}x}$$

Integrating both sides with respect to x gives:

$$\int u\frac{\mathrm{d}v}{\mathrm{d}x}\mathrm{d}x = \int \frac{\mathrm{d}}{\mathrm{d}x}(uv)\,\mathrm{d}x - \int v\frac{\mathrm{d}u}{\mathrm{d}x}\,\mathrm{d}x$$

Since integration is the reversal of the differentiation process this becomes:

$$\int u\,\mathrm{d}v = uv - \int v\,\mathrm{d}u$$

This formula enables products of certain simple functions to be integrated in cases where it is possible to evaluate $\int v\,\mathrm{d}u$. This is known as the **integration by parts formula** and is a useful method of integration enabling such integrals as $\int x\,e^x\,\mathrm{d}x$, $\int x\cos x\,\mathrm{d}x$, $\int t^2\sin t\,\mathrm{d}t$, $\int x^3\ln x\,\mathrm{d}x$, $\int \ln x\,\mathrm{d}x$, $\int e^{ax}\sin bx\,\mathrm{d}x$, etc., to be determined.

2 Application of the integration by parts formula

Problem 1. Find $\int x\, e^x\, dx$.

In the integration by parts formula we must let one function of our product be equal to u and the other be equal to dv.

Let $u = x$ and $dv = e^x\, dx$.

Then $\dfrac{du}{dx} = 1$, i.e. $du = dx$ and $v = \int e^x\, dx = e^x$.

There are now expressions for u, du, dv and v which are substituted into the formula:

$$\int u\ dv = u \cdot v - \int v\ du$$

$$\int x\ e^x\, dx = x \cdot e^x - \int e^x\, dx$$

$$= x \cdot e^x - e^x + c$$

Hence, $\int x\, e^x\, dx = e^x\, (x - 1) + c$.

The following four points should be noted:

(i) The above result may be checked by differentiation.

Thus $\dfrac{d}{dx} [e^x (x - 1) + c] = e^x (1) + (x - 1)\, e^x + 0$

$$= e^x + x e^x - e^x = x e^x$$

(ii) Given that $dv = e^x\, dx$ then $v = \int e^x\, dx$, which is strictly equal to $e^x + a$ constant. However, the constant is omitted at this stage. If a constant, say k, were included, then:

$\int x\, e^x\, dx = x(e^x + k) - \int(e^x + k)\, dx$

$\qquad\qquad = x\, e^x + x\, k - (e^x + kx + c)$

$\qquad\qquad = x\, e^x - e^x + c$, as before.

Thus the constant k is an unnecessary addition. A constant is added only after the final integration (i.e. c in the above problem).

(iii) If instead of choosing to let $u = x$ and $dv = e^x\, dx$ we let $u = e^x$ and $dv = x\, dx$ then $du = e^x\, dx$ and $v = \int x\, dx = \dfrac{x^2}{2}$.

Hence $\int x\, e^x\, dx = (e^x)\left(\dfrac{x^2}{2}\right) - \int\left(\dfrac{x^2}{2}\right) e^x\, dx$.

The integral on the far right-hand side is seen to be more complicated than the original, thus the original choice of letting $u = e^x$ and $dv = x\, dx$ was wrong. The choice must be such that the 'u part' becomes a constant after differentiation and the 'dv' part can be integrated easily. (It will be seen later that for the 'u part' to become a constant often requires more than one differentiation (see Problem 5).)

(iv) If a product to be integrated contains an 'x' term then this term is chosen as the '*u* part' and the other function as the '*d v* part' except where a logarithmic term is involved (see Problems 6 and 7).

Problem 2. Determine $\int x \sin x \, dx$.

Let $u = x$ and $dv = \sin x \, dx$
Then $du = dx$ and $v = \int \sin x \, dx = -\cos x$
Substituting into $\int u \, dv = uv - \int v \, du$ gives:

$$\int x \sin dx = (x)(-\cos x) - \int (-\cos x) \, dx$$

$$= -x \cos x + \int \cos x \, dx$$

$$= -x \cos x + \sin x + c.$$

This result can be checked by differentiating.

Thus $\dfrac{d}{dx}(-x \cos x + \sin x + c) = (-x)(-\sin x) + (\cos x)(-1) + \cos x$

$$= x \sin x - \cos x + \cos x = x \sin x$$

Problem 3. Evaluate $\displaystyle\int_0^1 2x \, e^{3x} \, dx$

Let $u = 2x$ and $dv = e^{3x} \, dx$
Then $\dfrac{du}{dx} = 2$, i.e. $du = 2 \, dx$ and $v = \int e^{3x} \, dx = \dfrac{e^{3x}}{3}$

Substituting into $\int u \, dv = uv - \int v \, du$ gives:

$$\int 2x \, e^{3x} \, dx = (2x)\left(\frac{e^{3x}}{3}\right) - \int \left(\frac{e^{3x}}{3}\right)(2 \, dx)$$

$$= \frac{2}{3} x \, e^{3x} - \frac{2}{3}\int e^{3x} \, dx$$

$$= \frac{2}{3} x \, e^{3x} - \frac{2}{3}\left(\frac{e^{3x}}{3}\right) + c$$

$$= \frac{2}{3} x \, e^{3x} - \frac{2}{9} e^{3x} + c$$

Hence $\displaystyle\int_0^1 2x \, e^{3x} \, dx = \left[\frac{2}{3} x \, e^{3x} - \frac{2}{9} e^{3x}\right]_0^1$

$$= \left(\frac{2}{3} e^3 - \frac{2}{9} e^3\right) - \left(0 - \frac{2}{9} e^0\right)$$

$$= \frac{4}{9} e^3 + \frac{2}{9} = 8.926\,9 + 0.222\,2$$

$$= \textbf{9.149 correct to three decimal places.}$$

Problem 4. Evaluate $\displaystyle\int_0^{\frac{\pi}{2}} 3t \cos 2t \, dt$.

Let $u = 3t$ and $dv = \cos 2t \, dt$

Then $du = 3dt$ and $v = \int\cos 2t \, dt = \dfrac{1}{2} \sin 2t$

Substituting into $\int u \, dv = uv - \int v \, du$ gives:

$$\int 3t \cos 2t \, dt = (3t)\left(\frac{\sin 2t}{2}\right) - \int \left(\frac{\sin 2t}{2}\right)(3 \, dt)$$

$$= \frac{3t}{2} \sin 2t - \frac{3}{2} \int \sin 2t \, dt$$

$$= \frac{3t}{2} \sin 2t - \frac{3}{2}\left(\frac{-\cos 2t}{2}\right) + c$$

$$= \frac{3t}{2} \sin 2t + \frac{3}{4} \cos 2t + c$$

Hence $\displaystyle\int_0^{\frac{\pi}{2}} 3t \cos 2t \, dt = \left[\frac{3t}{2} \sin 2t + \frac{3}{4} \cos 2t\right]_0^{\frac{\pi}{2}}$

$$= \left[\frac{3}{2}\left(\frac{\pi}{2}\right) \sin \pi + \frac{3}{4} \cos \pi\right] - \left[0 + \frac{3}{4} \cos 0\right]$$

$$= \left(0 + \frac{3}{4}(-1)\right) - \left(0 + \frac{3}{4}\right) = -\frac{3}{4} - \frac{3}{4} = -1\frac{1}{2}.$$

Problem 5. Determine $\int x^2 \cos x \, dx$.

Let $u = x^2$ and $dv = \cos x \, dx$

Then $du = 2x \, dx$ and $v = \int\cos x \, dx = \sin x$

Substituting into $\int u \, dv = uv - \int v \, du$ gives:

$$\int x^2 \cos x \, dx = (x^2)(\sin x) - \int (\sin x)(2x \, dx)$$

$$= x^2 \sin x - 2[\int x \sin x \, dx]$$

The integral on the right-hand side in the bracket is not a standard and cannot be determined 'on sight'. Since it is a product of two simple functions we may use integration by parts again.

Now $\int x \sin x \, dx = -x \cos x + \sin x$ (from Problem 2)

Thus $\int x^2 \cos x \, dx = x^2 \sin x - 2[-x \cos x + \sin x] + c$

$$= x^2 \sin x + 2x \cos x - 2 \sin x + c$$

$$= (x^2 - 2) \sin x + 2x \cos x + c$$

Generally, if the term in x is of power n then the integration by parts formula is applied n times, provided one of the functions is not $\ln x$, as in Problem 6 following.

Problem 6. $\int x^2 \ln x \, dx$.

Whenever a product consists of a term in x and a logarithmic function, as in this problem, it is always the logarithmic function that is chosen as the 'u part'. The reason for this is that $\dfrac{d}{dx}(\ln x)$ is $\dfrac{1}{x}$, but $\int \ln x \, dx$ is not normally remembered as a standard integral.

Thus if $u = \ln x$ and $dv = x^2 \, dx$

then $du = \dfrac{1}{x} \, dx$ and $v = \int x^2 \, dx = \dfrac{x^3}{3}$

Substituting into $\int u \, dv = uv - \int v \, du$ gives:

$$\int x^2 \ln x \, dx = (\ln x)\left(\frac{x^3}{3}\right) - \int \left(\frac{x^3}{3}\right)\left(\frac{1}{x} \, dx\right)$$

$$= \frac{x^3}{3}\ln x - \frac{1}{3}\int x^2 \, dx$$

$$= \frac{x^3}{3}\ln x - \frac{1}{3}\left(\frac{x^3}{3}\right) + c$$

$$= \frac{x^3}{3}\ln x - \frac{x^3}{9} + c$$

$$= \frac{x^3}{9}(3\ln x - 1) + c$$

Problem 7. Determine $\int \ln x \, dx$.

In each of the previous problems the components of the product have been obvious. However, $\int \ln x \, dx$ is a special case for initially it appears not to be a product. However, $\int \ln x \, dx$ is the same as $\int 1 \times \ln x \, dx$.

Let $u = \ln x$ and $dv = 1 \, dx$

Then $du = \dfrac{1}{x} \, dx$ and $v = \int 1 \, dx = x$.

Hence $\int \ln x \, dx = (\ln x)(x) - \int (x)\left(\frac{1}{x} \, dx\right)$

$$= x \ln x - \int dx$$

$$= x \ln x - x + c$$

$$= x(\ln x - 1) + c$$

Problem 8. Find $\int e^{ax} \sin bx \, dx$.

With an integral of a product of an exponential function and a sine or cosine function it does not matter which function is made equal to 'u'. Thus let $u = e^{ax}$ and $dv = \sin bx \, dx$

then $du = a \, e^{ax} \, dx$ and $v = \int \sin bx \, dx = \dfrac{-\cos bx}{b}$

Thus $\int e^{ax} \sin bx \, dx = (e^{ax}) \left(\dfrac{-\cos bx}{b} \right) - \int \left(\dfrac{-\cos bx}{b} \right) (a \, e^{ax} \, dx)$

$$= -\frac{1}{b} \, e^{ax} \cos bx + \frac{a}{b} \left[\int e^{ax} \cos bx \, dx \right] \qquad (1)$$

It would seem that we are no nearer a solution of the initial integral. However, the integration by parts formula may be applied to the integral in the bracket.

Let $u = e^{ax}$ and $dv = \cos bx \, dx$

Then $du = a \, e^{ax} \, dx$ and $v = \int \cos bx \, dx = \dfrac{\sin bx}{b}$

Thus $\int e^{ax} \cos bx \, dx = (e^{ax}) \left(\dfrac{\sin bx}{b} \right) - \int \left(\dfrac{\sin bx}{b} \right) (a \, e^{ax} \, dx)$

$$= \frac{1}{b} \, e^{ax} \sin bx - \frac{a}{b} \int e^{ax} \sin bx \, dx$$

Substituting this result into equation (1) gives:

$\int e^{ax} \sin bx \, dx = -\dfrac{1}{b} \, e^{ax} \cos bx + \dfrac{a}{b} \left[\dfrac{1}{b} \, e^{ax} \sin bx - \dfrac{a}{b} \int e^{ax} \sin bx \, dx \right]$

$$= \frac{-1}{b} \, e^{ax} \cos bx + \frac{a}{b^2} \, e^{ax} \sin bx - \frac{a^2}{b^2} \left[\int e^{ax} \sin bx \, dx \right]$$

The integral in the bracket on the right-hand side is the same as the integral on the left-hand side thus they may be combined on the left-hand side of the equation. Thus:

$\int e^{ax} \sin bx \, dx + \dfrac{a^2}{b^2} \int e^{ax} \sin bx \, dx = \dfrac{-1}{b} \, e^{ax} \cos bx + \dfrac{a}{b^2} \, e^{ax} \sin bx$

i.e. $\left(1 + \dfrac{a^2}{b^2} \right) \int e^{ax} \sin bx \, dx = \dfrac{e^{ax}}{b^2} \, (a \sin bx - b \cos bx)$

$\left(\dfrac{b^2 + a^2}{b^2} \right) \int e^{ax} \sin bx \, dx = \dfrac{e^{ax}}{b^2} \, (a \sin bx - b \cos bx)$

Hence $\int e^{ax} \sin bx \, dx = \left(\dfrac{b^2}{a^2 + b^2} \right) \dfrac{e^{ax}}{b^2} \, (a \sin bx - b \cos bx)$

$$= \left(\frac{e^{ax}}{a^2 + b^2} \right) (a \sin bx - b \cos bx) + c$$

A product of an exponential function and a sine or cosine function thus involves integration by parts twice. If, in the above problem, the exponential function is made equal to dv instead of u the same result is obtained. It is left as a student exercise to prove this.

By a similar method to above it may be shown that:

$$\int e^{ax} \cos bx \; dx = \left(\frac{e^{ax}}{a^2 + b^2}\right) (b \sin bx + a \cos bx) + c$$

Further problems on integration by parts may be found in the following Section (3) (Problems 1 to 27).

3 Further problems

Find the following integrals using integration by parts

1. $\int x \, e^{3x} \; dx \qquad \left[\frac{e^{3x}}{9} \, (3x - 1) + c\right]$

2. $\int 3x \, e^{2x} \; dx. \qquad \left[\frac{3}{4} \, e^{2x} \, (2x - 1) + c\right]$

3. $\displaystyle\int \frac{5x}{e^{4x}} \; dx. \qquad \left[\frac{-5}{16} \, e^{-4x} \, (4x + 1) + c\right]$

4. $\int x \cos x \; dx. \qquad [x \sin x + \cos x + c]$

5. $\int x \ln x \; dx. \qquad \left[\frac{x^2}{4} \, (2 \ln x - 1) + c\right]$

6. $\int 3x \sin 2x \; dx. \qquad \left[\frac{-3x}{2} \cos 2x + \frac{3}{4} \sin 2x + c\right]$

7. $\int \ln 4x \; dx. \qquad [x \, (\ln 4x - 1) + c]$

8. $\displaystyle\int \frac{\ln y \; dy}{y^2}. \qquad \left[-\frac{1}{y} \, (\ln y + 1) + c\right]$

9. $\int 2x \cos 5x \; dx. \qquad \left[\frac{2}{25} \, (5x \sin 5x + \cos 5x) + c\right]$

10. $\int 2x^2 \, e^x \; dx. \qquad [2e^x \, (x^2 - 2x + 2) + c]$

11. $\int x^2 \sin 2x \; dx. \qquad \left[\left(\frac{1}{4} - \frac{x^2}{2}\right) \cos 2x + \frac{x}{2} \sin 2x + c\right]$

12. $\int e^{2x} \cos x \; dx. \qquad \left[\frac{e^{2x}}{5} \, (\sin x + 2 \cos x) + c\right]$

13. $\int 4 \, \theta^2 \cos 3\theta \; d\theta. \qquad \left[4 \left(\frac{\theta^2}{3} - \frac{2}{27}\right) \sin 3\theta + \frac{8}{9} \, \theta \cos 3\theta + c\right]$

14. $\int 3e^x \sin 2x \, dx$. $\left[\dfrac{3}{5} e^x (\sin 2x - 2 \cos 2x) + c\right]$

15. $\int 4x \sec^2 x \, dx$. $[4[x \tan x - \ln(\sec x)] + c]$

Evaluate the following integrals correct to three decimal places.

16. $\displaystyle\int_0^2 x\, e^x \, dx$.   [8.389]

17. $\displaystyle\int_0^{\frac{\pi}{2}} x \cos x \, dx$.   [0.571]

18. $\displaystyle\int_0^1 t\, e^{2t} \, dt$.   [2.097]

19. $\displaystyle\int_0^{\frac{\pi}{4}} \phi \sin 2\phi \, d\phi$.   [0.250]

20. $\displaystyle\int_1^2 \ln x^2 \, dx$.   [0.773]

21. $\displaystyle\int_0^{\frac{\pi}{2}} 3x^2 \cos x \, dx$.   [1.402]

22. $\displaystyle\int_0^1 x^2\, e^x \, dx$.   [0.718]

23. $\displaystyle\int_1^4 \sqrt{t} \, \ln t \, dt$.   [4.282]

24. $\displaystyle\int_0^1 2e^t \cos 2t \, dt$.   [1.125]

25. $\displaystyle\int_0^{\frac{\pi}{2}} e^\theta \sin \theta \, d\theta$.   [2.905]

26. In the study of damped oscillations integrations of the following type are important:

$$C = \int_0^1 e^{-0.4\theta} \cos 1.2\theta \ d\theta$$

and $S = \int_0^1 e^{-0.4\theta} \sin 1.2\theta \ d\theta$

Determine C and S. $[C = 0.66, S = 0.41]$

27. If a string is plucked at a point $x = \dfrac{l}{3}$ with an amplitude a and released, the equation of motion is

$$K = \frac{2}{l}\left\{\int_0^{\frac{l}{3}} \frac{3a}{l} x \sin \frac{n\pi}{l} x \ dx + \int_{\frac{l}{3}}^{l} \frac{3a}{2l}(l-x) \sin \frac{n\pi}{l} x \ dx\right\}$$

where n is a constant.

Show that $K = \dfrac{9a}{\pi^2 n^2} \sin \dfrac{n\pi}{3}$.

Chapter 17

First order differential equations by separation of the variables

1 Introduction

Definition. A differential equation is one that contains differential coefficients. Examples include:

(i) $\dfrac{\mathrm{d}y}{\mathrm{d}x} = 5x$

(ii) $\dfrac{\mathrm{d}^2 y}{\mathrm{d}x^2} + 3\dfrac{\mathrm{d}y}{\mathrm{d}x} + 4y = 0$

(iii) $\left(\dfrac{\mathrm{d}^2 s}{\mathrm{d}t^2}\right)^3 + 2\left(\dfrac{\mathrm{d}s}{\mathrm{d}t}\right)^4 = 4.$

Order. Differential equations are classified according to the highest derivative which occurs in them. Hence example (i) above is a **first order differential equation** and examples (ii) and (iii) are both **second order differential equations**.

Degree. The degree of a differential equation is that of the highest power of the highest differential which the equation contains after any necessary simplification. Thus example (i) above is of first order, first degree, example (ii) is of second order, first degree and example (iii) is of second order, third degree.

Solutions of a differential equation

Starting with a differential equation it is possible, by integration and by being given enough data to determine the constants, to obtain the original function. The process of determining the original relationship between the two variables is called 'solving the differential equation'. A solution to a differential equation which contains one or more arbitrary constants of integration is called the **general solution** of the differential equation. When finding the general solution of a first order differential equation one arbitrary constant results; when finding the general solution of a second order differential equation two arbitrary constants result, and so on. When additional information is given so that constants may be calculated the **particular solution** of the differential equation is obtained. The additional information is called the **boundary conditions**.

In this text only differential equations of the first and second order and first degree are discussed and the method of solution of five types of first order and two types of second order are considered. The first order types discussed are:

(a) differential equations of the form $\dfrac{dy}{dx} = f(x)$,

(b) differential equations of the form $\dfrac{dy}{dx} = f(y)$,

(c) 'variable separable' types of differential equations,

(d) differential equations which are homogeneous in x and y, and

(e) differential equations of the form $\dfrac{dy}{dx} + Py = Q$, where P and Q are

functions of x.

Types (a), (b) and (c) are considered in this chapter, type (d) in Chapter 18 and type (e) in Chapter 19. Two second order types of differential equation are discussed in Chapters 20 and 21.

In Chapter 9, *Technician Mathematics level 3* by J.O. Bird and A.J.C. May, an introduction to differential equations is given, together with the method of solution of equations of the form $\dfrac{dy}{dx} = f(x)$ and $\dfrac{dQ}{dt} = kQ$. The former type is revised in the following section of this chapter.

2 Solution of differential equations of the form $\dfrac{dy}{dx} = f(x)$

An equation of the form $\dfrac{dy}{dx} = f(x)$ may be solved immediately by integration. The solution is $y = \int f(x)\, dx$.

Problem 1. Find the general solutions of the following differential equations.

(a) $\dfrac{dy}{dx} = 5x^2 + \cos 3x$ (b) $x\dfrac{dy}{dx} = 3 - 2x^2$.

(a) If $\dfrac{dy}{dx} = 5x^2 + \cos 3x$, then $y = \int(5x^2 + \cos 3x)\,dx$

$$\text{i.e. } y = \frac{5x^3}{3} + \frac{1}{3}\sin 3x + c$$

(b) If $x\dfrac{dy}{dx} = 3 - 2x^2$, then $\dfrac{dy}{dx} = \dfrac{3 - 2x^2}{x} = \dfrac{3}{x} - 2x$

$$\text{Hence } y = \int\left(\frac{3}{x} - 2x\right)dx$$

$$\text{i.e. } y = 3\ln x - x^2 + c$$

Problem 2. Find the particular solutions of the following differential equations satisfying the given boundary conditions:

(a) $3\dfrac{dy}{dx} + x = 6$ and $y = 5\frac{1}{2}$ when $x = 3$.

(b) $2\dfrac{dr}{d\theta} + \sin 2\theta = 0$ and $r = 2$ when $\theta = \dfrac{\pi}{2}$.

(a) If $3\dfrac{dy}{dx} + x = 6$, then $\dfrac{dy}{dx} = \dfrac{6 - x}{3} = 2 - \dfrac{x}{3}$

$$\text{Hence } y = \int\left(2 - \frac{x}{3}\right)dx$$

$$\text{i.e. } y = 2x - \frac{x^2}{6} + c$$

(This is the general solution.)

Substituting the boundary conditions, $y = 5\frac{1}{2}$ when $x = 3$ to evaluate c gives:

$$5\tfrac{1}{2} = 6 - 1\tfrac{1}{2} + c$$
$$\text{i.e. } c = 1.$$

Hence the particular solution is $y = 2x - \dfrac{x^2}{6} + 1$.

(b) If $2\dfrac{dr}{d\theta} + \sin 2\theta = 0$, then $\dfrac{dr}{d\theta} = \dfrac{-\sin 2\theta}{2}$

$$\text{Hence } r = \int\frac{-\sin 2\theta}{2}\,d\theta$$

$$\text{i.e. } r = \frac{1}{4}\cos 2\theta + c$$

Substituting the boundary conditions, $r = 2$ when $\theta = \dfrac{\pi}{2}$ to evaluate c

gives:

$$2 = \frac{1}{4} \cos \pi + c$$

i.e. $c = 2 - \left(-\frac{1}{4}\right) = 2\frac{1}{4}$

Hence the particular solution is $r = \frac{1}{4} \cos 2\theta + 2\frac{1}{4} = \frac{1}{4} (\cos 2\theta + 9)$.

Further problems on solving differentials equations of the form $\frac{dy}{dx} = f(x)$ *may be found in Section 5 (Problems 1 to 14), page 236.*

3 Solution of differential equations of the form $\frac{dy}{dx} = f(y)$

An equation of the form $\frac{dy}{dx} = f(y)$ may be rearranged to give:

$$\frac{dx}{dy} = \frac{1}{f(y)}$$

i.e. $dx = \frac{dy}{f(y)}$

Integrating both sides gives: $\int dx = \int \frac{dy}{f(y)}$

Hence the solution may be obtained by direct integration.

Problem 1. Solve the differential equations:

(a) $\frac{dy}{dx} = 2 + y.$ (b) $3\frac{dy}{dx} = \sec 2y.$

(a) Rearranging $\frac{dy}{dx} = 2 + y$ gives $dx = \frac{dy}{(2+y)}$

Integrating both sides gives: $\int dx = \int \frac{dy}{(2+y)}$

i.e. $x = \ln(2+y) + c$ (1)

The general solution of differential equations can sometimes be rearranged. In this case, for example, if $C = \ln D$, where D is a constant, then:

$$x = \ln(2+y) + \ln D$$
i.e. $x = \ln D(2+y)$, from the law of logarithms,
from which $e^x = D(2+y)$ (2)

Equations (1) and (2) are both acceptable general solutions of the differential equation $\frac{dy}{dx} = 2 + y$.

(b) Rearranging $3 \frac{dy}{dx} = \sec 2y$ gives $dx = \frac{3}{\sec 2y} dy = 3 \cos 2y \, dy$

Integrating both sides gives: $\int dx = \int 3 \cos 2y \, dy$

Hence the general solution is $x = \frac{3}{2} \sin 2y + c.$

Problem 2. The rate at which a body cools is given by the equation $\frac{d\theta}{dt} = -k\theta$, where θ is the temperature of the body above its surroundings and k is a constant. Solve the equation for θ given that at $t = 0, \theta = \theta_0$

$\frac{d\theta}{dt} = -k\theta$ is of the form $\frac{dy}{dx} = f(y)$.

Rearranging gives $dt = \frac{-1}{k\theta} d\theta$

Integrating both sides gives: $\int dt = \frac{-1}{k} \int \frac{d\theta}{\theta}$

i.e. $t = \frac{-1}{k} \ln \theta + c$ (1)

Substituting the boundary conditions $t = 0$ and $\theta = \theta_0$ to find c gives:

$0 = \frac{-1}{k} \ln \theta_0 + c$

i.e. $c = \frac{1}{k} \ln \theta_0$

Substituting $c = \frac{1}{k} \ln \theta_0$ in equation (1) gives:

$t = -\frac{1}{k} \ln \theta + \frac{1}{k} \ln \theta_0$

$t = \frac{1}{k} (\ln \theta_0 - \ln \theta) = \frac{1}{k} \ln \left(\frac{\theta_0}{\theta} \right)$

$kt = \ln \left(\frac{\theta_0}{\theta} \right)$

$e^{kt} = \frac{\theta_0}{\theta}$

$e^{-kt} = \frac{\theta}{\theta_0}$

Hence $\theta = \theta_0 e^{-kt}$

Further problems on solving differential equations of the form $\frac{dy}{dx} = f(y)$
may be found in Section 5 (Problems 15 to 29), page 237.

4 Solution of 'variable separable' type of differential equations

An equation of the form $\frac{dy}{dx} = f(x) \cdot g(y)$, where $f(x)$ is a function of x only
and $g(y)$ is a function of y only, may be rearranged thus:

$$\frac{dy}{g(y)} = f(x)\, dx$$

Integrating both sides gives $\int \frac{dy}{g(y)} = \int f(x)\, dx$, i.e. the left-hand side is
the integral of a function of y with respect to y and the right-hand side is the
integral of a function of x with respect to x.

When two variables can be rearranged into two separate groups as shown
above, each consisting of only one variable, the variables are said to be
separable.

The equations of the type $\frac{dy}{dx} = f(x)$ and $\frac{dy}{dx} = f(y)$ discussed in Sections
2 and 3 are, in fact, merely special simple cases of 'separating the variables'.

Problem 1. Solve the differential equations:

(a) $\frac{dy}{dx} = \frac{3x^2 - 2}{2y - 1}$ (b) $2xy\,\frac{dy}{dx} = 1 + y^2$

(a) $\frac{dy}{dx} = \frac{3x^2 - 2}{2y - 1}$.

Separating the variables gives $(2y - 1)\, dy = (3x^2 - 2)\, dx$.
Integrating both sides gives $\int(2y - 1)\, dy = \int(3x^2 - 2)\, dx$.
Hence the general solution is $y^2 - y = x^3 - 2x + C$.

Note that when integrating both sides of an equation there is no need
to put an arbitrary constant on both sides of the result. In this case, if
this was done, then:

$$y^2 - y + A = x^3 - 2x + B$$
$$\text{and} \quad y^2 - y = x^3 - 2x + C, \text{ where } C = B - A.$$

(b) $2xy\,\frac{dy}{dx} = 1 + y^2$.

Separating the variables gives: $\frac{2y}{1 + y^2}\, dy = \frac{1}{x}\, dx$.

Integrating both sides gives: $\int \dfrac{2y}{(1+y^2)}\ dy = \int \dfrac{1}{x}\ dx$

Hence the general solution is $\ln (1+y^2) = \ln x + C$ (1)

$$\text{or } \ln (1+y^2) - \ln x = C$$

from which $\ln \left(\dfrac{1+y^2}{x} \right) = C$

and $\dfrac{1+y^2}{x} = e^C$ (2)

Also, if in equation (1), $C = \ln A$, we have $\ln (1+y^2) = \ln x + \ln A$

$$\ln (1+y^2) = \ln (Ax)$$

i.e. $1 + y^2 = Ax$ (3)

Equations (1), (2) and (3) are all valid general solutions to the differential equation $2xy\ \dfrac{dy}{dx} = 1 + y^2$, none of them being any more correct than the others. Thus, by manipulation, it is possible to obtain several general solutions to a differential equation.

Problem 2. Find the particular solution of $\dfrac{dy}{dx} = 3e^{2x-3y}$ given that $y = 0$ when $x = 0$.

$\dfrac{dy}{dx} = 3e^{2x-3y} = 3(e^{2x})(e^{-3y})$ by the laws of indices.

Separating the variables gives $\dfrac{dy}{e^{-3y}} = 3e^{2x}\ dx$

i.e. $e^{3y}\ dy = 3e^{2x}\ dx$

Integrating both sides gives $\int e^{3y}\ dy = 3\int e^{2x}\ dx$

$$\frac{1}{3}\ e^{3y} = \frac{3}{2}\ e^{2x} + C.$$

(This is the general solution.)

When $y = 0, x = 0$, thus $\dfrac{1}{3}\ e^0 = \dfrac{3}{2}\ e^0 + C$

$$C = \frac{1}{3} - \frac{3}{2} = -\frac{7}{6}$$

Hence the particular solution is $\dfrac{1}{3}\ e^{3y} = \dfrac{3}{2}\ e^{2x} - \dfrac{7}{6}$

$$\text{or } 2e^{3y} = 9e^{2x} - 7.$$

Problem 3. An electrical circuit contains inductance L and resistance R connected to a constant voltage source E. The current i is given by the differential equation $E - L\, \dfrac{di}{dt} = Ri$, where L and R are constants. Find the current in terms of time t given that when $t = 0, i = 0$.

$E - L\, \dfrac{di}{dt} = Ri.$ Rearranging gives $\dfrac{di}{dt} = \dfrac{E - Ri}{L}$

and $\dfrac{di}{E - Ri} = \dfrac{dt}{L}$

Integrating both sides gives: $\displaystyle\int \dfrac{di}{E - Ri} = \int \dfrac{dt}{L}$

$$-\frac{1}{R}\, \ln\,(E - Ri) = \frac{t}{L} + C.$$

(This is the general solution.)

$t = 0$ when $i = 0$ hence $-\dfrac{1}{R}\, \ln E = C.$

Thus, $-\dfrac{1}{R}\, \ln\,(E - Ri) = \dfrac{t}{L} - \dfrac{1}{R}\, \ln E.$

This particular solution must now be transposed to find i.

$$-\frac{1}{R}\, \ln\,(E - Ri) + \frac{1}{R}\, \ln E = \frac{t}{L}$$

$$\frac{1}{R}\,(\ln E - \ln\,(E - Ri)) = \frac{t}{L}$$

$$\ln\left(\frac{E}{E - Ri}\right) = \frac{Rt}{L}$$

$$\frac{E}{E - Ri} = e^{\frac{Rt}{L}}$$

$$\frac{E - Ri}{E} = e^{-\frac{Rt}{L}}$$

$$Ri = E - E\,e^{-\frac{Rt}{L}}$$

Hence current $i = \dfrac{E}{R}\left(1 - e^{-\frac{Rt}{L}}\right)$

This expression for current represents the natural law of growth of current in an inductive circuit.

Further problems on solving 'variables separable' types of differential equations may be found in the following Section (5) (Problems 30 to 46), page 239.

5 Further problems

Differential equations of the form $\dfrac{dy}{dx} = f(x)$

In problems 1 to 9 solve the differential equations.

1. $\dfrac{dy}{dx} = 2x^4$. $\left[y = \dfrac{2}{5} x^5 + C \right]$

2. $\dfrac{dy}{dx} = 5x + \sin x$. $\left[y = \dfrac{5}{2} x^2 - \cos x + C \right]$

3. $x\dfrac{dy}{dx} = 4 - x^2$. $\left[y = 4 \ln x - \dfrac{x^2}{2} + C \right]$

4. $\dfrac{dy}{dx} - 2x^3 = e^{3x}$. $[6y = 2e^{3x} + 3x^4 + C]$

5. $x^2 \dfrac{dy}{dx} = 2 + x$. $\left[y = \ln x - \dfrac{2}{x} + C \right]$

6. $\dfrac{dy}{dx} + x = 2$ and $y = 3$ when $x = 2$. $\left[y = 2x - \dfrac{x^2}{2} + 1 \right]$

7. $2\dfrac{dr}{d\theta} + \cos \theta = 0$ and $r = \dfrac{5}{2}$ when $\theta = \dfrac{\pi}{2}$. $\left[r = 3 - \dfrac{1}{2} \sin \theta \right]$

8. $x(x - \dfrac{dy}{dx}) = 3$ and $y = 1$ when $x = 1$. $[2y = x^2 - 6 \ln x + 1]$

9. $\dfrac{1}{2e^t} + 4 = t - 3 \dfrac{d\theta}{dt}$ and $\theta = \dfrac{1}{6}$ when $t = 0$

$$\left[\theta = \dfrac{1}{3} \left(\dfrac{t^2}{2} + \dfrac{e^{-t}}{2} - 4t \right) \right]$$

10. The acceleration of a body a is equal to its rate of change of velocity, $\dfrac{dv}{dt}$. Determine an equation for v in terms of t given that the velocity is u when $t = 0$. $[v = u + at]$

11. The velocity of a body, v is equal to its rate of change of distance, $\dfrac{dx}{dt}$. Determine an equation for x in terms of t, given $v = u + at$, where u and a are constants and $x = 0$ when $t = 0$. $[x = ut + \frac{1}{2}at^2]$

12. The gradient of a curve is given by $\dfrac{dy}{dx} = 4x - \dfrac{x^3}{6}$. Determine the equation of the curve if it passes through the point $\left(2, 3\dfrac{1}{3}\right)$.

$$\left[y = 2x^2 - \frac{x^4}{24} - 4 \right]$$

13. The bending moment of a beam, M, and shear force F are related by the equation $\dfrac{dM}{dx} = F$, where x is the distance from one end of the beam. Determine M in terms of x when $F = -w\,(l - x)$ where w and l are constants and $M = \dfrac{1}{2}\,w\,l^2$ when $x = 0$. $\quad \left[M = \dfrac{1}{2}\,w(l - x)^2 \right]$

14. The angular velocity ω of a flywheel of moment of inertia I is given by $I\dfrac{d\omega}{dt} + N = 0$, where N is a constant. Determine ω in terms of t given that $\omega = \omega_0$ when $t = 0$. $\quad \left[\omega = \omega_0 - \dfrac{Nt}{I} \right]$

Differential equations of the form $\dfrac{dy}{dx} = f(y)$

In Problems 15 to 22 solve the differential equations.

15. $\dfrac{dy}{dx} = 3 + 2y$. $\quad \left[\dfrac{1}{2} \ln (3 + 2y) = x + c \right]$

16. $5\dfrac{dy}{dx} = \cot 2y$. $\quad \left[\dfrac{5}{2} \ln (\sec 2y) = x + c \right]$

17. $y\dfrac{dy}{dx} = 3 - y^2$. $\quad \left[-\dfrac{1}{2} \ln (3 - y^2) = x + c \right]$

18. $2\dfrac{dy}{dx} + 3y = 4$. $\quad \left[-\dfrac{2}{3} \ln (4 - 3y) = x + c \right]$

19. $\dfrac{dy}{dx} = 2 \tan y$. Find y. $\quad [y = \arcsin (e^{2x+c})]$

20. $y\dfrac{dy}{dx} = 1 - y$, and $y = 0$ when $x = 1$. $\quad [x + y + \ln (1 - y) = 1]$

21. $(y^2 + 1)\dfrac{dy}{dx} = 2y$ and $y = 1$ when $x = \dfrac{1}{4}$. $\quad [y^2 + 2 \ln y = 4x]$

22. $\sqrt{y}\,\dfrac{dy}{dx} - 1 = 0$ and $y = 4$ when $x = \dfrac{1}{3}$. $\quad \left[\dfrac{2}{3}\sqrt{y^3} = x + 5 \right]$

23. An equation of motion may be represented by the equation $\dfrac{dv}{dt} + kv^2 = 0$

where v is the velocity of a body travelling in a restraining medium. Show that $v = \dfrac{v_0}{1 + kt\,v_0}$ given that $v = v_0$ when $t = 0$.

24. The current in an electric circuit is given by $L\dfrac{di}{dt} + Ri = 0$ where L and R are constants. Solve for i given that $i = I$ when $t = 0$.

$$\left[i = I\,e^{-\frac{Rt}{L}} \right]$$

25. The difference in tension, T newtons, between two sides of a belt when in contact with a pulley over an angle of θ radians and when it is on the point of slipping, is given by $\dfrac{dT}{d\theta} = \mu\,T$, where μ is the coefficient of friction between the material of the belt and that of the pulley at the point of slipping. When $\theta = 0$ radians, the tension is 150 N and $\mu = 0.29$ as slipping starts. Find the tension at the point of slipping when $\theta = \dfrac{2\pi}{3}$ radians. Also determine the angle of lap correct to the nearest degree to give a tension of 300 N just before slipping starts. [275.3 N, 137°]

26. The charge Q coulomb at time t seconds for a capacitor of capacitance C farads when discharging through a resistance of R ohms is given by $R\dfrac{dQ}{dt} + \dfrac{Q}{C} = 0$. Solve for Q given that $Q = Q_0$ when $t = 0$. A circuit contains a resistance of 400 kilohms and a capacitance of 7.3 microfarads, and after 225 milliseconds the charge falls to 7.0 coulombs. Find the initial charge and the charge after 2 seconds, correct to three significant figures.

$$[Q = Q_0\,e^{-\frac{t}{CR}},\ 7.56\ \text{C};\ 3.81\ \text{C}]$$

27. In a chemical reaction in which x is the amount transformed in time t the velocity of the reaction is given by $\dfrac{dx}{dt} = k(a - x)$ where k is a constant and a is the concentration at time $t = 0$ when $x = 0$. Find x in terms of t. $[x = a(1 - e^{-kt})]$

28. The rate of decay of a radioactive substance is given by $\dfrac{dN}{dt} = -\lambda N$, where λ is the decay constant and λN is the number of radioactive atoms disintegrating per second. Determine the half-life of radium in years (i.e. the time for N to become one-half of its original value) taking the decay constant for radium as 1.36×10^{-11} atoms per second and assuming a '365 day' year. [1 616 years]

29. The variation of resistance R ohms, of a copper conductor with temperature, $\theta°C$, is given by $\dfrac{dR}{d\theta} = \alpha R$, where α is the temperature coefficient

of resistance of copper. If $R = R_0$ at $\theta = 0°C$, solve the equation for R. Taking α as 39×10^{-4} per $°C$, find the resistance of a copper conductor at $20°C$, correct to four significant figures, when its resistance at $80°C$ is 57.4 ohms.   [$R = R_0 e^{\alpha\theta}$; 45.42 ohms]

'Variables separable' types of differential equations

In Problems 30 to 36 solve the differential equations.

30. $\dfrac{dy}{dx} = (2y)(x^2)$.

$$\left[\frac{1}{2}\ln y = \frac{x^3}{3} + c \right.$$

$$\left. \text{or } \frac{1}{2}\ln 2y = \frac{x^3}{3} + k \right]$$

31. $\dfrac{dy}{dx} = y\cos x$. $[\ln y = \sin x + c]$

32. $(x+2)\dfrac{dy}{dx} = (1-y)$. $[\ln(x+2)(1-y) = c]$

33. $\dfrac{dy}{dx} = \dfrac{2x^2-1}{3y+2}$ and $x = 0$ when $y = 0$. $\left[\dfrac{3y^2}{2} + 2y = \dfrac{2x^3}{3} - x \right]$

34. $\dfrac{dy}{dx} = e^{x-2y}$ and $x = 0$ when $y = 0$. $[e^{2y} = 2e^x - 1]$

35. $\dfrac{1}{2}\dfrac{dy}{dx} = e^{x+3y}$ and $x = 0$ when $y = 0$. $[7 - e^{-3y} = 6e^x]$

36. $y(1+x) + x(1-y)\dfrac{dy}{dx} = 0$ and $x = 1$ when $y = 1$. $[\ln(xy) = y - x]$

37. Show that the solution of the equation $xy\dfrac{dy}{dx} = 1 + 2y^2$ may be of the form $y = \sqrt{\left(\dfrac{x^4 k - 1}{2}\right)}$, where k is a constant.

38. Show that the solution of the differential equation $\dfrac{y^2+2}{x^2+2} = \dfrac{y}{x}\dfrac{dy}{dx}$ is of the form $\sqrt{\left(\dfrac{x^2+2}{y^2+2}\right)} = $ constant.

39. Prove that $y = x$ is a solution of the equation $x\sqrt{(y^2-1)} - y\sqrt{(x^2-1)}\dfrac{dy}{dx} = 0$ when $x = 1$ and $y = 1$.

40. Solve $xy = (1+x^2)\dfrac{dy}{dx}$ for y. $\left[y = \dfrac{1}{k}\sqrt{(1+x^2)} \right]$

41. Find the curve which satisfies the equation $2xy\dfrac{dy}{dx} = x^2 + 1$ and which passes through the point $(1, 2)$. $[2y^2 = x^2 + 2 \ln x + 7]$

42. Solve the equation $y \cos^2 x \dfrac{dy}{dx} = \tan x + 2$ given that $y = 2$ when

$x = \dfrac{\pi}{4}$. $[y^2 = \tan^2 x + 4 \tan x - 1]$

43. A capacitor C is charged by applying a steady voltage E through a resistance R. The p.d. between the plates, V, is given by the differential equation $CR \dfrac{dV}{dt} + V = E$. Solve the equation for V given that $V = 0$ when $t = 0$ and evaluate V when $E = 20$ volts, $C = 25$ microfarads, $R = 300$ kilohms and $t = 2$ seconds.

$$[V = E\,(1 - e^{-\frac{t}{CR}}); 4.681 \text{ volts}]$$

44. For an adiabatic expansion of a gas $C_p \dfrac{dv}{V} + C_v \dfrac{dp}{P} = 0$, where C_p and C_v are constants. Show that $pv^n = $ constant, where $n = \dfrac{C_p}{C_v}$.

45. The streamlines of a cyclinder of radius a in a stream of liquid of ambient velocity v are given by the equation: $\dfrac{dr}{d\theta} = \dfrac{r(a^2 - r^2)}{(a^2 + r^2)} \cot \theta$.
r is the distance of the centre of the cyclinder from the applied force, the line joining them being at an angle θ to the direction of the liquid flow. Solve the equation. (Hint: let $a^2 + r^2 = a^2 - r^2 + 2r^2$.)

$$\left[\left(\dfrac{a^2}{r} - r\right) \sin \theta = C\right]$$

46. The equilibrium constant (K) of a chemical reaction varies with temperature (T) according to the equation: $\dfrac{d\,(\ln K)}{dT} = \dfrac{\Delta H}{RT^2}$. If $\Delta H = 10^4$, $R = 8.3$ and $K = 4$ when $T = 600$, solve the equation completely.

$$\left[\ln K = \dfrac{-\Delta H}{RT} + 3.39\right]$$

Chapter 18

Homogeneous first order differential equations

1 Solution of differential equations of the form $P \dfrac{dy}{dx} = Q$

Certain first order differential equations are not separable but can be made separable by a simple change of variables.

An equation of the form $P\dfrac{dy}{dx} = Q$, where P and Q are functions of both x and y of the same degree throughout, is said to be **homogeneous**. An example of a homogeneous function is $f(x,y) = x^3 + 2x^2 y + y^3$, since each of the three terms are of the degree 3. However, $f(x,y) = 2x^2 + xy^2 + y^2$ is not homogeneous since the second term is of degree 3 and the other two are of degree 2. Similarly, $f(x,y) = \dfrac{x + 2y}{3x - y}$ is homogeneous in x and y since each of the four terms are of degree 1.

Procedure to solve a differential equation of the form $P \dfrac{dy}{dx} = Q$

(i) Rearrange $P \dfrac{dy}{dx} = Q$ into the form $\dfrac{dy}{dx} = \dfrac{Q}{P}$.

(ii) Make the substitution $y = vx$, where v is a function of x. If $y = vx$ then, by the product rule, $\dfrac{dy}{dx} = v(1) + x \dfrac{dv}{dx}$. There is no particular reason for choosing the letter v, and in any case, it does not appear in the final solution.

(iii) Substitute for both y and $\dfrac{dy}{dx}$ in the equation $\dfrac{dy}{dx} = \dfrac{Q}{P}$. It will be found that the x terms will cancel leaving an equation in which the variables are separable.

(iv) Separate the variables and solve using the method shown in Chapter 17.

(v) The solution will be in terms of v and x. Thus, substitute $v = \dfrac{y}{x}$ to solve in terms of the original variables x any y.

Problem 1. Solve $xy \dfrac{dy}{dx} = x^2 + y^2$ given that $x = 1$ when $y = 2$.

Using the above procedure:

(i) Since $xy \dfrac{dy}{dx} = x^2 + y^2$ then $\dfrac{dy}{dx} = \dfrac{x^2 + y^2}{xy}$ which is homogeneous in x and y, since each of the three terms on the right-hand side are of the same degree (i.e. degree 2).

(ii) Let $y = vx$, then $\dfrac{dy}{dx} = v + x \dfrac{dv}{dx}$.

(iii) Substituting for y and $\dfrac{dy}{dx}$ in $\dfrac{dy}{dx} = \dfrac{x^2 + y^2}{xy}$ gives:

$$v + x \frac{dv}{dx} = \frac{x^2 + (vx)^2}{x(vx)} = \frac{x^2 + v^2 x^2}{x^2 v} = \frac{1 + v^2}{v}$$

Hence $\quad v + x \dfrac{dv}{dx} = \dfrac{1 + v^2}{v}$

(iv) Separating the variables gives: $x \dfrac{dv}{dx} = \dfrac{1 + v^2}{v} - v = \dfrac{1 + v^2 - v^2}{v} = \dfrac{1}{v}$

$$\text{i.e.} \quad v\,dv = \frac{dx}{x}$$

Integrating both sides gives: $\displaystyle \int v\,dv = \int \frac{dx}{x}$

$$\therefore \quad \frac{v^2}{2} = \ln x + c$$

(v) Replacing v by $\dfrac{y}{x}$ gives: $\dfrac{y^2}{2x^2} = \ln x + c$

When $x = 1, y = 2$ thus $\quad \dfrac{4}{2} = \ln 1 + c$

$$\text{i.e.} \quad c = 2$$

Thus the particular solution of $xy \dfrac{dy}{dx} = x^2 + y^2$ **is** $\dfrac{y^2}{2x^2} = \ln x + 2.$

Problem 2. Solve the equation $(4y + 3x)\dfrac{dy}{dx} = (3x - y)$.

(i) If $(4y + 3x)\dfrac{dy}{dx} = (3x - y)$ then $\dfrac{dy}{dx} = \dfrac{(3x - y)}{(4y + 3x)}$, which is homogeneous in x and y since each of the four terms on the right-hand side
is of the same degree (i.e. degree 1).

(ii) Let $y = vx$ then $\dfrac{dy}{dx} = v + x\dfrac{dv}{dx}$.

(iii) Substituting for y and $\dfrac{dy}{dx}$ gives: $v + x\dfrac{dv}{dx} = \dfrac{3x - vx}{4vx + 3x} = \dfrac{3 - v}{4v + 3}$

(iv) Separating the variables gives: $x\dfrac{dv}{dx} = \dfrac{3 - v}{4v + 3} - v$

$$= \dfrac{(3 - v) - v(4v + 3)}{4v + 3} = \dfrac{3 - 4v - 4v^2}{4v + 3}$$

Hence $\dfrac{4v + 3}{(3 + 2v)(1 - 2v)}\, dv = \dfrac{dx}{x}$

Integrating both sides gives: $\displaystyle\int \dfrac{4v + 3}{(3 + 2v)(1 - 2v)}\, dv = \int \dfrac{dx}{x}$

Resolving $\dfrac{4v + 3}{(3 + 2v)(1 - 2v)}$ into partial fractions gives:

$\dfrac{5}{4(1 - 2v)} - \dfrac{3}{4(3 + 2v)}$

Hence $\displaystyle\int \left\{ \dfrac{5}{4(1 - 2v)} - \dfrac{3}{4(3 + 2v)} \right\} dv = \int \dfrac{dx}{x}$

Thus $-\dfrac{5}{8}\ln(1 - 2v) - \dfrac{3}{8}\ln(3 + 2v) = \ln x + c$

(v) Replacing v by $\dfrac{y}{x}$ gives: $-\dfrac{5}{8}\ln\left(1 - \dfrac{2y}{x}\right) - \dfrac{3}{8}\ln\left(3 + \dfrac{2y}{x}\right) = \ln x + c$

or $\ln x + \dfrac{5}{8}\ln\left(\dfrac{x - 2y}{x}\right) + \dfrac{3}{8}\ln\left(\dfrac{3x + 2y}{x}\right) + c = 0$

Multiplying throughout by 8 gives:

$8\ln x + 5\ln\left(\dfrac{x - 2y}{x}\right) + 3\ln\left(\dfrac{3x + 2y}{x}\right) + 8c = 0.$

By the laws of logarithms: $\ln x^8 + \ln\left(\dfrac{x - 2y}{x}\right)^5 + \ln\left(\dfrac{3x + 2y}{x}\right)^3 = K$

(where $K = -8c$)

244

Hence $\ln \left\{ \dfrac{(x^8)(x-2y)^5\,(3x+2y)^3}{(x^5)(x^3)} \right\} = K$

i.e. $\ln \left\{ (x-2y)^5\,(3x+2y)^3 \right\} = K$ **is the general solution.**

Problem 3. Solve $(y-x)\dfrac{dy}{dx} = \dfrac{y^2}{x} - y + \dfrac{x^2}{y}$ given that $x=1$ when $y=3$.

(i) Multiplying throughout by xy gives: $xy\,(y-x)\dfrac{dy}{dx} = y^3 - xy^2 + x^3$

from which $\dfrac{dy}{dx} = \dfrac{y^3 - xy^2 + x^3}{xy(y-x)} = \dfrac{y^3 - xy^2 + x^3}{xy^2 - x^2 y}$,

which is homogeneous in x and y.

(ii) Let $y = vx$, then $\dfrac{dy}{dx} = v + x\dfrac{dv}{dx}$.

(iii) Substituting for y and $\dfrac{dy}{dx}$ gives:

$v + x\dfrac{dv}{dx} = \dfrac{(vx)^3 - x(vx)^2 + x^3}{x(vx)^2 - x^2(vx)} = \dfrac{v^3 - v^2 + 1}{v^2 - v}$

(iv) Separating the variables gives:

$x\dfrac{dv}{dx} = \dfrac{v^3 - v^2 + 1}{v^2 - v} - v = \dfrac{(v^3 - v^2 + 1) - v(v^2 - v)}{v^2 - v}$

$= \dfrac{v^3 - v^2 + 1 - v^3 + v^2}{v^2 - v} = \dfrac{1}{v^2 - v}$

Hence $(v^2 - v)\,dv = \dfrac{dx}{x}$

Integrating both sides gives: $\int (v^2 - v)\,dv = \int \dfrac{dx}{x}$

$\dfrac{v^3}{3} - \dfrac{v^2}{2} = \ln x + c$

(v) Replacing v by $\dfrac{y}{x}$ gives: $\dfrac{y^3}{3x^3} - \dfrac{y^2}{2x^2} = \ln x + c$.

(This is the general solution.)

When $x=1, y=3$, thus: $\dfrac{27}{3(1)} - \dfrac{9}{2(1)} = \ln 1 + c$

i.e. $c = \dfrac{9}{2}$

Hence the particular solution is $\dfrac{y^3}{3x^3} - \dfrac{y^2}{2x^2} = \ln x + \dfrac{9}{2}$

i.e. $2y^3 - 3xy^2 = 6x^3 \ln x + 27 x^3$

Further problems on homogeneous differential equations may be found in the following Section (2) (Problems 1 to 10).

2 Further problems

In Problems 1 to 5 find the general solution of the differential equations.

1. $x\dfrac{dy}{dx} + \dfrac{y^2}{x} = y.$ $\left[\dfrac{x}{y} = \ln x + c \text{ or } x = y \ln k x \right]$

2. $(x + y)\dfrac{dy}{dx} = \dfrac{y}{x}\,(y - x).$

$\left[-\dfrac{1}{2} \ln \dfrac{y}{x} - \ln x = \dfrac{y}{2x} + c \text{ or } \dfrac{y}{2x} + \ln \sqrt{(xy)} = K \right]$

3. $y^2 = x^2 \dfrac{dy}{dx}.$ $[y = x(1 + ky)]$

4. $xy \dfrac{dy}{dx} = \dfrac{x^2 + y^2}{2}.$ $[x^2 - y^2 = kx]$

5. $(x + y) = (x - y)\dfrac{dy}{dx}.$ $\left[\arctan \dfrac{y}{x} = \ln \sqrt{(x^2 + y^2)} + c \right]$

6. If $x + y + x \dfrac{dy}{dx} = 0$, show that $x(x + 2y) = k$, where k is a constant.

In Problems 7 to 9 find the particular solution of the differential equations.

7. $\dfrac{dy}{dx} = \dfrac{xy}{x^2 + y^2}$ given that $x = 1$ when $y = 1$. $\left[y = e^{\frac{x^2 - y^2}{2y^2}} \right]$

8. $(x + y)\dfrac{dy}{dx} = x + \dfrac{y^2}{x}$ given that $x = 4$ when $y = 2$.

$\left[2x \ln \left\{ \dfrac{(x - y)^2}{x} \right\} = x - 2y \right]$

9. $\dfrac{xy^2}{(x^3 + y^3)}\dfrac{dy}{dx} = 1$ given that $x = 1$ when $y = 3$. $[y^3 = 3x^3 (\ln x + 9)]$

10. Given that $\dfrac{4xy}{(x^2 - y^2)}\dfrac{dy}{dx} = 1$, and $y = 0$ when $x = 1$, show that $(\sqrt{x})(x^2 - 5y^2) = 1$.

Chapter 19

Linear first order differential equations

1 Solution of differential equations of the form $\dfrac{dy}{dx} + Py = Q$

An equation of the form $\dfrac{dy}{dx} + Py = Q$, where P and Q are functions of x only, is called a **linear differential equation** since y and its derivative (i.e. $\dfrac{dy}{dx}$) are of the first degree.

The solution of $\dfrac{dy}{dx} + Py = Q$ is obtained by multiplying throughout by what is called an **integrating factor**.

Multiplying $\dfrac{dy}{dx} + Py = Q$ throughout by some function of x alone, say R, gives:

$$R\,\frac{dy}{dx} + RPy = RQ \tag{1}$$

Now the differential coefficient of a product Ry is given by the product rule and is:

$$\frac{d}{dx}(Ry) = R\,\frac{dy}{dx} + y\,\frac{dR}{dx}$$

$R\,\dfrac{dy}{dx} + y\,\dfrac{dR}{dx}$ is the same as the left hand side of equation (1) when R is chosen such that RP is equal to $\dfrac{dR}{dx}$.

If $\dfrac{dR}{dx} = Rp$, separating the variables gives:

$\dfrac{dR}{R} = P\,dx$

Integrating both sides gives: $\displaystyle\int \dfrac{dR}{R} = \int P\,dx$

i.e. $\ln R = \int P\,dx + c$

and $R = e^{\int P\,dx + c} = e^{\int P\,dx}\, e^c$

i.e. $R = A e^{\int P\,dx}$ (where $A = e^c$, an arbitrary constant)

Substituting $R = A e^{\int P\,dx}$ in equation (1) gives:

$$A e^{\int P\,dx} \left(\dfrac{dy}{dx}\right) + A e^{\int P\,dx} Py = A e^{\int P\,dx} Q$$

and by cancelling the constant A:

$$e^{\int P\,dx} \left(\dfrac{dy}{dx}\right) + e^{\int P\,dx} Py = e^{\int P\,dx} Q \qquad (2)$$

The left-hand side of equation (2) is $\dfrac{d}{dx}\left(y e^{\int P\,dx}\right)$, which may be checked by differentiating $y e^{\int P\,dx}$ with respect to x, using the product rule.

Hence from equation (2):

$$\dfrac{d}{dx}\left[y e^{\int P\,dx}\right] = e^{\int P\,dx}\, \dot{Q}$$

Integrating both sides gives:

$$y e^{\int P\,dx} = \int e^{\int P\,dx} Q\, dx \qquad (3)$$

Procedure to solve differential equations of the form $\dfrac{dy}{dx} + Py = Q$

(i) Rearrange the differential equation, if necessary, into the form $\dfrac{dy}{dx} + Py = Q$, where P and Q are functions of x.

(ii) Find $\int P\,dx$. Since the arbitrary constant of integration disappears (as shown in equation (2) above) it is unnecessary to include the constant for this indefinite integral.

(iii) Find the integrating factor $e^{\int P\,dx}$

(iv) Substitute $e^{\int P\,dx}$ into equation (3).

(v) Integrate the right-hand side of equation (3) to give the general solution. If boundary conditions are given the arbitrary constant of integration may be determined and the particular solution of the differential equation may be found.

Problem 1. Solve the equation $\dfrac{1}{x}\dfrac{dy}{dx} + 2y = 1$ given that when $x = 0$, $y = 1$.

Following the above procedure:

(i) Multiplying throughout by x gives $\dfrac{dy}{dx} + 2xy = x$, which is of the form $\dfrac{dy}{dx} + Py = Q$, where $P = 2x$ and $Q = x$.

(ii) $\int P\,dx = \int 2x\,dx = x^2$.
(iii) Integrating factor $= e^{x^2}$.
(iv) Substituting into equation (3) gives: $ye^{x^2} = \int e^{x^2}(x)\,dx$
The right-hand side may be integrated by using the substitution $u = x^2$.

Hence $ye^{x^2} = \dfrac{e^{x^2}}{2} + c.$ (This is the general solution.)

When $x = 0$, $y = 1$, thus $(1)e^0 = \dfrac{e^0}{2} + c$

i.e. $c = \tfrac{1}{2}$

Hence the particular solution is $ye^{x^2} = \dfrac{e^{x^2}}{2} + \dfrac{1}{2}$ or $2y = 1 + e^{-x^2}$

Problem 2. Solve: $x \cos x \dfrac{dy}{dx} + (x \sin x + \cos x)y = 1.$

(i) Dividing throughout by $x \cos x$ gives:

$$\dfrac{dy}{dx} + \left(\dfrac{x \sin x + \cos x}{x \cos x}\right) y = \dfrac{1}{x \cos x}$$

i.e. $\dfrac{dy}{dx} + \left(\tan x + \dfrac{1}{x}\right) y = \dfrac{\sec x}{x}$

which is of the form $\dfrac{dy}{dx} + Py = Q$, where $P = \left(\tan x + \dfrac{1}{x}\right)$

and $Q = \left(\dfrac{\sec x}{x}\right)$

(ii) $\int P\,dx = \int \left(\tan x + \dfrac{1}{x}\right) dx = \ln (\sec x) + \ln x = \ln (x \sec x)$

(iii) Integrating factor $= e^{\int P\,dx} = e^{\ln (x \sec x)} = x \sec x.$

(iv) Substituting in equation (3) gives:

$$y (x \sec x) = \int (x \sec x)\left(\dfrac{\sec x}{x}\right) dx$$

i.e. $xy \sec x = \int \sec^2 x\,dx$

(v) Integrating gives: $xy \sec x = \tan x + c$,

and dividing by $\sec x$: $xy = \dfrac{\tan x}{\sec x} + \dfrac{c}{\sec x}$

i.e. the general solution is $xy = \sin x + c \cos x$.

Problem 3. Find the particular solution of the differential equation $\dfrac{dy}{dx} + 2x = y$, given that $x = 0$ when $y = 2$.

(i) Rearranging gives $\dfrac{dy}{dx} - y = -2x$, which is of the form $\dfrac{dy}{dx} + Py = Q$, where $P = -1$ and $Q = -2x$.

(ii) $\int P \, dx = \int -1 \, dx = -x$

(iii) Integrating factor $= e^{\int P \, dx} = e^{-x}$.

(iv) Substituting into equation (3) gives

$$ye^{-x} = \int e^{-x}(-2x) \, dx$$

i.e. $ye^{-x} = -2 \int x e^{-x} \, dx$ \hfill (4)

(v) $\int x e^{-x} \, dx$ is determined using integration by parts.

Let $u = x$ and $dv = e^{-x} \, dx$.

Then $du = dx$ and $v = \int e^{-x} \, dx = -e^{-x}$.

Thus $\int x e^{-x} \, dx = x(-e^{-x}) - \int(-e^{-x}) \, dx$

$$= -xe^{-x} - e^{-x} + c.$$

Hence from equation (4):

$$ye^{-x} = -2(-xe^{-x} - e^{-x} + c)$$

which is the general solution.

When $x = 0, y = 2$, thus $2(1) = -2(0 - 1 + c)$

from which $c = 0$

Thus the particular solution is $ye^{-x} = 2xe^{-x} + 2e^{-x}$

or $y = 2x + 2$

or $y = 2(x + 1)$.

Problem 4. Solve: $(x + 2) \dfrac{dy}{dx} = 3 - \dfrac{2y}{x}$

(i) Rearranging gives: $\dfrac{dy}{dx} + \dfrac{2}{x(x+2)} \, y = \dfrac{3}{(x+2)}$, which is of the form $\dfrac{dy}{dx} + Py = Q$, where $P = \dfrac{2}{x(x+2)}$ and $Q = \dfrac{3}{(x+2)}$.

(ii) $\int P \, dx = \int \dfrac{2}{x \, (x + 2)} \, dx$ which may be integrated using partial fractions.

Let $\dfrac{2}{x \, (x + 2)} \equiv \dfrac{A}{x} + \dfrac{B}{(x + 2)} \equiv \dfrac{A \, (x + 2) + Bx}{x \, (x + 2)}$

Equating the numerators gives: $2 \equiv A \, (x + 2) + Bx$

When $x = 0$, $\qquad\qquad\qquad\qquad 2 = 2A$, i.e. $A = 1$

When $x = -2$, $\qquad\qquad\qquad\; 2 = -2B$, i.e. $B = -1$.

Hence $\displaystyle\int \dfrac{2}{x \, (x + 2)} \, dx = \int \left\{ \dfrac{1}{x} - \dfrac{1}{(x + 2)} \right\} \, dx = \ln x - \ln (x + 2)$

$$= \ln \left(\dfrac{x}{x + 2} \right)$$

(iii) Integrating factor $= e^{\int P \, dx} = e^{\ln (x / x + 2)} = \dfrac{x}{x + 2}$

(iv) Substituting in equation (3) gives:

$$y \left(\dfrac{x}{x + 2} \right) = \int \left(\dfrac{x}{x + 2} \right) \left(\dfrac{3}{x + 2} \right) \, dx = \int \dfrac{3x}{(x + 2)^2} \, dx$$

(v) $\displaystyle\int \dfrac{3x}{(x + 2)^2} \, dx$ cannot be determined 'on sight'. It is thus split into two fractions, the first fraction being specifically chosen to give a logarithmic solution.

Hence $\dfrac{xy}{x + 2} = \displaystyle\int \left\{ \dfrac{\frac{3}{2} (2x + 4)}{(x + 2)^2} - \dfrac{6}{(x + 2)^2} \right\} \, dx$

$$= \tfrac{3}{2} \ln (x + 2)^2 + \dfrac{6}{(x + 2)} + c$$

$\dfrac{xy}{x + 2} = \ln [(x + 2)^2]^{3/2} + \dfrac{6}{(x + 2)} + \ln k \quad$ (where $\ln k = c$)

Hence the general solution is $xy = (x + 2) \ln \{ k \, (x + 2)^3 \} + 6$.

Further problems on solving linear differential equations may be found in the following Section (2) (Problems 1 to 18).

2 Further problems

In Problems 1 to 10 solve the differential equations.

1. $x\dfrac{dy}{dx} + y = 2. \quad [xy = 2x + c]$

2. $y = x(3 - \dfrac{dy}{dx})$ and $x = 1$ when $y = 2$. $[2xy = 3x^2 + 1]$

3. $x\dfrac{dy}{dx} = y + x^3 - 2x$ and $x = 1$ when $y = 3$. $[2y = x^3 + 5x - 4x \ln x]$

4. $\dfrac{dy}{dx} + 1 = \dfrac{-y}{x}$ and $x = 2$ when $y = 1$. $[2xy + x^2 = 8]$

5. $\dfrac{dy}{dx} = x - \dfrac{2}{x} y$ and $x = 2$ when $y = 1$. $[4y = x^2]$

6. $\cos x \dfrac{dy}{dx} = (\sin x)(1 - 2y)$. $[2y = 1 + k \cos^2 x]$

7. $\tan x \dfrac{dy}{dx} + y = x^2 \tan x$. $[(y - 2x) \sin x = (2 - x^2) \cos x + C]$

8. $\dfrac{dy}{dx} + x = y$ and $x = 0$ when $y = 2$. $[y = x + 1 + e^x]$

9. $(x + 1)\dfrac{dy}{dx} + \dfrac{y}{x} = 2$. $[xy = 2(x + 1) \ln (x + 1) + 2 + C(x + 1)]$

10. $2(x + 1)\dfrac{dy}{dx} - (1 + 2x)y = x^2 \sqrt{(1 + x)}$.

$$[2y \sqrt{(x + 1)} = C e^x - (x^2 + 2x + 2)]$$

11. In an electrical alternating current circuit containing resistance R and inductance L the current i is given by the equation $Ri + L\dfrac{di}{dt} = E$. If $E = E_0 \sin \omega t$ and $i = 0$ when $t = 0$ show that:

$$i = \dfrac{E_0}{R^2 + \omega^2 L^2} (R \sin \omega t - \omega L \cos \omega t) + \dfrac{E_0 \omega L}{R^2 + \omega^2 L^2} e^{-Rt/L}.$$

12. Solve the following equation for θ, given that when $t = 0, \theta = \frac{1}{2}$:

$\cosh t \dfrac{d\theta}{dt} + \theta \sinh t = \sinh 2t$. $[\theta = \frac{1}{2} \cosh 2t \operatorname{sech} t]$

13. An equation of motion when a particle moves in a resisting medium is given by: $\dfrac{dv}{dt} = - (kv + bt)$, where k and b are constants. Solve the equation for v given that $v = u$ when $t = 0$.

$$\left[v = \dfrac{b}{k^2} - \dfrac{bt}{k} + \left(u - \dfrac{b}{k^2} \right) e^{-kt} \right]$$

14. A train of total mass m is moved from rest by the engine which exerts a time-dependent force $m k (1 - e^{-t})$ on the train. The resistance to motion is mcv, where v is the speed of the train and c is a constant. Find

the subsequent speed of the train if the equation of motion is given by
$$m \frac{dv}{dt} = m k (1 - e^{-t}) - mcv.$$

$$\left[v = k \left(\frac{1}{c} - \frac{e^{-t}}{(c-1)} + \frac{e^{-ct}}{c(c-1)} \right) \right]$$

15. In a unimolecular chemical reaction a substance α changes into β. If x is the number of molecules of β present at any time t, and a the initial number of molecules of α, then $\frac{dx}{dt} = k_1 (a - x) - k_2 x$, where k_1 and k_2 are constants. Show that if $x = 0$ when $t = 0$, then

$$x = \frac{k_1 a}{k_1 + k_2} \left[1 - e^{-(k_1 + k_2)t} \right]$$

16. The instantaneous current (i) passing through a solution, in a circuit of resistance R and inductance L, whose dielectric constant is to be measured is given by $\frac{di}{dt} + \frac{R}{L} i = \frac{V_0}{L} \sin pt$, where t is time and V_0 and p are constants. Show that the solution of this equation is

$$i = \left\{ \frac{V_0}{R^2 + p^2 L^2} \right\} \left\{ R \sin pt - pL \cos pt \right\} + C e^{-Rt/L}.$$

17. In the drain tank of an oil purifier in a ship the concentration (c) of impurities varies with time (t) according to the equation

$$M \frac{dc}{dt} = p + km - cm,$$

where M, p, m and k are constants.
Solve this equation given $c = c_0$ when $t = 0$.

$$\left[c = c_0 e^{-mt/M} + M \left(k + \frac{p}{m} \right) \left(1 - e^{-mt/M} \right) \right]$$

18. The angular velocity of a flywheel (ω) at time t is given by
$$I \frac{d\omega}{dt} + K \omega = A + B \sin^2 pt,$$
where I, K, A, B and p are constants. If $\omega = 0$ when $t = 0$ show that:

$$\omega = \frac{2A + B}{2K} - \frac{B(K \cos 2pt + 2 pI \sin 2pt)}{2(K^2 + 4p^2 I^2)}$$

$$- \frac{1}{K} \left(A + \frac{2 p^2 I^2 B}{K^2 + 4p^2 I^2} \right) e^{-Kt/I}.$$

Chapter 20

The solution of linear second order differential equations of the form

$$a\frac{d^2y}{dx^2} + b\frac{dy}{dx} + cy = 0$$

1 Introduction

An equation of the form:

$$a\frac{d^2y}{dx^2} + b\frac{dy}{dx} + cy = 0, \tag{1}$$

where a, b and c are constants, is called a **linear second order differential equation with constant coefficients**. It is termed 'linear' since y and its derivatives are all of the first degree, and it is of 'second order' since its highest derivative is $\frac{d^2y}{dx^2}$ (see Chapter 17, Section 1).

An alternative way of stating equation (1) is:

$$(a D^2 + b D + c)y = 0, \tag{2}$$

where D indicates the operation $\frac{d}{dx}$ and D^2 symbolises $\frac{d^2}{dx^2}$. When an equation is stated as in equation (2) it is said to be in **'D operator'** form. Equations of the form stated in equations (1) and (2) have important engineering applications, especially in electrical and mechanical work. Two examples include:

(i) $L\frac{d^2q}{dt^2} + R\frac{dq}{dt} + \frac{1}{c}q = 0$, representing an equation for charge q in an electrical circuit containing resistance R, inductance L and capacitance C in series, and

(ii) $m \dfrac{d^2 s}{dt^2} + a \dfrac{ds}{dt} + ks = 0$ defines a mechanical system, where s is the distance from a fixed point after t seconds, m is a mass, a the damping factor and k the spring stiffness.

In this chapter the solution of differential equations of the type $a \dfrac{d^2 y}{dx^2} + b \dfrac{dy}{dx} + cy = 0$ are dealt with, and in Chapter 21 the more general type, $a \dfrac{d^2 y}{dx^2} + b \dfrac{dy}{dx} + cy = f(x)$ are dealt with.

2 Types of solution of second order differential equations with constant coefficients

If in equation (1) we let $a = 0$, then $b \dfrac{dy}{dx} + cy = 0$

$$\text{i.e.} \quad \dfrac{dy}{dx} + \dfrac{c}{b} y = 0$$

Hence $\quad \dfrac{dy}{dx} = - \dfrac{c}{b} y$

and $\quad \dfrac{dy}{y} = - \dfrac{c}{b} dx$ by separating the variables

Therefore $\displaystyle\int \dfrac{dy}{y} = - \dfrac{c}{b} \int dx$

$\ln y = - \dfrac{c}{b} x + k$, where k is a constant.

Hence, $\quad y = e^{\left(- \frac{c}{b} x + k\right)} = \left(e^{-\frac{c}{b} x}\right) (e^k)$

If we let $e^k = A = $ a constant, then

$y = A e^{-\frac{c}{b} x}$ or $y = A e^{mx}$, where m is a constant.

Let us see if the solution $y = A e^{mx}$ also satisfies equation (1), m being a constant to be found by substituting $y = A e^{mx}$ into equation (1).

If $y = A e^{mx}$ then $\dfrac{dy}{dx} = A m e^{mx}$

and $\dfrac{d^2 y}{dx^2} = A m^2 e^{mx}$

Substituting for y, $\dfrac{dy}{dx}$ and $\dfrac{d^2 y}{dx^2}$ in equation (1) gives:

$$a\frac{d^2y}{dx^2} + b\frac{dy}{dx} + cy = a\,A\,m^2\,e^{mx} + b\,A\,m\,e^{mx} + c\,A\,e^{mx} = 0$$

i.e. $A\,e^{mx}\,(am^2 + bm + c) = 0$ \hfill (3)

Equation (3) is true for all values of x. Since e^{mx} is not equal to zero and A is not zero (for this would make $y = 0$ always) then:

$$am^2 + b\,m + c = 0 \hfill (4)$$

which is a quadratic equation in m.

Equation (4) is called the **auxiliary equation** (or the characteristic equation).

From equation (4), m may be obtained by factorisation or by using the quadratic formula, $m = \dfrac{-b \pm \sqrt{[b^2 - 4ac]}}{2a}$

Since a, b and c are real the auxiliary equation may have either:

    (i) two different real roots   (when $b^2 > 4ac$),
    (ii) two real equal roots   (when $b^2 = 4ac$),
or   (iii) two complex roots   (when $b^2 < 4ac$).

These three cases will now be discussed separately.

(i) Two different real roots

Let the solutions to equation (4) be $m = \alpha$ and $m = \beta$. Then either $y = A\,e^{\alpha x}$ or $y = B\,e^{\beta x}$ is a solution of equation (1). However, it is shown below that a general property of linear equations, such as equation (4), is that the sum of the two solutions is also a solution.

Hence: $\quad y = A\,e^{\alpha x} + B\,e^{\beta x}$ \hfill (5)

is a solution of $a\,\dfrac{d^2y}{dx^2} + b\dfrac{dy}{dx} + cy = 0$, and since this expression contains two arbitrary constants, A and B, it is the **general solution**. If sufficient data is given the constants A and B may be determined and the **particular solution** found.

The general solution $y = A\,e^{\alpha x} + B\,e^{\beta x}$, stated in equation (5), may be derived directly as follows:

Let α and β be the roots of the auxiliary equation $am^2 + bm + c = 0$

i.e. $m^2 + \dfrac{b}{a}\,m + \dfrac{c}{a} = 0$.

From the theory of quadratic equations, $\alpha + \beta = -\dfrac{b}{a}$

$$\text{and} \quad \alpha\beta = \frac{c}{a}$$

$a\dfrac{d^2y}{dx^2} + b\dfrac{dy}{dx} + cy = 0$ may be written as

$$\frac{d^2y}{dx^2} + \frac{b}{a}\frac{dy}{dx} + \frac{c}{a}\,y = 0,$$

or $\quad \dfrac{d^2y}{dx^2} - (\alpha + \beta)\dfrac{dy}{dx} + \alpha\beta y = 0$, and by rearranging,

$$\frac{d}{dx}\left(\frac{dy}{dx} - \beta y\right) = \alpha\left(\frac{dy}{dx} - \beta y\right) \tag{6}$$

Let $\quad \dfrac{dy}{dx} - \beta y = Z$

then $\qquad \dfrac{dZ}{dx} = \alpha Z$ from equation (6)

Hence $\displaystyle\int \frac{dZ}{Z} = \int \alpha\,dx$, by separation of variables.

$\qquad\qquad \ln Z = \alpha x + k$, where k is a constant

$\qquad\qquad$ and $Z = e^{\alpha x + k} = (e^{\alpha x})(e^{k}) = C\,e^{\alpha x}$ where $e^{k} = C =$ a constant,

i.e. $Z = \dfrac{dy}{dx} - \beta y = C\,e^{\alpha x}$ $\hfill (7)$

Equation (7) is of the form of the linear first order differential equations discussed in Chapter 19 solved by introducing an integrating factor, and the solution of this equation is:

$$y\,e^{\int -\beta\,dx} = \int e^{\int -\beta\,dx}\,(C\,e^{\alpha x})\,dx$$

i.e. $\quad y\,e^{-\beta x} = \int C\,e^{(\alpha-\beta)x}\,dx$ $\hfill (8)$

$$y\,e^{-\beta x} = \frac{C}{\alpha - \beta}\,e^{(\alpha-\beta)x} + B, \text{ where } B \text{ is a constant}$$

Multiplying throughout by $e^{\beta x}$ gives:

$$y = \frac{C}{\alpha - \beta}\,e^{\alpha x}\,e^{-\beta x}\,e^{\beta x} + B\,e^{\beta x}$$

$$y = A\,e^{\alpha x} + B\,e^{\beta x}, \text{ where } A = \frac{C}{\alpha - \beta} = \text{ a constant.}$$

This is the same as equation (5) showing that the sum of the two solutions is also a solution.

Thus, given the differential equation $2\dfrac{d^2y}{dx^2} - \dfrac{dy}{dx} - 6y = 0$

$$\text{then } (2D^2 - D - 6)y = 0.$$

Letting $y = A\,e^{mx}$ we obtain $A\,e^{mx}\,(2m^2 - m - 6) = 0$, and since $A\,e^{mx} \neq 0$ then $2m^2 - m - 6 = 0$. This is the auxiliary equation. (Note that the auxiliary equation may be obtained from the original equation by replacing D by m.)

If $2m^2 - m - 6 = 0$, then $(2m + 3)(m - 2) = 0$,

i.e. $m = -\dfrac{3}{2}$ or $m = 2$,

i.e. two different real roots.

Hence the general solution of the differential equation

$2\dfrac{d^2y}{dx^2} - \dfrac{dy}{dx} - 6y = 0$ is $y = A e^{-\frac{3}{2}x} + B e^{2x}$.

Such a solution may always be checked:

if $y = A e^{-\frac{3}{2}x} + B e^{2x}$ then $\dfrac{dy}{dx} = -\dfrac{3}{2} A e^{-\frac{3}{2}x} + 2B e^{2x}$,

and $\dfrac{d^2y}{dx^2} = \dfrac{9}{4} A e^{-\frac{3}{2}x} + 4B e^{2x}$.

Substituting for y, $\dfrac{dy}{dx}$ and $\dfrac{d^2y}{dx^2}$ into $2\dfrac{d^2y}{dx^2} - \dfrac{dy}{dx} - 6y$ gives:

$2\left(\dfrac{9}{4} A e^{-\frac{3}{2}x} + 4B e^{2x}\right) - \left(-\dfrac{3}{2} A e^{-\frac{3}{2}x} + 2B e^{2x}\right)$

$-6\left(A e^{-\frac{3}{2}x} + B e^{2x}\right) = \left(\dfrac{9}{2} + \dfrac{3}{2} - 6\right) A e^{-\frac{3}{2}x} + (8 - 2 - 6) B e^{2x} = 0$.

Hence $y = A e^{-\frac{3}{2}x} + B e^{2x}$ is a solution of the differential equation
$2\dfrac{d^2y}{dx^2} - \dfrac{dy}{dx} - 6y = 0$.

(ii) Two real equal roots

If from the auxiliary equation $\alpha = \beta$ then the general solution is given by
$y = A e^{\alpha x} + B e^{\alpha x}$.

i.e. $y = (A + B) e^{\alpha x} = D e^{\alpha x}$, where $D = A + B = $ a constant.

However this is no longer the general solution of equation (1) since the solution **must** contain **two** arbitrary constants. The analysis used in (i) can be used as far as equation (8), which becomes:

$y e^{-\alpha x} = \int C \, dx$, since $\alpha = \beta$,
i.e. $y e^{-\alpha x} = Cx + B$.

Multiplying throughout by $e^{\alpha x}$ gives: $y = Cx e^{\alpha x} + B e^{\alpha x}$

Replacing C by A gives:

$y = A x e^{\alpha x} + B e^{\alpha x}$

i.e. $y = (Ax + B) e^{\alpha x}$.

This is the general solution for the case of two real equal roots. Thus, given the differential equation $\dfrac{d^2y}{dx^2} - 6\dfrac{dy}{dx} + 9y = 0$

i.e. $(D^2 - 6D + 9)y = 0$

The auxiliary equation is $m^2 - 6m + 9 = 0$

i.e. $(m - 3)(m - 3) = 0$

Hence $m = 3$, twice.

The general solution of $\dfrac{d^2y}{dx^2} - 6\dfrac{dy}{dx} + 9y = 0$ is $y = (Ax + B)\,e^{3x}$.

This solution may be checked:

If $y = (Ax + B)\,e^{3x}$ then $\dfrac{dy}{dx} = (Ax + B)(3e^{3x}) + (e^{3x})(A)$

and $\dfrac{d^2y}{dx^2} = (Ax + B)(9e^{3x}) + (3e^{3x})(A) + 3A\,e^{3x}$

Substituting for y, $\dfrac{dy}{dx}$ and $\dfrac{d^2y}{dx^2}$ into $\dfrac{d^2y}{dx^2} - 6\dfrac{dy}{dx} + 9y$ gives:

$(9Ax\,e^{3x} + 9Be^{3x} + 3A\,e^{3x} + 3A\,e^{3x}) - 6(3Ax\,e^{3x} + 3B\,e^{3x} + A\,e^{3x})$
$$+ 9(Ax\,e^{3x} + B\,e^{3x}) = 0$$

Hence $y = (Ax + B)e^{3x}$ is a solution of the differential equation $\dfrac{d^2y}{dx^2} - 6\dfrac{dy}{dx} + 9y = 0$.

(iii) Two complex roots

When solving the auxiliary equation $am^2 + bm + c = 0$, then

$$m = \frac{-b \pm \sqrt{[b^2 - 4ac]}}{2a}$$

If the coefficients of the differential equation are such that $b^2 < 4ac$ then the values of m are complex.

For example, to solve $m^2 - 6m + 13 = 0$

$$m = \frac{-(-6) \pm \sqrt{[(-6)^2 - 4(1)(13)]}}{2(1)} = \frac{6 \pm \sqrt{(-16)}}{2}$$

Now $\sqrt{(-16)} = \sqrt{[(-1)(16)]} = (\sqrt{-1})(\sqrt{16})$

By definition, $\sqrt{-1} = j$

Hence $\sqrt{(-16)} = \pm j4$

Thus $m = \dfrac{6 \pm j4}{2} = 3 \pm j2$

Hence the solution of the differential equation:

$$\frac{d^2y}{dx^2} - 6\frac{dy}{dx} + 13y = 0$$

259

is $y = A e^{(3+j2)x} + B e^{(3-j2)x}$

i.e. $y = e^{3x} \{ A e^{j2x} + B e^{-j2x} \}$

It was shown in Chapter 6 that $e^{jx} = \cos x + j \sin x$

and $e^{-jx} = \cos x - j \sin x$

Hence $y = e^{3x} \{ A(\cos 2x + j \sin 2x) + B(\cos 2x - j \sin 2x) \}$

$y = e^{3x} \{ (A + B) \cos 2x + j (A - B) \sin 2x \}$

It would appear at first sight that this is a complex solution. However, y (which represents a real voltage or current or displacement, etc.) cannot be equal to a complex expression. In practice it is always found that A and B are complex conjugate numbers. For example, let $A = 3 + j4$ and $B = 3 - j4$. Then $A + B = 6$, a real number and $j(A - B) = j(j8) = -8$, also a real number.

Hence, given that the solution of the auxiliary equation is $m = 3 \pm j2$ then the solution may be written down directly as $y = e^{3x} \{ C \cos 2x + D \sin 2x \}$, where C and D are two real unknown constants. This solution may be checked by substituting y, $\dfrac{dy}{dx}$ and $\dfrac{d^2 y}{dx^2}$ into the left hand side of the differential equation $\dfrac{d^2 y}{dx^2} - 6\dfrac{dy}{dx} + 13y = 0$.

Generally, if the roots of the auxiliary equation are $\alpha \pm j \beta$, then the general solution is $y = e^{\alpha x} \{ C \cos \beta x + D \sin \beta x \}$.

3 Summary of the procedure used to solve differential equations of the form $a\dfrac{d^2 y}{dx^2} + b\dfrac{dy}{dx} + cy = 0$

(a) Rewrite the given differential equation $a \dfrac{d^2 y}{dx^2} + b\dfrac{dy}{dx} + cy = 0$ as $(a D^2 + b D + c)y = 0$.

(b) Substitute m for D and solve the auxiliary equation $am^2 + bm + c = 0$ for m.

(c)   (i) If the roots of the auxiliary equation are **real and different**, say, $m = \alpha$ and $m = \beta$, then the general solution is:

$y = A e^{\alpha x} + B e^{\beta x}$

(ii) If the roots of the auxiliary equation are **real and equal**, say $m = \alpha$, twice, then the general solution is:

$y = (Ax + B) e^{\alpha x}$.

(iii) If the roots of the auxiliary equation are **complex**, say $m = \alpha \pm j \beta$, then the general solution is:

$y = e^{\alpha x} \{ C \cos \beta x + D \sin \beta x \}$

(d) If the **particular solution** of a differential equation is required then substitute the given boundary conditions to find the unknown constants (i.e. to find A and B in (c, i) and (c, ii) or C and D in (c, iii)).

Worked problems on solving differential equations of the form
$$a\frac{d^2 y}{dx^2} + b\frac{dy}{dx} + cy = 0$$

Problem 1. (a) Find the general solution of $6\frac{d^2 y}{dx^2} + 5\frac{dy}{dx} - 4y = 0$

          (b) Find the particular solution of the differential equation in (a) given that when $x = 0, y = 11$ and $\frac{dy}{dx} = 0$.

(a) In D operator form the differential equation is $(6D^2 + 5D - 4)y = 0$ where $D \equiv \dfrac{d}{dx}$

The auxiliary equation is $\quad 6m^2 + 5m - 4 = 0$
Factorising gives: $\qquad (3m + 4)(2m - 1) = 0$
$$\text{i.e. } m = -\frac{4}{3} \text{ or } m = \frac{1}{2}$$

Since the two roots are real and different **the general solution is**

$$y = A e^{-\frac{4}{3}x} + B e^{\frac{1}{2}x}.$$

(b) When $x = 0, y = 11$, then
$$11 = A + B \tag{1}$$

Since $y = A e^{-\frac{4}{3}x} + B e^{\frac{1}{2}x}$ then

$$\frac{dy}{dx} = -\frac{4}{3} A e^{-\frac{4}{3}x} + \frac{1}{2} B e^{\frac{1}{2}x}$$

When $x = 0, \dfrac{dy}{dx} = 0$.

Thus $0 = -\dfrac{4}{3} A + \dfrac{1}{2} B$ $\tag{2}$

$2 \times (2)$ gives: $0 = -\dfrac{8}{3} A + B$ $\tag{3}$

(1)–(3) gives: $11 = 3\dfrac{2}{3} A$, i.e. $A = 3$

Substituting $A = 3$ in equation (1) gives $B = 8$.

Hence the particular solution is $y = 3 e^{-\frac{4}{3}x} + 8 e^{\frac{1}{2}x}$.

Problem 2. (a) Determine the general solution of $4\dfrac{d^2y}{dt^2} - 12\dfrac{dy}{dt} + 9y = 0$.

      (b) Find the particular solution of the equation given in (a)
which makes $y = 2$ at $t = 0$ and $\dfrac{dy}{dt} = 4$ at $t = 0$.

(a) In D operator form the differential equation is $(4D^2 - 12D + 9)y = 0$
where $D \equiv \dfrac{d}{dt}$.

The auxiliary equation is $\quad 4m^2 - 12m + 9 = 0$

        i.e. $\quad (2m - 3)(2m - 3) = 0$

        i.e. $\quad m = \dfrac{3}{2}$ twice.

Since the two roots are real and equal **the general solution is**

$$y = (At + B)\, e^{\frac{3}{2}t}.$$

(b) When $y = 2, t = 0$.

Thus $2 = (0 + B)\, e^0$, i.e. $B = 2$

$\dfrac{dy}{dt} = (At + B)\left(\dfrac{3}{2}\, e^{\frac{3}{2}t}\right) + (e^{\frac{3}{2}t})\,(A)$ by the product rule of
                                      differentiation.

When $\dfrac{dy}{dt} = 4, t = 0$.

Thus $4 = (0 + B)\left(\dfrac{3}{2}\, e^0\right) + (e^0)(A)$

i.e. $\quad 4 = \dfrac{3}{2}\, B + A$

Since $B = 2, A = 1$.
Hence the particular solution is $y = (t + 2)\, e^{\frac{3}{2}t}$.

Problem 3. (a) Determine the general solution of the differential equation
$\dfrac{d^2y}{dx^2} + 2\dfrac{dy}{dx} + 5y = 0$.

      (b) Given the boundary conditions that when $x = 0, y = 1$ and
when $x = 0, \dfrac{dy}{dx} = 5$, find the particular solution of the
equation given in (a).

(a) In D operator form the differential equation is $(D^2 + 2D + 5)y = 0$
where $D \equiv \dfrac{d}{dx}$.
The auxiliary equation is $\quad m^2 + 2m + 5 = 0$.

Using the quadratic formula, $m = \dfrac{-2 \pm \sqrt{[(2)^2 - 4(1)(5)]}}{2(1)}$

$= \dfrac{-2 \pm \sqrt{-16}}{2}$

i.e. $m = \dfrac{-2 \pm j\,4}{2} = -1 \pm j\,2$

Since the roots are complex **the general solution is**
$y = e^{-x}\{\, C \cos 2x + D \sin 2x \,\}$

(b) When $x = 0, y = 1.$
   Thus $1 = e^0\{C \cos 0 + D \sin 0\}$
   Hence $1 = C$

$\dfrac{dy}{dx} = e^{-x}(-2C \sin 2x + 2D \cos 2x) - e^{-x}(C \cos 2x + D \sin 2x),$
   by the product rule,

$= e^{-x}\{\,(2D - C)\cos 2x - (2C + D)\sin 2x\,\}$

When $x = 0$, $\dfrac{dy}{dx} = 5.$

   Hence $5 = e^0\{(2D - C)\cos 0 - (2C + D)\sin 0\}$
         $5 = 2D - C.$
   Since $C = 1, D = 3.$
   Hence the particular solution is $y = e^{-x}\{\cos 2x + 3 \sin 2x\}.$

Since $a \cos \omega t + b \sin \omega t \equiv R \sin(\omega t + \alpha)$, where $R = \sqrt{(a^2 + b^2)}$ and

$$\alpha = \arctan \dfrac{a}{b},$$

$\cos 2x + 3 \sin 2x \equiv \sqrt{(1^2 + 3^2)}\sin(2x + \arctan \tfrac{1}{3})$
$\equiv \sqrt{10}\sin(2x + 18° 26')$
$\equiv \sqrt{10}\sin(2x + 0.321\,8)$
Thus the particular solution may also be expressed as

$$y = \sqrt{10}\,e^{-x}\sin(2x + 0.321\,8).$$

Problem 4. Solve the differential equations.

(a) $\dfrac{d^2y}{dx^2} + n^2 y = 0$

(b) $\dfrac{d^2y}{dx^2} - n^2 y = 0$, where n is a constant.

(a) $\dfrac{d^2y}{dx^2} + n^2 y = 0$ is a differential equation representing simple harmonic motion (S.H.M.) and has many practical applications.

In D operator form the equation is $(D^2 + n^2)y = 0$

The auxiliary equation is $\qquad m^2 + n^2 \qquad = 0$

$$\text{i.e.} \qquad m^2 = -n^2$$
$$m = \sqrt{(-n^2)} = \pm j\, n$$

Since the two roots are complex the general solution is
$y = e^0 \{ C \cos nx + D \sin nx \}$, i.e. $y = C \cos nx + D \sin nx$ or

$y = R \sin (nx + \alpha)$, where $R = \sqrt{(C^2 + D^2)}$ and $\alpha = \arctan \dfrac{C}{D}$.

(b) In D operator form the differential equation $\dfrac{d^2 y}{dx^2} - n^2 y = 0$ is

$(D^2 - n^2)y = 0$.

The auxiliary equation is $m^2 - n^2 = 0$.

$$\text{Hence} \qquad m^2 = n^2$$
$$m = \pm n.$$

Since the two roots are real and different **the general solution is**
$y = A\, e^{nx} + B\, e^{-nx}$.

Since $\sinh nx = \dfrac{1}{2}\,(e^{nx} - e^{-nx})$ and $\cosh nx = \dfrac{1}{2}\,(e^{nx} + e^{-nx})$

then $\sinh nx + \cosh nx = e^{nx}$
and $\cosh nx - \sinh nx = e^{-nx}$.

Hence the general solution may also be written as:

$$y = A(\sinh nx + \cosh nx) + B(\cosh nx - \sinh nx)$$
i.e. $y = (A + B) \cosh nx + (A - B) \sinh nx$
i.e. $y = C \cosh nx + D \sinh nx$.

Problem 5. A circuit containing resistance R, inductance L and capacitance C in series has an equation for current i given by:

$L\, \dfrac{d^2 i}{dt^2} + R\, \dfrac{di}{dt} + \dfrac{1}{C}\, i = 0$.

If $L = 0.25$ henry, $C = 25 \times 10^{-6}$ farads and $R = 200$ ohms, solve the equation given that when time $t = 0, i = 0$ and $\dfrac{di}{dt} = 50$.

In D operator form the differential equation is $\left(L\, D^2 + R\, D + \dfrac{1}{C} \right) i = 0$.

The auxiliary equation is $L\, m^2 + R\, m + \dfrac{1}{C} = 0$

$$\text{Hence } m = \dfrac{-R \pm \sqrt{(R^2 - \dfrac{4L}{C})}}{2L}$$

Substituting $L = 0.25$, $C = 25 \times 10^{-6}$ and $R = 200$ gives:

$$m = \frac{-200 \pm \sqrt{\left[(200)^2 - \frac{4(0.25)}{25.10^{-6}}\right]}}{2(0.25)} = \frac{-200 \pm \sqrt{0}}{0.5}$$

i.e. $m = \dfrac{-200}{0.5} = -400$.

For a second order differential equation there must be two solutions. Hence $m = -400$ twice. **Hence the general solution is $i = (At + B)e^{-400t}$.**

When $t = 0$, $i = 0$.
  Hence $0 = B$

$$\frac{di}{dt} = (At + B)(-400 e^{-400t}) + (e^{-400t})(A)$$

When $t = 0$, $\dfrac{di}{dt} = 50$.

  Hence $50 = -400 B + A$
  Since $B = 0$, $A = 50$.

Hence the particular solution is $i = 50\, t\, e^{-400t}$.

Further problems on solving differential equations of the form
$a\dfrac{d^2y}{dx^2} + b\dfrac{dy}{dx} + cy = 0$ *may be found in the following section (4)*
(Problems 1 to 21).

4 Further problems

In Problems 1 to 6 find the general solution of the given differential equations.

1. (a) $\dfrac{d^2y}{dx^2} - 5\dfrac{dy}{dx} + 6y = 0$. $[y = A\, e^{3x} + B\, e^{2x}]$

   (b) $\dfrac{d^2y}{dx^2} - 2\dfrac{dy}{dx} + y = 0$. $[y = (Ax + B)\, e^x]$

2. (a) $\dfrac{d^2y}{dx^2} + 6\dfrac{dy}{dx} + 13y = 0$. $[y = e^{-3x}(C \cos 2x + D \sin 2x)]$

   (b) $\dfrac{d^2x}{dt^2} + 3\dfrac{dx}{dt} + 2x = 0$. $[x = A\, e^{-t} + B\, e^{-2t}]$

3. (a) $\dfrac{d^2y}{d\theta^2} - 4\dfrac{dy}{d\theta} + 4y = 0$ $[y = (A\theta + B)\, e^{2\theta}]$

   (b) $\dfrac{d^2y}{dt^2} + y = 0$. $[y = C \cos t + D \sin t]$

4. (a) $6\dfrac{d^2\theta}{dt^2} + 4\dfrac{d\theta}{dt} - 2\theta = 0$ $\quad[\theta = A\,e^{\frac{1}{3}t} + B\,e^{-t}]$

   (b) $4\dfrac{d^2y}{dx^2} + 4\dfrac{dy}{dx} + y = 0$ $\quad[y = (Ax+B)\,e^{-\frac{1}{2}x}]$

5. (a) $(3D^2 - 2D + 5)y = 0$ where $D \equiv \dfrac{d}{dx}$.

$$\left[y = e^{\frac{1}{3}x}\left(C\cos\frac{\sqrt{14}}{3}x + D\sin\frac{\sqrt{14}}{3}x\right)\right]$$

   (b) $(4D^2 - 7D - 15)y = 0$ where $D \equiv \dfrac{d}{dx}$.

$$\left[y = A\,e^{-\frac{5}{4}x} + B\,e^{3x}\right]$$

6. (a) $(16D^2 + 8D + 1)y = 0$ where $D \equiv \dfrac{d}{dx}$.

$$\left[y = (Ax+B)\,e^{-\frac{1}{4}x}\right]$$

   (b) $(D^2 + D + 1)\theta = 0$ where $D \equiv \dfrac{d}{dt}$.

$$\left[\theta = e^{-\frac{1}{2}t}\left(C\cos\frac{\sqrt{3}}{2}t + D\sin\frac{\sqrt{3}}{2}t\right)\right]$$

7. Show that the solution to the differential equation
$\dfrac{d^2y}{dx^2} + (2a\cos\theta)\dfrac{dy}{dx} + a^2y = 0$, where a and θ are constants, may be
may be expressed as: $y = A\,e^{-ax\,\cos\theta}\cos(B + ax\sin\theta)$.

In Problems 8 to 13, find the particular solution of the given differential
equations for the stated boundary conditions.

8. $12\dfrac{d^2y}{dx^2} - 3y = 0$ $\quad y = 3$ when $x = 0$

$\qquad\qquad \dfrac{dy}{dx} = \dfrac{1}{2}$ when $x = 0$ $\quad\left[y = 2e^{\frac{1}{2}x} + e^{-\frac{1}{2}x}\right]$

9. $9\dfrac{d^2y}{dt^2} - 12\dfrac{dy}{dt} + 4y = 0.$ $\begin{cases} y = 3 \text{ when } t = 0 \\ \\ \dfrac{dy}{dt} = 4 \text{ when } t = 0 \end{cases}$ $\left[y = (2t+3)\,e^{\frac{2}{3}t}\right]$

10. $\dfrac{d^2y}{dx^2} + 2\dfrac{dy}{dx} + 6y = 0.$ $\begin{cases} y = 2 \text{ when } x = 0 \\ \\ \dfrac{dy}{dx} = 3 \text{ when } x = 0 \end{cases}$

$$[y = e^{-x}(2\cos\sqrt{5}x + \sqrt{5}\sin\sqrt{5}x)]$$

11. $(35D^2 - 11D - 6)x = 0$ where $D \equiv \dfrac{d}{dt}$ $\begin{cases} x = 5 \text{ when } t = 0 \\[2mm] \dfrac{dx}{dt} = \dfrac{12}{35} \text{ when } t = 0 \end{cases}$

$$\left[x = 3e^{-\frac{2}{7}t} + 2e^{\frac{3}{5}t} \right]$$

12. $(2D^2 + 2D + 1)y = 0$ where $D \equiv \dfrac{d}{dx}$. $\begin{cases} y = 4 \text{ when } x = 0 \\[2mm] \dfrac{dy}{dx} = 5 \text{ when } x = 0 \end{cases}$

$$\left[y = 2e^{-\frac{1}{2}x} \left(2 \cos \frac{x}{2} + 7 \sin \frac{x}{2} \right) \right]$$

13. $(25D^2 - 20D + 4)y = 0$ where $D \equiv \dfrac{d}{dx}$. $\begin{cases} y = 5 \text{ when } x = 0 \\[2mm] \dfrac{dy}{dx} = 3 \text{ when } x = 0 \end{cases}$

$$\left[y = (x + 5) e^{\frac{2}{5}x} \right]$$

14. The oscillations of a heavily damped pendulum satisfy the differential equation $\dfrac{d^2 x}{dt^2} + 7\dfrac{dx}{dt} + 12x = 0$, where x cm is the displacement of the bob at time t seconds. The initial displacement is equal to $+3$ cm and the initial velocity (i.e. $\dfrac{dx}{dt}$) is 6 cm/s. Solve the equation for x.

$$[x = 3 (6e^{-3t} - 5 e^{-4t})]$$

15. If $\dfrac{d^2 V}{dx^2} - w^2 V = 0$ where w is a constant show that:

$$V = V_0 \cosh wx + \frac{2}{w} \sinh wx,$$

given that when $x = 0$, $V = V_0$ and $\dfrac{dV}{dx} = 2$.

16. The charge q on a capacitor in a certain electrical circuit satisfies the differential equation $\dfrac{d^2 q}{dt^2} + 3\dfrac{dq}{dt} + 4q = 0$. Initially (i.e. when $t = 0$), $q = Q$ and $\dfrac{dQ}{dt} = 0$. Show that the charge in the circuit may be expressed as:

$$q = \frac{4}{\sqrt{7}} Q e^{-\frac{3}{2}t} \sin \left(\frac{\sqrt{7}}{2} t + 0.723 \right).$$

17. The differential equation $I\dfrac{d^2 \theta}{dt^2} + K \dfrac{d\theta}{dt} + F\theta = 0$ represents the motion of the pointer of a galvanometer about its position of equilibrium. I is the moment of inertia of the pointer about its pivot, K is the resistance due to friction at unit angular velocity and F is the force on the spring necessary to produce unit displacement. If $I = 0.006, K = 0.03$ and $F = 0.187\,5$ solve the equation for θ in terms of t given that when $t = 0$,

$\theta = 0.2$ and $\dfrac{d\theta}{dt} = 0$. $[\theta = e^{-2.5t}\,(0.2\cos 5t + 0.1\sin 5t)]$

18. A body moves in a straight line so that its distance S metres from the origin after time t seconds is given by $\dfrac{d^2 S}{dt^2} + n^2\,S = 0$, where n is a constant. Solve the equation for S given that $S = k$ and $\dfrac{dS}{dt} = 0$ when

$t = \dfrac{2\pi}{n}$. $[S = k\cos nt]$

19. The equation $\dfrac{d^2 i}{dt^2} + \dfrac{R}{L}\dfrac{di}{dt} + \dfrac{1}{LC}i = 0$ represents a current i flowing in an electrical circuit containing resistance R, inductance L and capacitance C connected in series. Current $i = 0$ when time $t = 0$ and $\dfrac{di}{dt} = 100$ when $t = 0$. Inductance $L = 0.20$ henry and capacitance $C = 80 \times 10^{-6}$ farads. Solve the equation for i when (a) $R = 100$ ohms and (b) $R = 223.6$ ohms.

(a) $\left[i = 100\,t\,e^{-250t} \right]$ (b) $\left[i = \dfrac{1}{10}\left(e^{-59t} - e^{-1059t} \right) \right]$

20. The equation of motion of a body oscillating on the end of a spring is: $\dfrac{d^2 x}{dt^2} + 225x = 0$, where x is the displacement in metres of the body from its equilibrium position after time t seconds. Find x in terms of t given that at time $t = 0, x = 1$ and $\dfrac{dx}{dt} = 0$. $[x = \cos 15t]$

21. The equation: $\dfrac{d^2 y}{dx^2} = k\sqrt{\left\{ 1 + \left(\dfrac{dy}{dx} \right)^2 \right\}}$ on solution will give that for the catenary. Solve it for y given that $\dfrac{dy}{dx} = 0$ when $x = 0$.

$$\left[y = \frac{1}{k}\cosh(kx) + B \right]$$

Chapter 21

The solution of linear second order differential equations of the form

$$a\,\frac{d^2y}{dx^2} + b\,\frac{dy}{dx} + cy = f(x)$$

1 Complementary function and particular integral

The differential equation $a\dfrac{d^2y}{dx^2} + b\dfrac{dy}{dx} + cy = f(x)$ (1)

where a, b and c are constants, is a linear, second order differential equation with constant coefficients.

In order to solve this type of equation a substitution is made as follows. Let $u + v$ be substituted for y in equation (1). The equation then becomes:

$$a\,\frac{d^2}{dx^2}(u+v) + b\,\frac{d}{dx}(u+v) + c(u+v) = f(x)$$

Hence $a\dfrac{d^2u}{dx^2} + a\dfrac{d^2v}{dx^2} + b\dfrac{du}{dx} + b\dfrac{dv}{dx} + cu + cv = f(x)$

i.e. $\left(a\dfrac{d^2u}{dx^2} + b\dfrac{du}{dx} + cu\right) + \left(a\dfrac{d^2v}{dx^2} + b\dfrac{dv}{dx} + cv\right) = f(x)$

Let v be any solution of the equation (1) such that:

$$a\,\frac{d^2v}{dx^2} + b\,\frac{dv}{dx} + cv = f(x) \qquad (2)$$

This would mean that $\quad a\dfrac{d^2u}{dx^2} + b\dfrac{du}{dx} + cu = 0 \qquad (3)$

The general solution, u, of equation (3) will contain two unknown constants as required for the general solution of equation (1). (The method of solution of this type of equation was discussed in the previous chapter.)

If, also, the particular solution, v, of equation (2) can be found without containing any unknown constants then $y = u + v$ will give the general solution of equation (1). Section 2 following discusses methods whereby v may be determined.

The function u is called the **complementary function (C.F.)**.

The function v is called the **particular integral (P.I.)**.

The general solution y of a linear differential equation such as equation (1) is given by the sum of the complementary function and the particular integral.

$$y = \textbf{C.F.} + \textbf{P.I.}$$
or $\quad y = u + v.$

2 Methods of finding the particular integral

The function $f(x)$ on the right-hand side of equation (1) may be any non-zero expression, the more common expressions being a constant, a polynomial (i.e. of the form $p + qx + rx^2 + \ldots$, where p, q and r are constants), an exponential function, a sine or cosine function, or a sum of product of these functions. The method of finding the particular integral differs for each expression of $f(x)$ and it is usual to assume a form of particular integral which is suggested by $f(x)$. The following examples will make this clear.

(a) $f(x) =$ **a constant**
(i) Straightforward case

If $f(x)$ is a constant then firstly try $v = k$, where k is a constant, as the particular integral. $v = k$ is substituted into equation (2) and k may be determined.

Problem 1. Solve $\dfrac{d^2y}{dx^2} + 4\dfrac{dy}{dx} + 3y = 6.$

Let $v = k$ then from equation (2): $(D^2 + 4D + 3)k = 6.$
Since $D^2(k) = 0$ and $D(k) = 0$, then $3k = 6$, i.e. $k = 2.$
Hence the particular integral, $v = 2.$

The complementary function, u is obtained as in the previous chapter.

The auxiliary equation is $\quad m^2 + 4m + 3 = 0$
$$\text{i.e.} \ (m + 3)(m + 1) = 0$$
$$\text{i.e.} \ m = -3 \text{ and } m = -1.$$
Hence the complementary function $u = A\,e^{-3x} + B\,e^{-x}.$
The general solution is given by $\quad y = u + v,$
$$\text{i.e.} \ y = A\,e^{-3x} + B\,e^{-x} + 2.$$

(ii) 'Snag' case
When $f(x)$ is a constant the substitution of $y = k$ will not always work. If the complementary function is found to contain a constant term (as in Problem 2) then try $y = kx$ as the particular integral.

270

Problem 2. (a) Find the general solution of $\dfrac{d^2 y}{dx^2} + 4\dfrac{dy}{dx} = 6$.

    (b) Find the particular solution for (a) given that when $x = 0$,
$y = 0$ and $\dfrac{dy}{dx} = 0$.

(a) In D operator form the differential equation is $(D^2 + 4D)y = 6$.
    The auxiliary equation is $m^2 + 4m = 0$
                  i.e. $m(m + 4) = 0$
         Hence $m = 0$ or $m = -4$.
    Hence the complementary function $u = A\,e^0 + B\,e^{-4x} = A + B\,e^{-4x}$.
    If we assume a particular integral of $v = k$ in this case, we obtain
    $(D^2 + 4D)k = 6$ from equation (2) from which $0 = 6$ which is impossible.
    This occurs because the particular integral assumed, namely a constant,
    already occurs in the complementary function (i.e. as A).
    In this case let $v = kx$, then $(D^2 + 4D)kx \equiv 6$,
      i.e. $D^2\,(kx) + 4D\,(kx) \equiv 6$.
      $D^2\,(kx) = 0$ and $4D\,(kx) = 4k$

      Hence $4k = 6$, i.e. $k = \dfrac{3}{2}$

      Hence the particular integral, $v = \dfrac{3}{2}\,x$.

The general solution, $y = u + v = A + B\,e^{-4x} + \dfrac{3}{2}x$.

(b) When $x = 0, y = 0$ and since $y = A + B\,e^{-4x} + \dfrac{3}{2}\,x$, then

$0 = A + B$ (1)

$\dfrac{dy}{dx} = 0 - 4B\,e^{-4x} + \dfrac{3}{2}$

$\dfrac{dy}{dx} = 0$ when $x = 0$, thus $0 = -4B + \dfrac{3}{2}$

                    i.e. $B = \dfrac{3}{8}$

From equation (1) $A = -\dfrac{3}{8}$

Hence the particular solution is $y = -\dfrac{3}{8} + \dfrac{3}{8}\,e^{-4x} + \dfrac{3}{2}\,x$

            i.e. $y = \dfrac{3}{8}\,(e^{-4x} - 1) + \dfrac{3}{2}\,x$.

Typical practical examples of the type of equation solved in Problems

1 and 2 are the equations for the charge q in a series-connected electrical circuit and the force equation for a 'mass, spring, damped' mechanical system, i.e.

$$L\frac{dq^2}{dt^2} + R\frac{dq}{dt} + \frac{1}{c}q = E \text{ and } m\frac{d^2s}{dt^2} + a\frac{ds}{dt} + ks = F$$

(b) $f(x) = $ a polynomial

If $f(x) = p + qx + rx^2 + \ldots$, where p, q and r are constants, then try $v = a + bx + cx^2 + \ldots$, where a, b and c are constants, as the particular integral.

For example, if $f(x) = 4 + 3x$ then try $v = a + bx$
if $f(x) = 4x^2 + 2x - 1$ then try $v = ax^2 + bx + c$,
if $f(x) = 2x^3 - 3$ then try $v = ax^3 + bx^2 + cx + d$,

and so on.

After substitution into equation (2), the coefficients of similar terms are equated, as shown in Problem 3.

Problem 3. Solve the equation $2\frac{d^2y}{dx^2} - \frac{dy}{dx} - 6y = 3x + 2$.

In D operator form, the equation is $(2D^2 - D - 6)y = 3x + 2$.
The auxiliary equation is $2m^2 - m - 6 = 0$
i.e. $(2m + 3)(m - 2) = 0$
i.e. $m = -\frac{3}{2}$ or $m = 2$

The complementary function, $u = A e^{-\frac{3}{2}x} + B e^{2x}$.
Let the particular integral $v = ax + b$, then substituting into equation (2) gives:

$(2D^2 - D - 6)(ax + b) \equiv 3x + 2$
i.e. $2D^2[ax + b] - D[ax + b] - 6[ax + b] \equiv 3x + 2$
$2D^2(ax + b) = 0, \quad D(ax + b) = a$
Hence $0 - a - 6ax - 6b \equiv 3x + 2$

Equating the coefficients of x gives: $-6a = 3$

i.e. $a = -\frac{1}{2}$

Equating the constant terms gives: $-a - 6b = 2$

i.e. $-\left(-\frac{1}{2}\right) - 6b = 2$

$-6b = \frac{3}{2}$

$b = -\frac{1}{4}$

Hence the particular integral $v = -\dfrac{1}{2}x - \dfrac{1}{4}$

The general solution $y = u + v = A\,e^{-\frac{3}{2}x} + B\,e^{2x} - \dfrac{1}{2}x - \dfrac{1}{4}$

(c) $f(x)$ = an exponential function
(i) Straightforward case
If $f(x) = C\,e^{px}$, where C and p are constants, then try as the particular integral $v = k\,e^{px}$. The reason for such a substitution is that all the differential coefficients of e^{px} are multiples of e^{px}.

Problem 4. Find the general solution of the equation

$$3\dfrac{d^2 y}{dx^2} + \dfrac{dy}{dx} - 4y = 2\,e^{-3x}$$

Given the boundary conditions that when $x = 0, y = \dfrac{3}{5}$ and $\dfrac{dy}{dx} = -6\dfrac{4}{5}$, find the particular solution.

In D operator form the equation is $(3D^2 + D - 4)y = 2\,e^{-3x}$.
The auxiliary equation is $3m^2 + m - 4 = 0$
$$\text{i.e. } (m - 1)(3m + 4) = 0$$
$$\text{i.e. } m = 1 \text{ or } m = -\dfrac{4}{3}$$

Hence the complementary function, $u = A\,e^{x} + B\,e^{-\frac{4}{3}x}$
Let the particular integral, $v = k\,e^{-3x}$ then

$(3D^2 + D - 4)\,[k\,e^{-3x}] \equiv 2\,e^{-3x}$ from equation (2)
$D(k\,e^{-3x}) = -3\,k\,e^{-3x}$
$D^2\,(k\,e^{-3x}) = 9\,k\,e^{-3x}$
Hence $27\,k\,e^{-3x} - 3\,k\,e^{-3x} - 4\,k\,e^{-3x} \equiv 2\,e^{-3x}$
$$\text{i.e. } 20\,k\,e^{-3x} \equiv 2\,e^{-3x}$$
$$\text{Hence } k = \dfrac{1}{10}$$

Hence the particular integral, $v = \dfrac{1}{10}e^{-3x}$

The general solution, $y = u + v = A\,e^{x} + B\,e^{-\frac{4}{3}x} + \dfrac{1}{10}e^{-3x}$

When $x = 0, y = \dfrac{3}{5}$, and substituting in the general solution gives:

$$\dfrac{3}{5} = A + B + \dfrac{1}{10}$$

i.e. $\dfrac{1}{2} = A + B$ (1)

$$\dfrac{dy}{dx} = A\,e^x - \dfrac{4}{3}B\,e^{-\frac{4}{3}x} - \dfrac{3}{10}\,e^{-3x}$$

When $x = 0$, $\dfrac{dy}{dx} = -\,6\dfrac{4}{5}$. Therefore $-\,6\dfrac{4}{5} = A - \dfrac{4}{3}B - \dfrac{3}{10}$

i.e. $-\,6\dfrac{1}{2} = A - \dfrac{4}{3}B$ (2)

$(1) - (2)$ gives $\dfrac{1}{2} - -6\,\dfrac{1}{2} = \left(1 - -\dfrac{4}{3}\right)B$

i.e. $7 = \dfrac{7}{3}B$

i.e. $B = 3$

From (1), when $B = 3$, $A = -\,2\dfrac{1}{2}$

Hence the particular solution is $y = 3\,e^{-\frac{4}{3}x} - \dfrac{5}{2}\,e^x + \dfrac{1}{10}\,e^{-3x}$.

(ii) 'Snag' cases

If $f(x) = C\,e^{px}$, where C and p are constants, and e^{px} occurs in the complementary function the particular integral cannot be found by assuming that $v = k\,e^{px}$, for on substituting k becomes zero. In this case try $v = kx\,e^{px}$ for the particular integral (see Problem 5). If e^{px} and $x\,e^{px}$ both appear in the complementary function then try $v = k\,x^2\,e^{px}$ as the particular integral, and so on (see Problem 6).

Problem 5. Solve the differential equation $(D^2 - D - 2)y = 6\,e^{-x}$.

The auxiliary equation is $m^2 - m - 2 = 0$
i.e. $(m - 2)(m + 1) = 0$
i.e. $m = 2$ and $m = -1$
Hence the complementary function, $u = A\,e^{2x} + B\,e^{-x}$

Since e^{-x} appears in the complementary function **and** in the right-hand side of the differential equation, let the particular integral $v = kx\,e^{-x}$.

Then $(D^2 - D - 2)[kx\,e^{-x}] \equiv 6\,e^{-x}$

$D(kx\,e^{-x}) = (kx)(-e^{-x}) + (e^{-x})(k) = k\,e^{-x}\,(1 - x)$, by the product rule

$D^2\,(kx\,e^{-x}) = D(k\,e^{-x}\,(1 - x)) = (k\,e^{-x})(-1) + (1 - x)(-k\,e^{-x})$ by the
product rule

$$= -k\,e^{-x}\,(2 - x)$$

Hence $(D^2 - D - 2)[kx\,e^{-x}] = -k\,e^{-x}(2-x) - k\,e^{-x}(1-x) - 2\,kx\,e^{-x}$
$$\equiv 6\,e^{-x}$$

Thus $-2\,ke^{-x} + kx\,e^{-x} - ke^{-x} + kx\,e^{-x} - 2\,kx\,e^{-x} \equiv 6\,e^{-x}$

i.e. $\qquad -3k\,e^{-x} = 6\,e^{-x}$

i.e. $\qquad k = -2$

Hence the particular integral, $v = -2x\,e^{-x}$

The general solution $y = u + v = A\,e^{2x} + B\,e^{-x} - 2x\,e^{-x}$

Problem 6. Solve $\dfrac{d^2 y}{dx^2} - 2\dfrac{dy}{dx} + y = 5\,e^x$

In D operator form the differential equation is $(D^2 - 2D + 1)y = 5\,e^x$
The auxiliary equation is $m^2 - 2m + 1 = 0$
$\qquad$ i.e. $\quad (m-1)(m-1) = 0$
$\qquad$ i.e. $\quad m = 1$ twice
Hence the complementary function, $u = (Ax + B)\,e^x$

Since e^x **and** $x\,e^x$ both appear in the complementary function, let the particular integral, $v = kx^2\,e^x$

$\qquad$ Thus $\quad (D^2 - 2D + 1)[kx^2\,e^x] \equiv 5\,e^x$

$D(kx^2\,e^x) = (kx^2)(e^x) + (e^x)(2\,kx) \equiv k\,e^x\,(x^2 + 2x)$

$D^2(kx^2\,e^x) = D(ke^x\,(x^2 + 2x)) = (ke^x)(2x + 2) + (x^2 + 2x)(ke^x)$
$\qquad\qquad\qquad = ke^x\,(x^2 + 4x + 2)$

Hence $(D^2 - 2D + 1)[kx^2\,e^x] = ke^x\,(x^2 + 4x + 2) - 2ke^x(x^2 + 2x)$
$\qquad\qquad\qquad\qquad\qquad\qquad + kx^2\,e^x \equiv 5e^x$

i.e. $2\,k\,e^x = 5\,e^x$

i.e. $\qquad k = \dfrac{5}{2}$

Hence the particular integral, $v = \dfrac{5}{2}\,x^2\,e^x$.

The general solution, $y = u + v = (Ax + B)\,e^x + \dfrac{5}{2}\,x^2\,e^x$

i.e. $y = e^x\left(Ax + B + \dfrac{5}{2}\,x^2\right).$

(d) $f(x) =$ **a sine or cosine**
(i) Straightforward case
If $f(x) = k_1 \sin px + k_2 \cos px$, where p, k_1 and k_2 are constants and either k_1 or k_2 may be zero, then try $v = A \sin px + B \cos px$ as the particular

integral. The reason for this is that all the differential coefficients of either
$\sin px$ or $\cos px$ are multiples of either $\sin px$ or $\cos px$. (Note that in the
suggested substitution for v it is insufficient to just let $v = A \sin pt$ or $v = B \cos pt$ even if either k_1 or k_2 are zero). Substituting for v in the differential
equation and comparing coefficients of $\sin px$ and $\cos px$ on both sides
produces two equations to determine A and B.

Problem 7. Solve $2\dfrac{d^2y}{dx^2} + 5\dfrac{dy}{dx} - 3y = 4 \sin 2x$.

In D operator form the differential equation is $(2D^2 + 5D - 3)y = 4 \sin 2x$.
The auxiliary equation is $\quad 2m^2 + 5m - 3 = 0$

$$\text{i.e.} \quad (2m - 1)(m + 3) = 0$$

$$\text{i.e.} \quad m = \frac{1}{2} \text{ or } m = -3.$$

Hence the complementary function, $u = A\,e^{\frac{1}{2}x} + B\,e^{-3x}$

Let the particular integral be $v = A \sin 2x + B \cos 2x$

then $(2D^2 + 5D - 3)\,[A \sin 2x + B \cos 2x] \equiv 4 \sin 2x$

$D\,(A \sin 2x + B \cos 2x) = 2A \cos 2x - 2B \sin 2x$

$D^2(A \sin 2x + B \cos 2x) = D(2A \cos 2x - 2B \sin 2x)$
$\qquad\qquad\qquad\qquad = -4A \sin 2x - 4B \cos 2x.$

Hence $(2D^2 + 5D - 3)\,[A \sin 2x + B \cos 2x]$
$= -8A \sin 2x - 8B \cos 2x + 10A \cos 2x - 10B \sin 2x - 3A \sin 2x$
$\qquad\qquad\qquad\qquad\qquad\qquad - 3B \cos 2x \equiv 4 \sin 2x$

Equating coefficients of $\sin 2x$ gives: $-11A - 10B = 4$ (1)
Equating coefficients of $\cos 2x$ gives: $\;\;10A - 11B = 0$ (2)

$$10 \times (1) \text{ gives:} -110A - 100B = 40 \qquad (3)$$
$$11 \times (2) \text{ gives:} \;\; 110A - 121B = 0 \qquad (4)$$
$$(3) + (4) \text{ gives:} \qquad -221B = 40$$

$$\text{i.e.} \qquad B = \frac{-40}{221}$$

Substituting $B = \dfrac{-40}{221}$ into equation (1) or (2) gives:

$$A = \frac{-44}{221}$$

Hence the particular integral, $v = \dfrac{4}{221}\,(-11 \sin 2x - 10 \cos 2x)$

The general solution,

$$y = u + v = A\,e^{\frac{1}{2}x} + B\,e^{-3x} - \frac{4}{221}\,(11 \sin 2x + 10 \cos 2x)$$

(ii) 'Snag' case

If $f(x) = k_1 \sin px + k_2 \cos px$ and $\sin px$ and/or $\cos px$ occur in the complementary function then let the particular integral, $v = x (A \sin px + B \cos px)$. This case is demonstrated in Problem 8.

Problem 8. Find the general solution of $\dfrac{d^2 y}{dx^2} + 9y = 12 \cos 3x$. If $y = 2$

and $\dfrac{dy}{dx} = 3$ when $x = 0$, determine the particular solution.

In D operator form the differential equation is $(D^2 + 9)y = 12 \cos 3x$.
The auxiliary equation is $m^2 + 9 = 0$
$$\text{i.e. } m = \sqrt{-9} = \pm j\,3$$
Hence the complementary function $u = e^0 \, (C \cos 3x + D \sin 3x)$
$$= C \cos 3x + D \sin 3x.$$

Since $\cos 3x$ occurs in the complementary function and in the right-hand side of the differential equation, let the particular integral
$v = x (A \sin 3x + B \cos 3x)$.

$$\text{Hence } (D^2 + 9) \, [x \, (A \sin 3x + B \cos 3x)] \equiv 12 \cos 3x.$$

$$D\,[x(A \sin 3x + B \cos 3x)] = (x)(3A \cos 3x - 3B \sin 3x)$$
$$+ (A \sin 3x + B \cos 3x)(1), \text{ by the product rule}$$

$$D^2\,[x(A \sin 3x + B \cos 3x)] = (x)(-9A \sin 3x - 9B \cos 3x)$$
$$+ (3A \cos 3x - 3B \sin 3x)\,(1) + (3A \cos 3x - 3B \sin 3x)$$

Hence $(D^2 + 9)\,[x(A \sin 3x + B \cos 3x)] = -9Ax \sin 3x - 9Bx \cos 3x$
$+ 6A \cos 3x - 6B \sin 3x + 9Ax \sin 3x + 9Bx \cos 3x \equiv 12 \cos 3x$

$$\text{i.e. } 6A \cos 3x - 6B \sin 3x \equiv 12 \cos 3x$$

Equating coefficients of $\cos 3x$ gives: $6A = 12$
$$\text{i.e.} \quad A = 2$$

Equating coefficients of $\sin 3x$ gives: $-6B = 0$
$$\text{i.e.} \quad B = 0$$

Hence the particular integral, $v = x(2 \sin 3x)$
The general solution, $y = u + v = C \cos 3x + D \sin 3x + 2x \sin 3x$.

When $x = 0, y = 2$. Hence $2 = C \cos 0 + D \sin 0 + 0$
$$\text{i.e.} \quad C = 2$$

$$\frac{dy}{dx} = -3\,C \sin 3x + 3\,D \cos 3x + (2x)(3 \cos 3x) + (2 \sin 3x)$$

When $x = 0$, $\dfrac{dy}{dx} = 3$. Hence $3 = -3\,C \sin 0 + 3\,D \cos 0 + 0 + 2 \sin 0$
$$\text{i.e.} \quad 3 = 3D$$
$$D = 1$$

Hence the particular solution is $y = 2 \cos 3x + \sin 3x + 2x \sin 3x$
i.e. $y = 2 \cos 3x + (1 + 2x) \sin 3x$.

Typical practical examples of the type of equation solved in Problems 7 and 8 include:

(i) the equation of motion of a body: $\dfrac{d^2 y}{dt^2} + n^2 y = k \cos \omega t$, and

(ii) the equation for variation of charge q in an alternating current circuit containing L, R and C in series:

$$L \frac{d^2 q}{dt^2} + R \frac{dq}{dt} + \frac{1}{C} q = V_0 \sin \omega t.$$

(e) $f(x) = $ a sum or product

In all of the straightforward cases of finding particular integrals it is noticed that a form of particular integral is assumed which is suggested by $f(x)$ but which contains undetermined coefficients.

If $f(x)$ consists of a **sum** of terms the particular integral is the sum of the particular integrals corresponding to the separate terms. Thus, for example, if $f(x) = 2x + 3 \sin 4x$ then let the particular integral, $v = (ax + b) + (C \sin 2x + D \cos 2x)$.

If $f(x)$ is a **product** of two terms then the particular integral assumed is that suggested by $f(x)$. Thus, for example, if $f(x) = e^{\alpha x} \cos \beta x$ then let the particular integral, $v = e^{\alpha x} (A \cos \beta x + B \sin \beta x)$.

Problem 9. Solve $(D^2 - 4D + 4)y = 4x + 3 \cos 2x$.

The auxiliary equation is $m^2 - 4m + 4 = 0$
i.e. $(m - 2)(m - 2) = 0$
i.e. $m = 2$ twice.
The complementary function, $u = (Ax + B) e^{2x}$.

Let the particular integral, $v = ax + b + C \cos 2x + D \sin 2x$,

then $(D^2 - 4D + 4) [v] = 4x + 3 \cos 2x$.

$D(v) = a - 2 C \sin 2x + 2 D \cos 2x$
$D^2 (v) = -4 C \cos 2x - 4 D \sin 2x$.

Hence $(D^2 - 4D + 4) [v] = -4 C \cos 2x - 4 D \sin 2x - 4a + 8 C \sin 2x$
$- 8 D \cos 2x + 4ax + 4b + 4 C \cos 2x$
$+ 4 D \sin 2x \equiv 4x + 3 \cos 2x$.

Equating constant terms gives: $-4a + 4b = 0$
Equating the coefficients of x gives: $4a = 4$
i.e. $a = 1$ and thus $b = 1$.

Equating coefficients of $\cos 2x$ gives: $-4C - 8D + 4C = 3$

$$\text{i.e.} \qquad 8D = 3$$
$$D = \frac{-3}{8}$$

Equating coefficients of sin 2x gives: $-4D + 8C + 4D = 0$
$$\text{i.e.} \qquad 8C = 0$$
$$C = 0$$

Hence the particular integral, $v = x + 1 - \dfrac{3}{8} \sin 2x.$

The general solution, $y = u + v = (Ax + B)\,e^{2x} + x + 1 - \dfrac{3}{8} \sin 2x.$

Problem 10. Solve $\dfrac{d^2 y}{dx^2} + 2\dfrac{dy}{dx} + 2y = 6\,e^x \sin 2x.$

In D operator form the differential equation is $(D^2 + 2D + 2)y = 6e^x \sin 2x.$
The auxiliary equation is $m^2 + 2m + 2 = 0$

$$\text{i.e.} \quad m = \frac{-2 \pm \sqrt{[2^2 - 4(1)(2)]}}{2} = -1 \pm j\,1$$

The complementary function, $u = e^{-x}\,(C \cos x + D \sin x)$
Let the particular integral, $v = e^x\,(A \cos 2x + B \sin 2x)$
then $(D^2 + 2D + 2)\,[v] \equiv 6\,e^x \sin 2x$

$D(v) = (e^x)(-2A \sin 2x + 2B \cos 2x) + e^x\,(A \cos 2x + B \sin 2x)$
$\qquad\qquad [= e^x\,(-2A + B) \sin 2x + e^x\,(2B + A) \cos 2x]$

$D^2\,(v) = (e^x)(-4A \cos 2x - 4B \sin 2x) + (e^x)(-2A \sin 2x + 2B \cos 2x)$
$\qquad + (e^x)(-2A \sin 2x + 2B \cos 2x) + (e^x)(A \cos 2x + B \sin 2x)$
$\qquad = e^x\,[(-3A + 4B) \cos 2x + (-3B - 4A) \sin 2x]$

Hence $(D^2 + 2D + 2)\,[v] = e^x(-3A + 4B) \cos 2x + e^x(-3B - 4A) \sin 2x$
$\qquad\qquad + 2\,e^x\,(-2A + B) \sin 2x + 2\,e^x\,(2B + A) \cos 2x$
$\qquad\qquad + 2\,e^x\,A \cos 2x + 2\,e^x\,B \sin 2x \equiv 6\,e^x \sin 2x$

Equating coefficients of $e^x \sin 2x$ gives: $-3B - 4A - 4A + 2B + 2B = 6$
$$\text{i.e.} \quad -8A + B = 6 \qquad\qquad (1)$$

Equating coefficients of $e^x \cos 2x$ gives: $-3A + 4B + 4B + 2A + 2A = 0$
$$\text{i.e.} \quad A + 8B = 0 \qquad\qquad (2)$$

Solving equations (1) and (2) gives $A = \dfrac{-48}{65}$ and $B = \dfrac{6}{65}$

Hence the particular integral, $v = e^x \left(\dfrac{-48}{65} \cos 2x + \dfrac{6}{65} \sin 2x \right)$

The general solution is

$$y = u + v = e^{-x}\,(C \cos x + D \sin x) + \frac{e^x}{65}\,(6 \sin 2x - 48 \cos 2x)$$

Further problems on solving differential equations of the type
$a\dfrac{d^2y}{dx^2} + b\dfrac{dy}{dx} + cy = f(x)$ *may be found in Section 4 (Problems 1 to 26),*
page 280.

3 Summary of procedure to solve differential equations of the type $a\dfrac{d^2y}{dx^2} + b\,\dfrac{dy}{dx} + cy = f(x)$

1. Rewrite the given differential equation as $(a\,D^2 + b\,D + c)y = f(x)$.
2. Substitute m for D and solve the **auxiliary equation** $am^2 + bm + c = 0$ for m.
3. Obtain the **complementary function, u**. (This is exactly the same procedure discussed in Section 3, Chapter 20, although it was not referred to at that stage as the complementary function.)
4. To find the **particular integral, v**, firstly assume a particular integral which is suggested by $f(x)$ but which contains undetermined coefficients. Below are a list of suggested substitutions.

Type	Straightforward cases Try as particular integral:	'Snag' cases Try as particular integral:
(a) $f(x) =$ a constant	$v = k$	$v = kx$ (used when C.F. contains a constant)
(b) $f(x) =$ polynomial ($f(x) = L + Mx + Nx^2$ $+ \ldots$ where any of the coefficients may be zero)	$v = a + bx + cx^2 + \ldots$	
(c) $f(x) =$ an exponential function. ($f(x) = A\,e^{\alpha x}$)	$v = k\,e^{\alpha x}$	$v = k\,x\,e^{\alpha x}$ (used when $e^{\alpha x}$ appears in the C.F.) $v = kx^2\,e^{\alpha x}$ (used when $e^{\alpha x}$ **and** $xe^{\alpha x}$ both appear in the C.F.), and so on
(d) $f(x) =$ a sine or cosine ($f(x) = a\sin px +$ $b\cos px$, where a or b may be zero.)	$v = A\sin px +$ $B\cos px$	$v = x\,(A\sin px +$ $B\cos px)$ (used when $\sin px$ and/or $\cos px$ appears in the C.F.)

Type	Straightforward cases	'Snag' cases
	Try as particular integral:	Try as particular integral:

(e) $f(x)$ = a sum
    e.g. (i) $f(x) =$ (i) $v = ax^2 + bx + c$
         $2x^2 + 5 \cos 3x$ $+ d \cos 3x$
                                           $+ e \sin 3x$
       (ii) $f(x) = x+1 - e^{-x}$ (ii) $v = ax + b + ce^{-x}$

     $f(x)$ = a product $v = e^{2x} (A \cos 4x +$
     e.g. $f(x) = 3e^{2x} \sin 4x$ $B \sin 4x)$

5. Substitute the suggested particular integral into the differential equation
   $a\dfrac{d^2v}{dx^2} + b\dfrac{dv}{dx} + cv = f(x)$ and equate relevant coefficients to find the
   constants introduced.
6. The **general solution** is given by y = complementary function + particular
   integral, i.e. $y = u + v$.
7. Given sufficient **boundary conditions** the arbitrary constants in the complementary function may be determined to give the **particular solution**.

4 Further problems

In Problems 1 to 12 find the general solution of the given differential
equations.

1. (a) $\dfrac{d^2y}{dx^2} - 5\dfrac{dy}{dx} + 6y = 3$ $[y = A\,e^{3x} + B\,e^{2x} + \tfrac{1}{2}]$

   (b) $\dfrac{d^2y}{dt^2} + 2\dfrac{dy}{dt} + 2y = 5$ $[y = e^{-t}\,(C \cos t + D \sin t) + \tfrac{5}{2}]$

2. (a) $(D^2 + 9D)y = 2$ where $D \equiv \dfrac{d}{dx}$. $[y = A + B\,e^{-9x} + \tfrac{2}{9}x]$

   (b) $\dfrac{d^2x}{d\theta^2} + 7\dfrac{dx}{d\theta} = 7$ $[x = A + B\,e^{-7\theta} + \theta]$

3. (a) $4\dfrac{d^2y}{dx^2} - 12\dfrac{dy}{dx} + 9y = x$ $\left[y = (A + Bx)\,e^{\frac{3}{2}x} + \dfrac{1}{27}\,(3x + 4) \right]$

   (b) $(D^2 + 3D - 4)y = 2x + 3$ $[y = Ae^x + Be^{-4x} - \tfrac{1}{8}\,(4x + 9)]$

4. (a) $2\dfrac{d^2y}{dx^2} + 3\dfrac{dy}{dx} + 2y = x^2 + 2x + 1.$

$$\left[y = e^{-\frac{3}{4}x} \left(C \cos \frac{\sqrt{7}}{4} x + D \sin \frac{\sqrt{7}}{4} x \right) + \frac{1}{2} x^2 - \frac{1}{2} x + \frac{1}{4} \right]$$

(b) $(2D^2 + 5D - 3)y = 2x^3 - x + 7.$

$$\left[y = Ae^{\frac{1}{2}x} + Be^{-3x} - \frac{1}{27} (18x^3 + 90x^2 + 363x + 788) \right]$$

5. (a) $\dfrac{d^2 y}{dx^2} + 3\dfrac{dy}{dx} - 10y = 2e^{3x}$ $[y = Ae^{2x} + Be^{-5x} + \frac{1}{4} e^{3x}]$

(b) $9\dfrac{d^2 x}{dt^2} - 6\dfrac{dx}{dt} + x = 5e^{2t}$ $[x = (A + Bt)e^{\frac{1}{3}t} + \frac{1}{5} e^{2t}]$

6. (a) $\dfrac{d^2 y}{dx^2} + \dfrac{dy}{dx} - 6y = e^{-3x}$ $[y = Ae^{2x} + Be^{-3x} - \frac{1}{5} x e^{-3x}]$

(b) $(2D^2 - D - 3)y = 5e^{-x}$ $[y = Ae^{-x} + Be^{\frac{3}{2}x} - x e^{-x}]$

7. (a) $\dfrac{d^2 y}{dx^2} + 4\dfrac{dy}{dx} + 4y = 4e^{-2x}$ $[y = (A + Bx)e^{-2x} + 2x^2 e^{-2x}]$

(b) $(4D^2 + 20D + 25)y = 8e^{-\frac{5}{2}x}$ $[y = (A + Bx)e^{-\frac{5}{2}x} + x^2 e^{-\frac{5}{2}x}]$

8. (a) $2\dfrac{d^2 x}{dt^2} + 5\dfrac{dx}{dt} - 3x = 2 \sin 3t.$

$$\left[x = Ae^{\frac{1}{2}t} + Be^{-3t} - \frac{1}{111} (7 \sin 3t + 5 \cos 3t) \right]$$

(b) $(9D^2 - 6D + 1)y = 40 \cos 2x.$

$$\left[y = (A + Bx)e^{\frac{1}{3}x} - (0.350\,6 \sin 2x + 1.022\,6 \cos 2x) \right]$$

9. (a) $\dfrac{d^2 y}{dx^2} + 4y = 2 \sin 2x.$ $[y = C \cos 2x + D \sin 2x - \frac{1}{2} x \cos 2x]$

(b) $(D^2 + 6D + 13)y = \cos 2x.$

$$\left[y = e^{-3x} (C \cos 2x + D \sin 2x) + \frac{x}{26} (2 \sin 2x + 3 \cos 2x) \right]$$

10. (a) $15\dfrac{d^2 y}{dx^2} - 2\dfrac{dy}{dx} - y = 3x + 65 \sin x.$

$$[y = Ae^{-\frac{1}{5}x} + Be^{\frac{1}{3}x} - 3x + 6 - 4 \sin x + \frac{1}{2} \cos x]$$

(b) $(4D^2 + 12D + 9)y = e^x \cos x.$

$$\left[y = (A + Bx)e^{-\frac{3}{2}x} + \frac{e^x}{841} (20 \sin x + 21 \cos x) \right]$$

11. (a) $\dfrac{d^2x}{dt^2} - 2\dfrac{dx}{dt} + 2x = e^t \sin t.$

$$[x = e^t (C \cos t + D \sin t) - \tfrac{1}{2} e^t t \cos t]$$

(b) $(2D^2 - 5D - 7)y = 3 \sin x + 4 \cos x.$

$$\left[y = A\, e^{\frac{7}{2}x} + Be^{-x} - \dfrac{1}{106} \ (47 \sin x + 21 \cos x) \right]$$

12. (a) $\dfrac{d^2y}{dx^2} - 3\dfrac{dy}{dx} + 2y = x^2 + e^x.$

$$\left[y = Ae^x + Be^{2x} + \dfrac{1}{2}\,x^2 + \dfrac{3}{2}\,x + \dfrac{7}{4} - x\,e^x \right]$$

(b) $(D^2 - 1)y = \sinh 2x.$ $\left[y = A\,e^x + B\,e^{-x} + \dfrac{1}{3}\,\sinh 2x \right]$

In Problems 13 to 19 find the particular solution of the given differential equations for the stated boundary conditions.

13. $2\dfrac{d^2y}{dx^2} + 5\dfrac{dy}{dx} - 3y = 9$; when $x = 0, y = 0$ and $\dfrac{dy}{dx} = 2.$

$$\left[y = \dfrac{1}{7} \left(22e^{\frac{1}{2}x} - e^{-3x} \right) - 3 \right]$$

14. $(D^2 + 1)y = 4$; when $x = 0, y = 0$ and $D(y) = 3.$ $\left(D \equiv \dfrac{d}{dx} \right).$

$$[y = 3 \sin x - 4 \cos x + 4]$$

15. $3\dfrac{d^2x}{dt^2} + \dfrac{dx}{dt} - 4x = 2e^{-t}$; when $t = 0, x = 0$ and $\dfrac{dx}{dt} = -5.$

$$\left[x = 3e^{-\frac{4}{3}t} - 2e^t - e^{-t} \right]$$

16. $(D^2 - 2D + 1)y = 5e^x$; when $x = 0, y = 2$ and $D(y) = 3.$

$$\left[y = (2 + x)e^x + \dfrac{5}{2}\,x^2\,e^x \right]$$

17. $(D^2 - 5D + 6)y = -3 \sin x$; when $x = 0, y = 0$ and $D(y) = \dfrac{2}{5}$

$$\left[y = \dfrac{1}{10}\left\{ e^{3x} + 2e^{2x} - 3(\sin x + \cos x) \right\} \right]$$

18. $\dfrac{d^2\theta}{dt^2} - 6\dfrac{d\theta}{dt} + 10\theta = 20 - e^{2t}$; when $t = 0, \theta = 4$ and $\dfrac{d\theta}{dt} = \dfrac{25}{2}.$

$$\left[\theta = e^{3t}\left(\dfrac{5}{2}\ \cos t + 6 \sin t \right) + 2 - \dfrac{1}{2}\ e^{2t} \right]$$

19. $2\dfrac{d^2x}{dt^2} - \dfrac{dx}{dt} - 6x = 12\,e^t \cos t$; when $t = 0$, $x = \dfrac{-13}{29}$ and $\dfrac{dx}{dt} = \dfrac{-169}{29}$.

$$\left[x = 2e^{-\frac{3}{2}t} - e^{2t} + \dfrac{e^t}{29}\,(18 \sin t - 42 \cos t) \right]$$

20. The charge q in an electrical circuit of time t satisfies the differential equation $L\,\dfrac{d^2q}{dt^2} + R\,\dfrac{dq}{dt} + \dfrac{1}{C}\,q = E$, where L, R, C and E are constants. $L = 0.9$, $C = 40 \times 10^{-6}$ and $E = 100$. Solve the equation for q when (a) $R = 300$ and (b) R is negligible. Assume that when $t = 0$, $q = 0$ and $\dfrac{dq}{dt} = 0$.

$$\text{(a)} \left[q = \dfrac{1}{250} - \left(\dfrac{1}{250} + \dfrac{2}{3}\,t \right) e^{-\frac{500}{3}t} \right]$$

$$\text{(b)} \left[q = \dfrac{1}{250} \left(1 - \cos \dfrac{500}{3}t \right) \right]$$

21. Solve the following differential equation representing the motion of a body: $\dfrac{d^2y}{dt^2} + n^2 y = a \sin pt$,

where $n \neq 0$ and $p^2 \neq n^2$, given that when $t = 0$, $y = 0$ and $\dfrac{dy}{dt} = 0$

$$\left[y = \dfrac{a}{n^2 - p^2}\,\left(\sin pt - \dfrac{p}{n} \sin nt \right) \right]$$

22. In a galvanometer the deflection θ satisfies the differential equation: $\dfrac{d^2\theta}{dt^2} + 2\dfrac{d\theta}{dt} + \theta = 4$.

Solve the equation for θ given that when $t = 0$, $\theta = 0$ and $\dfrac{d\theta}{dt} = 0$.

$$[\theta = 4 - 4e^{-t}\,(1 + t)]$$

23. Solve the equation $\dfrac{d^2x}{dt^2} + n^2 x = k \cos \omega t$ representing the equation of motion of a body, given that when $t = 0$, $x = 0$ and $\dfrac{dx}{dt} = 0$.

$$\left[x = \dfrac{k}{n^2 - \omega^2}\,(\cos \omega t - \cos nt) \right]$$

24. The differential equation describing the variation of capacitor charge in an alternating current circuit containing inductance L, resistance R and capacitance C in series is:

$$L\,\dfrac{d^2q}{dt^2} + R\,\dfrac{dq}{dt} + \dfrac{1}{C}\,q = V_0 \sin \omega t.$$

Find an expression for the charge q of the capacitor in the circuit at any time t seconds, given that when $t = 0, q = 0$ and $\dfrac{dq}{dt} = -44.29$, in the following case: $R = 20$ ohms, $L = 10 \times 10^{-3}$ henry, $C = 100 \times 10^{-6}$ farads, $V_0 = 500$ volts and $\omega = 250$.

$$[q = (0.022\ 15 - 32.53t)e^{-1\ 000t} + 0.0415\ 2 \sin 250t$$
$$- 0.022\ 15 \cos 250t]$$

25. The equation of motion of a particle moving in a straight line under damping is: $2\dfrac{d^2x}{dt^2} + 2\dfrac{dx}{dt} + x = k \sin t$, where k is a constant. Show that if when $t = 0, x = n$ and $\dfrac{dx}{dt} = 0$, x is given by:

$$x = e^{-\frac{1}{2}t}\left\{ n \cos \frac{t}{2} + (n - 2k) \sin \frac{t}{2} \right\} + k \sin t - \frac{2k}{5} \cos t$$

26. The motion of a mass vibrating in a given way is shown by the equation:

$$\frac{d^2y}{dt^2} + 10\frac{dy}{dt} + 81y = 4 \sin 3t$$

Solve the equation for y.

$$[y = e^{-5} (c \cos \sqrt{56}t + D \sin \sqrt{56}t) + \frac{2}{507} (12 \sin 3t - 5 \cos 3t)]$$

Chapter 22

The Fourier series for periodic functions of period 2π

1 Introduction

There are many phenomena that are studied in engineering and science which are periodic in nature. Some examples include the current and voltage in an alternating current circuit, the displacement, velocity and acceleration of a slider-crank mechanism of a reciprocating engine, acoustic waves and vibrating systems in general.

The best known of mathematical periodic functions are the sine and cosine functions and it thus seems logical that these functions be used to model such practical situations as mentioned above. Fourier series is a mode of analysing a periodic function into its constituent components (called its fundamental and harmonics — see Section 2) and is a technique developed and named after the French mathematician and physicist Jean Baptiste Fourier (1768–1830). Fourier used the series initially to solve problems in the theory of heat conduction. However, Fourier series have found application in many other fields, such as analysis work in electrical and mechanical vibrations, bending of beams, and in radio waves. It is an important function of many electronic devices to send signals without distortion. The result of amplification of a waveform can be studied by Fourier analysis. The original wave may be analysed into its harmonics and the amplification of each harmonic calculated from the characteristics of the amplifying device. It is possible to determine from a Fourier analysis how much of each harmonic is present in the amplified wave and hence the degree of distortion produced. The range of freedom from such distortion is termed the bandwidth of the device. It is important to know how much the amplitude and phase of the harmonics can be changed (many harmonics can be discarded) before the quality of reproduction suffers noticeably. The effects of filters can also be studied by first analysing the input wave into its constituent components.

Fourier series can be applied to many more functions than can Taylor's and Maclaurin's series (see Chapter 5), and while the theory of the analysis is complicated the application of these series is quite simple. To appreciate Fourier series it is firstly essential to

(a) understand what a periodic function is (see Section 2, following),
(b) be able to evaluate definite integrals, such as

$$\int_{-\pi}^{\pi} \sin mx \cos nx \, dx \text{ (see Section 3, following), and sometimes to}$$

(c) use integration be parts (see Chapter 16).

This chapter, together with Chapters 23 to 26, attempt to present the basic theory of Fourier series, the understanding of which leads to the application of the analysis to the behaviour of the physical systems mentioned above.

2 Periodic functions

A function $f(x)$ is said to be periodic if $f(x + l) = f(x)$, for all values of x, where l is some positive number. l is the interval between two successive repetitions and is called the **period** of the function $f(x)$. Figure 1 shows a periodic function $f(x)$ having a period l.

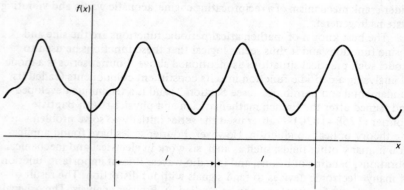

Figure 1

Familiar examples of periodic functions are the sine and cosine functions and graphs of $y = \sin x$ and $y = \cos x$, as shown in Fig. 2(a) and (b). Each of the graphs repeat themselves at regular intervals of 2π, i.e. $\sin x = \sin (x + 2\pi)$ $= \sin (x + 4\pi)$, etc. and $\cos x = \cos (x + 2\pi) = \cos (x + 4\pi)$, etc.

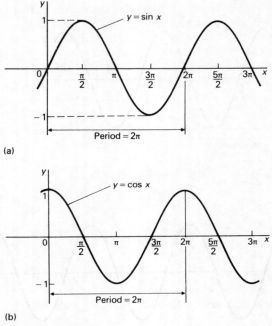

(a)

(b)

Figure 2

It was shown in *Technician Mathematics level 3*, by J.O. Bird and A.J.C. May, Chapter 1, that generally, if $y = \sin \omega t$ or $y = \cos \omega t$ then the period of the waveform is $\dfrac{2\pi}{\omega}$. A graph of $y = \sin 2x$ is shown in Fig. 3(*a*) and is of period $\dfrac{2\pi}{2}$, i.e. π. A graph of $y = 2 \cos 3x$ is shown in Fig. 3(*b*) and is of period $\dfrac{2\pi}{3}$.

If a function is a constant, such as $y = 3$, then this is also considered as a periodic function, the period being infinite.

Continuity

If a graph of a function has no sudden jumps or breaks it is called a **continuous function**, examples being the graphs shown in Figs 2 and 3 of sine and cosine functions. It is possible for a function to have what is called a **finite discontinuity**, which means that the function makes a finite jump at some point, or points in the interval. A typical square wave function is shown in Fig. 4 where finite discontinuities are found at $x = \pi, 2\pi, 3\pi$ and so on. The function has a period of 2π and is defined by $f(x) = \begin{cases} + k, & 0 < x < \pi \\ - k, & \pi < x < 2\pi \end{cases}$

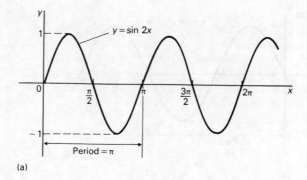

(a)

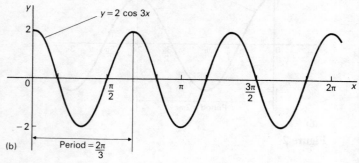

(b)

Figure 3

It was shown in Chapter 9 that the period (or *T*) by I.O. Div. and A. C. Max, T. Linear) that generally, if *y* = sin *ax* or *y* = cos *ax* the period of the waveform is $\dfrac{2\pi}{a}$. A graph of *y* = sine 2x, i.e. in Fig. 3(a) and a set of

period $\dfrac{2\pi}{...}$, i.e. a A graph of *y* = cos 3x are shown in Fig. 3(b) and is of

period $\dfrac{2\pi}{3}$.

If a function is a constant, such as *y* = *C*, then this can be consider'd as a periodic function, the period being infinite.

Continuity

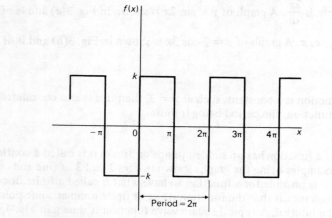

Figure 4

Trigonometric series

If, in the equation $y = x$, for example, we let x be any given value than there is only one corresponding value of y. The function y is then called a **one-valued function**. In contrast, in the equation $x = \sin y$, i.e. $y = \arcsin x$, for example, to each value of x between -1 and $+1$ there corresponds an infinite number of values of y. Thus the function y is then called a **many-valued function**.

If a finite, one-valued function $f(x)$ recurs periodically over successive intervals of 2π it is possible to represent it as a series of the form:

$$
\begin{aligned}
f(x) = \ &a_0 + a_1 \cos x + a_2 \cos 2x + a_3 \cos 3x + \ldots \\
&+ b_1 \sin x + b_2 \sin 2x + b_3 \sin 3x + \ldots
\end{aligned} \tag{1}
$$

where $a_0, a_1, a_2, \ldots b_1, b_2 \ldots$ are real constants.

Such a series is called a **trigonometric series** and a_n and b_n are called the **coefficients** of the series. If the coefficient can be determined, the series is then called the **Fourier series** corresponding to $f(x)$ and its coefficients are called the **Fourier coefficients**.

Alternatively equation (1) may be expressed as:

$$f(x) = a_0 + (a_1 \cos x + b_1 \sin x) + (a_2 \cos 2x + b_2 \sin 2x) + \ldots$$

i.e. $f(x) = a_0 + c_1 \sin(x + \alpha_1) + c_2 \sin(2x + \alpha_2) + \ldots$

$$+ c_n \sin(nx + \alpha_n) \tag{2}$$

where $c_1 = \sqrt{(a_1^2 + b_1^2)}$, $c_n = \sqrt{(a_n^2 + b_n^2)}$ and $\alpha_n = \arctan \dfrac{a_n}{b_n}$.

[Note that if $R \sin(\omega t + \alpha) = b \sin \omega t + a \cos \omega t$

then $b = R \cos \alpha$, $a = R \sin \alpha$, $R = \sqrt{(a^2 + b^2)}$ and $\alpha = \arctan \dfrac{a}{b}$.

See *Technician Mathematics level 3*, Chapter 1, Section 9; or *Mathematics for Electrical Technicians level 3*, Chapter 2, Section 2.]

Periodic functions that occur in practice are often rather complicated and it is therefore desirable to represent these functions in terms of simple periodic functions. Almost any periodic function $f(x)$ of period 2π that appears in applications can be represented by a trigonometric series such as that of equation (1) or equation (2).

A great advantage of a Fourier series is that it can be applied to functions which are discontinuous, that is, functions that have a finite discontinuity.

Harmonics

In the trigonometric series of equations (1) and (2), $(a_1 \cos x + b_1 \sin x)$ or $c_1 \sin(x + \alpha_1)$ is called the **first harmonic** or the **fundamental**, $(a_2 \cos 2x + b_2 \sin 2x)$ or $c_2 \sin(2x + \alpha_2)$ is called the **second harmonic**, and so on. $c_1, c_2, \ldots$ are the **amplitudes** of the various harmonics.

3 Evaluation of definite integrals involving trigonometric functions

The following definite integrals, discussed in Chapter 15, are used when calculating the coefficients $a_0, a_1, a_2, \ldots b_1, b_2, \ldots$ of the trigonometric series of the form stated in equation (1). If m and n are any integers (except zero) then:

(a) $\displaystyle\int_{-\pi}^{\pi} \cos nx \, dx = 0$ (3)

(b) $\displaystyle\int_{-\pi}^{\pi} \sin nx \, dx = 0$ (4)

(c) $\displaystyle\int_{-\pi}^{\pi} \cos^2 nx \, dx \quad = \int_{-\pi}^{\pi} \tfrac{1}{2}\,(1 + \cos 2\,nx) \, dx$

$$= \frac{1}{2}\left[x + \frac{\sin 2\,nx}{2n} \right]_{-\pi}^{\pi}$$

$$= \pi \quad \text{if } n \neq 0 \tag{5}$$

(d) $\displaystyle\int_{-\pi}^{\pi} \sin^2 nx \, dx \quad = \int_{-\pi}^{\pi} \tfrac{1}{2}\,(1 - \cos 2\,nx) \, dx$

$$= \frac{1}{2}\left[x - \frac{\sin 2\,nx}{2n} \right]_{-\pi}^{\pi}$$

$$= \pi \quad \text{if } n \neq 0 \tag{6}$$

(e) $\displaystyle\int_{-\pi}^{\pi} \cos mx \cos nx \, dx \; = \int_{-\pi}^{\pi} \tfrac{1}{2}\,[\cos (m+n)x + \cos (m-n)x] \, dx$

$$= \frac{1}{2}\left[\frac{\sin (m+n)x}{(m+n)} + \frac{\sin (m-n)x}{(m-n)} \right]_{-\pi}^{\pi}$$

$$\left. \begin{aligned} &= 0 \quad \text{if } m \neq n \\ &= \pi \quad \text{if } m = n \end{aligned} \right\} \qquad \begin{aligned} &\text{(7(a))} \\ &\text{(7(b))} \end{aligned}$$

[Note that if $m = n$, the initial integral becomes $\displaystyle\int_{-\pi}^{\pi} \cos^2 nx \, dx = \pi$ if $n \neq 0$ (see (c) above).]

(f) $\displaystyle\int_{-\pi}^{\pi} \sin mx \sin nx \, dx \quad = \int_{-\pi}^{\pi} -\tfrac{1}{2}\,[\cos (m+n)x - \cos (m-n)x] \, dx$

$$= -\frac{1}{2}\left[\frac{\sin{(m+n)x}}{(m+n)} - \frac{\sin{(m-n)x}}{(m-n)}\right]_{-\pi}^{\pi}$$

$$= 0 \quad \text{if } m \neq n \left.\right\} \tag{8(a)}$$
$$= \pi \quad \text{if } m = n \left.\right\} \tag{8(b)}$$

[Note that if $m = n$, the initial integral becomes $\displaystyle\int_{-\pi}^{\pi} \sin^2 nx \, dx = \pi$ if $n \neq 0$

(see (d) above).]

(g) $\displaystyle\int_{-\pi}^{\pi} \sin mx \cos nx \, dx = \int_{-\pi}^{\pi} \tfrac{1}{2}\left[\sin{(m+n)x} + \sin{(m-n)x}\right] dx$

$$= -\frac{1}{2}\left[\frac{\cos{(m+n)x}}{(m+n)} + \frac{\cos{(m-n)x}}{(m-n)}\right]_{-\pi}^{\pi}$$

$$= 0 \tag{9}$$

4 Fourier series

Let $f(x)$ be a periodic function with period 2π such that:

$$f(x) = a_0 + a_1 \cos x + a_2 \cos 2x + a_3 \cos 3x + \ldots$$
$$+ b_1 \sin x + b_2 \sin 2x + b_2 \sin 3x + \ldots$$

i.e. $f(x) = a_0 + \displaystyle\sum_{n=1}^{\infty} (a_n \cos nx + b_n \sin nx) \tag{10}$

To find the coefficient a_0

Integrating both sides of equation (10) from $-\pi$ to π gives:

$$\int_{-\pi}^{\pi} f(x) \, dx = \int_{-\pi}^{\pi} a_0 \, dx, \text{ all the other terms becoming zero}$$
$$\text{(from equations (3) and (4)).}$$

$$\therefore \quad \int_{-\pi}^{\pi} f(x) \, dx = a_0 \, [x]_{-\pi}^{\pi} = 2\pi a_0 \tag{11}$$

Hence $\quad a_0 = \dfrac{1}{2\pi} \displaystyle\int_{-\pi}^{\pi} f(x) \, dx$ (i.e. a_0 = the area under the curve $f(x)$ from $-\pi$ to π, divided by 2π).

To find the coefficient a_n $(n \neq 0)$

Multiplying both sides of equation (10) by $\cos mx$ and integrating from $-\pi$ to π gives:

292

$$\int_{\pi}^{\pi} f(x)\cos mx\,dx = \int_{-\pi}^{\pi} a_0 \cos mx\,dx + \sum_{n=1}^{\infty}\left\{\int_{-\pi}^{\pi}(a_n\cos nx\cos mx + b_n\sin nx\cos mx)\,dx\right\}$$

All the terms on the right-hand side become zero (from equations (3), (7(a))

and (9)) with the exception of $\int_{-\pi}^{\pi} a_n\cos nx\cos mx\,dx$, when $m=n$. This equals

π from equation 7(b).

Hence $\int_{-\pi}^{\pi} f(x)\cos mx\,dx = a_m(\pi)$

i.e. $a_m = \dfrac{1}{\pi}\int_{-\pi}^{\pi} f(x)\cos mx\,dx$

i.e. $a_n = \dfrac{1}{\pi}\left\{\int_{-\pi}^{\pi} f(x)\cos nx\,dx,\ \text{where } n=1,2,3,\ldots\right\}$ (12)

To find the coefficient b_n
Multiplying both sides of equation (10) by $\sin mx$ and integrating from $-\pi$ to π gives:

$$\int_{-\pi}^{\pi} f(x)\sin mx\,dx = \int_{-\pi}^{\pi} a_0 \sin mx\,dx + \sum_{n=1}^{\infty}\left\{\int_{-\pi}^{\pi}(a_n\cos nx\sin mx + b_n\sin nx\sin mx)\,dx\right\}$$

All the terms on the right-hand side become zero (from equations (4), (8(a))

and (9)) with the exception of $\int_{-\pi}^{\pi} b_n\sin nx\sin mx\,dx$ when $m=n$. This

equals π from equation (8(b)).

Hence $\int_{-\pi}^{\pi} f(x)\sin mx\,dx = b_m(\pi)$

i.e. $b_m = \dfrac{1}{\pi}\int_{-\pi}^{\pi} f(x)\sin mx\,dx$

or $b_n = \dfrac{1}{\pi}\int_{-\pi}^{\pi} f(x)\sin nx\,dx,\ \text{where } n=1,2,3,\ldots$ (13)

Hence, given a periodic function $f(x)$ with period 2π, a_n and b_n may be determined and the trigonometric series of equation (10) may be found. This series is then called the **Fourier series** corresponding to $f(x)$ and its coefficients (from equations (11), (12) and (13) are called the **Fourier coefficients**. Since

$f(x)$ is periodic the same results as in equations (11), (12) and (13) would be achieved if the range is taken between say 0 to 2π instead of $-\pi$ to π, or any other interval of length 2π.

Summary

If $f(x)$ is a periodic function of period 2π then its Fourier series is given by:

$$f(x) = a_0 + \sum_{n=1}^{\infty} (a_n \cos nx + b_n \sin nx)$$

where, for range $-\pi$ to $+\pi$

$$a_0 = \frac{1}{2\pi} \int_{-\pi}^{\pi} f(x)\, dx$$

$$a_n = \frac{1}{\pi} \int_{-\pi}^{\pi} f(x) \cos nx\, dx \ (n = 1, 2, 3, \ldots)$$

$$b_n = \frac{1}{\pi} \int_{-\pi}^{\pi} f(x) \sin nx\, dx \ (n = 1, 2, 3, \ldots)$$

Worked problems on Fourier series of periodic functions of period 2π

Problem 1. Obtain a Fourier series for the periodic function $f(x)$ defined as follows:

$$f(x) = \begin{cases} -1 & \text{when } -\pi < x < 0 \\ 1 & \text{when } 0 < x < \pi \end{cases}$$

The function is periodic outside of that range with period 2π (i.e. $f(x) = f(x + 2\pi)$).

The square wave function defined above is shown in Fig. 5 and such functions may occur on external forces acting on mechanical systems, or as electromotive forces in electrical circuits.

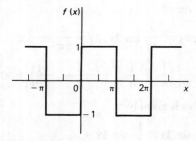

Figure 5

Since $f(x)$ is given by two different expressions in the two halves of the range the integration is divided into two parts, one from $-\pi$ to 0 and the other from 0 to π.

$$a_0 = \frac{1}{2\pi}\int_{-\pi}^{\pi} f(x)\,\mathrm{d}x = \frac{1}{2\pi}\left\{\int_{-\pi}^{0} -1\,\mathrm{d}x + \int_{0}^{\pi} 1\,\mathrm{d}x\right\}$$

$$= \frac{1}{2\pi}\left\{\left[-x\right]_{-\pi}^{0} + \left[x\right]_{0}^{\pi}\right\} = 0$$

(This step was really unnecessary since from the graph in Fig. 5 it is seen that the area under the curve $f(x)$ between $-\pi$ and π is zero and hence $a_0 = 0$.)

$$a_n = \frac{1}{\pi}\int_{-\pi}^{\pi} f(x)\cos nx\,\mathrm{d}x = \frac{1}{\pi}\left\{\int_{-\pi}^{0} (-1)\cos nx\,\mathrm{d}x + \int_{0}^{\pi} (1)\cos nx\,\mathrm{d}x\right\}$$

$$= \frac{1}{\pi}\left\{\left[\frac{-\sin nx}{n}\right]_{-\pi}^{0} + \left[\frac{\sin nx}{n}\right]_{0}^{\pi}\right\} = 0$$

Hence $a_1, a_2, a_3, \ldots$ are all zero and no cosine terms will appear in the Fourier series.

$$b_n = \frac{1}{\pi}\int_{-\pi}^{\pi} f(x)\sin nx\,\mathrm{d}x = \frac{1}{\pi}\left\{\int_{-\pi}^{0} (-1)\sin nx\,\mathrm{d}x + \int_{0}^{\pi} (1)\sin nx\,\mathrm{d}x\right\}$$

$$= \frac{1}{\pi}\left\{\left[\frac{\cos nx}{n}\right]_{-\pi}^{0} + \left[\frac{-\cos nx}{n}\right]_{0}^{\pi}\right\}$$

When n is odd: $b_n = \dfrac{1}{\pi}\left\{\left[\dfrac{1}{n} - \left(-\dfrac{1}{n}\right)\right] + \left[-\left(-\dfrac{1}{n}\right) - \left(-\dfrac{1}{n}\right)\right]\right\} = \dfrac{1}{\pi}\left(\dfrac{2}{n} + \dfrac{2}{n}\right)$

i.e. $b_n = \dfrac{4}{n\pi}$

Hence $b_1 = \dfrac{4}{\pi}\sin x,\; b_3 = \dfrac{4}{3\pi}\sin 3x,\; b_5 = \dfrac{4}{5\pi}\sin 5x,\;\ldots$

When n is even: $b_n = \dfrac{1}{\pi}\left\{\left(\dfrac{1}{n} - \dfrac{1}{n}\right) + \left(-\dfrac{1}{n} - -\dfrac{1}{n}\right)\right\} = 0$

Hence the Fourier series is given by:

$$f(x) = \frac{4}{\pi}\left(\sin x + \frac{1}{3}\sin 3x + \frac{1}{5}\sin 5x + \ldots\right)$$

$\frac{4}{\pi}$ sin x is termed the first **partial sum** of the Fourier series of $f(x)$,

$\frac{4}{\pi}$ (sin x + $\frac{1}{3}$ sin $3x$) is termed the second partial sum of the Fourier series,

$\frac{4}{\pi}$ (sin x + $\frac{1}{3}$ sin $3x$ + $\frac{1}{5}$ sin $5x$) is termed the third partial sum of the Fourier series, and so on.

Problem 2.   (i) Show, by plotting the first three partial sums of the Fourier series for Problem 1, that, as the Fourier series is added together term by term, the result approximates more and more closely to the function it represents.

(ii) Deduce from the Fourier series of Problem 1 that

$$\frac{\pi}{4} = 1 - \frac{1}{3} + \frac{1}{5} - \frac{1}{7} + \dots$$

(i) Let $P_1 = \frac{4}{\pi}$ sin x, $P_2 = \frac{4}{\pi}$ (sin x + $\frac{1}{3}$ sin $3x$),

$P_3 = \frac{4}{\pi}$ (sin x + $\frac{1}{3}$ sin $3x$ + $\frac{1}{5}$ sin $5x$), and so on.

Graphs of P_1, P_2 and P_3, obtained by drawing up tables of values, are shown in Fig. 6(a), (b) and (c) and they indicate that the series is convergent, that is, continually approximating towards a definite limit, as more and more partial sums are taken, and in the limit will have the sum $f(x)$.

[From the graphs of Fig. 6 it is seen that at the points of discontinuity of $f(x)$, i.e. at $-\pi, 0, \pi, \dots$ all the partial sums have the value zero. This is, in fact, the **arithmetic mean of the two limiting values of $f(x)$** as x approaches the point of discontinuity from the two sides (i.e. $\frac{1 + -1}{2} = 0$).

This is true for any point of discontinuity of a periodic function. Figure 7 shows two functions, one a square wave and the other a saw-tooth waveform. For Fig. 7(a) the sum of the Fourier series at the point of discontinuity (i.e. at $0, \pi, 2\pi, \dots$) is given by $\frac{15 + 0}{2}$, i.e. $7\frac{1}{2}$. For Fig. 7(b) the sum of the Fourier series at the point of discontinuity is given by $\frac{8 + 0}{2}$, i.e. 4.]

(ii) $f(x) = \frac{4}{\pi}$ (sin x + $\frac{1}{3}$ sin $3x$ + $\frac{1}{5}$ sin $5x$ + $\frac{1}{7}$ sin $7x$ + ...)

From Fig. 5, when $x = \frac{\pi}{2}$, $f(x) = 1$ and sin $\frac{\pi}{2} = 1$, sin $\frac{3\pi}{2} = -1$,

sin $\frac{5\pi}{2} = 1$, sin $\frac{7\pi}{2} = -1$, and so on.

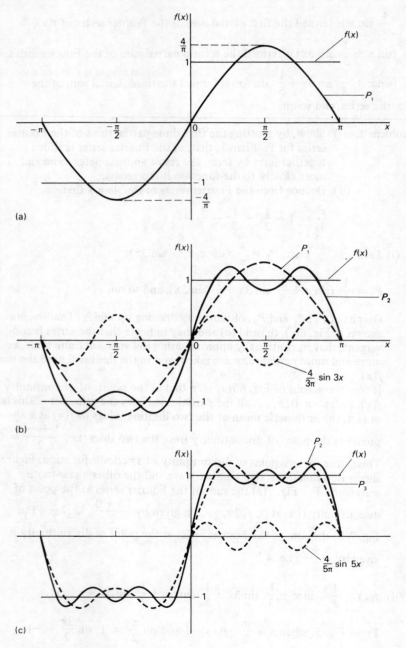

Figure 6

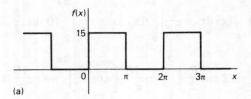

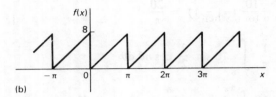

Figure 7

Hence $1 = \frac{4}{\pi} \left\{ 1 + \frac{1}{3} (-1) + \frac{1}{5} (1) + \frac{1}{7} (-1) + \dots \right\}$

Hence $\frac{\pi}{4} = 1 - \frac{1}{3} + \frac{1}{5} - \frac{1}{7} + \dots$

Thus the values of various series with constant terms can be obtained by evaluating Fourier series at specific points.

Problem 3. Find the Fourier series for the rectified sine wave $v = 10 \sin \frac{\theta}{2}$, shown in Fig. 8.

v is a periodic function of period 2π.

Thus $v = f(\theta) = a_0 + \sum\limits_{n=1}^{\infty} (a_n \cos n\theta + b_n \sin n\theta)$

In this case it is better to take the range from 0 to 2π instead of $-\pi$ to $+\pi$ since the waveform is continuous between 0 and 2π.

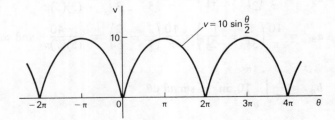

Figure 8

Thus $a_0 = \dfrac{1}{2\pi}\displaystyle\int_{-\pi}^{\pi} f(x)\,dx = \dfrac{1}{2\pi}\int_{0}^{2\pi} f(x)\,dx = \dfrac{1}{2\pi}\int_{0}^{2\pi} 10\sin\dfrac{\theta}{2}\,d\theta$

$$= \frac{10}{2\pi}\left[-2\cos\frac{\theta}{2}\right]_{0}^{2\pi} = \frac{-10}{\pi}\left[\cos\frac{\theta}{2}\right]_{0}^{2\pi} = \frac{-10}{\pi}\left(\cos\pi - \cos 0\right)$$

$$= \frac{-10}{\pi}\,(-1-1) = \frac{20}{\pi}$$

$$a_n = \frac{1}{\pi}\int_{0}^{2\pi} 10\sin\frac{\theta}{2}\,\cos n\theta\,d\theta$$

$$= \frac{10}{\pi}\int_{0}^{2\pi}\frac{1}{2}\left[\sin\left(\frac{\theta}{2}+n\theta\right) + \sin\left(\frac{\theta}{2}-n\theta\right)\right]d\theta$$

$$= \frac{10}{2\pi}\left[\frac{-\cos\left[\theta\left(\frac{1}{2}+n\right)\right]}{\left(\frac{1}{2}+n\right)} - \frac{\cos\left[\theta\left(\frac{1}{2}-n\right)\right]}{\left(\frac{1}{2}-n\right)}\right]_{0}^{2\pi}$$

$$= \frac{5}{\pi}\left[\left\{\frac{-\cos\left[2\pi\left(\frac{1}{2}+n\right)\right]}{\left(\frac{1}{2}+n\right)} - \frac{\cos\left[2\pi\left(\frac{1}{2}-n\right)\right]}{\left(\frac{1}{2}-n\right)}\right\} - \left\{\frac{-\cos 0}{\left(\frac{1}{2}+n\right)} - \frac{\cos 0}{\left(\frac{1}{2}-n\right)}\right\}\right]$$

When n is both odd and even,

$$a_n = \frac{5}{\pi}\left[\left\{\frac{1}{\left(\frac{1}{2}+n\right)} + \frac{1}{\left(\frac{1}{2}-n\right)}\right\} - \left\{\frac{-1}{\left(\frac{1}{2}+n\right)} - \frac{1}{\left(\frac{1}{2}-n\right)}\right\}\right]$$

$$= \frac{5}{\pi}\left[\frac{2}{\left(\frac{1}{2}+n\right)} + \frac{2}{\left(\frac{1}{2}-n\right)}\right] = \frac{10}{\pi}\left[\frac{1}{\left(\frac{1}{2}+n\right)} + \frac{1}{\left(\frac{1}{2}-n\right)}\right]$$

Hence, $a_1 = \dfrac{10}{\pi}\left(\dfrac{1}{\frac{3}{2}} + \dfrac{1}{-\frac{1}{2}}\right) = \dfrac{10}{\pi}\left(\dfrac{2}{3} - \dfrac{2}{1}\right) = \dfrac{-40}{3\pi}$

$a_2 = \dfrac{10}{\pi}\left(\dfrac{1}{2\frac{1}{2}} + \dfrac{1}{-1\frac{1}{2}}\right) = \dfrac{10}{\pi}\left(\dfrac{2}{5} - \dfrac{2}{3}\right) = \dfrac{-40}{(3)(5)\pi}$

$a_3 = \dfrac{10}{\pi}\left(\dfrac{1}{3\frac{1}{2}} + \dfrac{1}{-2\frac{1}{2}}\right) = \dfrac{10}{\pi}\left(\dfrac{2}{7} - \dfrac{2}{5}\right) = \dfrac{-40}{(5)(7)\pi}$ and so on.

$$b_n = \frac{1}{\pi}\int_{0}^{2\pi} 10\sin\frac{\theta}{2}\,\sin n\theta\,d\theta$$

$$= \frac{10}{\pi}\int_{0}^{2\pi} -\left[\frac{1}{2}\quad\cos\theta\left(\frac{1}{2}+n\right) - \cos\theta\left(\frac{1}{2}-n\right)\right]d\theta$$

$$= \frac{10}{2\pi} \left[\frac{\sin \left[\theta \left(\frac{1}{2}-n\right)\right]}{\left(\frac{1}{2}-n\right)} - \frac{\sin \left[\theta \left(\frac{1}{2}+n\right)\right]}{\left(\frac{1}{2}+n\right)} \right]_0^{2\pi}$$

$$= \frac{5}{\pi} \left[\left\{ \frac{\sin 2\pi\left(\frac{1}{2}-n\right)}{\left(\frac{1}{2}-n\right)} - \frac{\sin 2\pi\left(\frac{1}{2}+n\right)}{\left(\frac{1}{2}+n\right)} \right\} - \left\{ \frac{\sin 0}{\left(\frac{1}{2}-n\right)} - \frac{\sin 0}{\left(\frac{1}{2}+n\right)} \right\} \right]$$

When n is both odd and even, $b_n = 0$ since $\sin 0$, $\sin \pi$, $\sin 3\pi$, ... are all zero.

Hence the Fourier series for the rectified sine wave $v = 10 \sin \dfrac{\theta}{2}$ is given by:

$$v = f(\theta) = \frac{20}{\pi} - \frac{40}{3\pi} \cos \theta - \frac{40}{(3)(5)\pi} \cos 2\theta - \frac{40}{(5)(7)\pi} \cos 3\theta - \ldots$$

i.e. $v = \dfrac{20}{\pi} \left\{ 1 - \dfrac{2}{3} \cos \theta - \dfrac{2}{(3)(5)} \cos 2\theta - \dfrac{2}{(5)(7)} \cos 3\theta - \ldots \right\}$

Further problems on Fourier series of periodic functions of period 2π may be found in the following section (5) (Problems 1 to 6).

5 Further problems

In Problems 1 to 3 find the Fourier series for the given periodic functions of period 2π.

1. $f(x) = \begin{cases} 0 \text{ when } -\pi < x < 0 \\ 1 \text{ when } 0 < x < \pi. \end{cases}$

$$\left[f(x) = \frac{1}{2} + \frac{2}{\pi} \left(\sin x + \frac{1}{3} \sin 3x + \frac{1}{5} \sin 5x + \ldots \right) \right]$$

2. $f(x) = \begin{cases} 0 \text{ when } -\pi < x < 0 \\ \sin x \text{ when } 0 < x < \pi. \end{cases}$

$$\left[f(x) = \frac{1}{\pi} - \frac{2}{\pi} \left(\frac{\cos 2x}{3} + \frac{\cos 4x}{(3)(5)} + \frac{\cos 6x}{(5)(7)} + \ldots \right) \right]$$

3. $f(t) = \begin{cases} 0 \text{ when } -\pi < t < -\dfrac{\pi}{2} \\ 1 \text{ when } -\dfrac{\pi}{2} < t < \dfrac{\pi}{2} \\ 0 \text{ when } \dfrac{\pi}{2} < t < \pi. \end{cases}$

$$\left[f(t) = \frac{1}{2} + \frac{2}{\pi} \left(\cos t - \frac{1}{3} \cos 3t + \frac{1}{5} \cos 5t - \ldots \right) \right]$$

4. For the waveform shown in Fig. 9 show that:

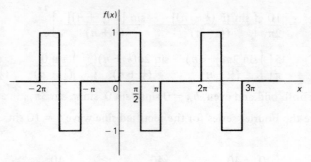

Figure 9

$$f(x) = \frac{2}{\pi} \left(\cos x - \frac{1}{3} \cos 3x + \frac{1}{5} \cos 5x \ldots + \sin 2x + \frac{1}{3} \sin 6x + \right.$$
$$\left. + \frac{1}{5} \sin 10x \ldots \right)$$

5. Obtain the Fourier series for the periodic function defined by:

$$f(t) = \begin{cases} -2, & -\pi < t < -\frac{\pi}{2} \\ 2, & -\frac{\pi}{2} < t < \frac{\pi}{2} \\ -2, & \frac{\pi}{2} < t < \pi \end{cases}$$

The function has a period of 2π.

$$\left[f(t) = \frac{8}{\pi} \left(\cos t - \frac{1}{3} \cos 3t + \frac{1}{5} \cos 5t - \ldots \right) \right]$$

6. (a) For Problems 1, 3 and 5 draw graphs of the first three partial sums of the Fourier series and show that as the series is added together term by term the result approximates more and more closely to the function it represents.

   (b) For Problem 3, find the sum of the Fourier series at the point of discontinuity at $\frac{\pi}{2}$. $\left[\frac{1}{2} \right]$

Chapter 23

The Fourier series for a non-periodic function over range 2π

1 Expansion of non-periodic functions

If a function $f(x)$ is not periodic then it cannot be expanded in a Fourier series for **all** values of x. However, it is possible to find a Fourier series to represent the function over any range of width 2π, such as from $-\pi$ to $+\pi$ or from 0 to 2π. Thus a new function, say $g(x)$, is constructed by taking the values of $f(x)$ in the given range and then repeating them outside of the given range at intervals of 2π. Since, by construction, the function $g(x)$ is periodic with period 2π it may then be expanded in a Fourier series for all values of x. Since $g(x) = f(x)$ in the given range, the sum of the series is equal to $f(x)$ at all points in the given range, but it is **not** equal to $f(x)$ at points outside of the given range. We use exactly the same formulae for the Fourier series and its coefficients as used in the previous chapter. The following worked problems show how Fourier series for non-periodic functions over a range of 2π may be determined.

Worked Problems on Fourier series of non-periodic functions over a range of 2π.

Problem 1. (a) Obtain a Fourier series to represent $f(x) = x$ in the range $-\pi$ to $+\pi$.

(b) In the Fourier series obtained in (a) let $f(x) = \dfrac{\pi}{2}$ and deduce a series for $\dfrac{\pi}{4}$.

302

(a) The function $f(x) = x$ is not periodic. The function is shown in the range $-\pi$ to $+\pi$ in Fig. 1 and is then purposely constructed outside of that range so that it is periodic with period 2π. This is known as a saw-tooth waveform. Points of discontinuity occur at $-\pi, \pi, 3\pi, \dots$ and so on.

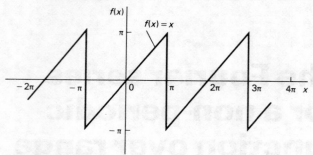

Figure 1

$$f(x) = a_0 + \sum_{n=1}^{\infty} (a_n \cos nx + b_n \sin nx)$$

$$a_0 = \frac{1}{2\pi} \int_{-\pi}^{\pi} f(x)\,dx = \frac{1}{2\pi} \int_{-\pi}^{\pi} x\,dx = \frac{1}{2\pi} \left[\frac{x^2}{2} \right]_{-\pi}^{\pi} = 0$$

$$a_n = \frac{1}{\pi} \int_{-\pi}^{\pi} f(x) \cos nx\,dx = \frac{1}{\pi} \int_{-\pi}^{\pi} x \cos nx\,dx$$

$$= \frac{1}{\pi} \left[\frac{x \sin nx}{n} - \int \frac{\sin nx}{n}\,dx \right]_{-\pi}^{\pi} \text{ by parts}$$

$$= \frac{1}{\pi} \left[\frac{x \sin nx}{n} + \frac{\cos nx}{n^2} \right]_{-\pi}^{\pi}$$

$$= \frac{1}{\pi} \left[\left(0 + \frac{\cos n\pi}{n^2} \right) - \left(0 + \frac{\cos n(-\pi)}{n^2} \right) \right]$$

$$= 0$$

$$b_n = \frac{1}{\pi} \int_{-\pi}^{\pi} f(x) \sin nx\,dx$$

$$= \frac{1}{\pi} \int_{-\pi}^{\pi} x \sin nx\,dx$$

$$= \frac{1}{\pi} \left[\frac{-x \cos nx}{n} - \int \left(\frac{-\cos nx}{n} \right) dx \right]_{-\pi}^{\pi} \text{ by parts}$$

$$= \frac{1}{\pi} \left[\frac{-x \cos nx}{n} + \frac{\sin nx}{n^2} \right]_{-\pi}^{\pi}$$

$$= \frac{1}{\pi} \left[\left(\frac{-\pi \cos n\pi}{n} + \frac{\sin n\pi}{n^2} \right) - \left(\frac{-(-\pi) \cos n(-\pi)}{n} + \frac{\sin n(-\pi)}{n^2} \right) \right]$$

$$= \frac{1}{\pi} \left[\frac{-\pi \cos n\pi}{n} - \frac{\pi \cos n\pi}{n} \right]$$

$$= -\frac{2}{n} \cos n\pi$$

When n is odd, $b_n = \frac{2}{n}$. Thus $b_1 = 2, b_3 = \frac{2}{3}, b_5 = \frac{2}{5}$, and so on.

When n is even, $b_n = -\frac{2}{n}$. Thus $b_2 = -1, b_4 = -\frac{1}{2}, b_6 = -\frac{1}{3}$, and so on.

Thus $f(x) = x = 2 \sin x - \sin 2x + \frac{2}{3} \sin 3x - \frac{1}{2} \sin 4x + \frac{2}{5} \sin 5x$

$$- \frac{1}{3} \sin 6x + \ldots$$

i.e. $x = 2(\sin x - \frac{1}{2} \sin 2x + \frac{1}{3} \sin 3x - \frac{1}{4} \sin 4x + \frac{1}{5} \sin 5x$

$$- \frac{1}{6} \sin 6x + \ldots)$$

for values of x between $-\pi$ and $+\pi$.
For values of x outside of the range $-\pi$ to $+\pi$ the sum of the series is not equal to x.

(b) If $x = \frac{\pi}{2}$ then $f(x) = \frac{\pi}{2}$.

Thus $\frac{\pi}{2} = 2 \left(\sin \frac{\pi}{2} - \frac{1}{2} \sin \pi + \frac{1}{3} \sin \frac{3\pi}{2} - \frac{1}{4} \sin 2\pi \right.$

$$\left. + \frac{1}{5} \sin \frac{5\pi}{2} - \frac{1}{6} \sin 3\pi + \ldots \right)$$

i.e. $\frac{\pi}{4} = 1 - \frac{1}{3} + \frac{1}{5} - \frac{1}{7} + \ldots$

Problem 2. Expand $f(x) = x^2$ in the range $0 < x < 2\pi$ in a Fourier series. By letting $x = \pi$ in the series obtained, find a series for $\frac{\pi^2}{12}$.

$f(x) = x^2$ is not a periodic function but is constructed in the range 0 to 2π and periodic with period 2π outside that range, as shown in Fig. 2.

304

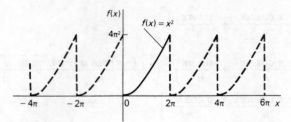

Figure 2

$$f(x) = a_0 + \sum_{n=1}^{\infty} (a_n \cos nx + b_n \sin nx)$$

$$a_0 = \frac{1}{2\pi} \int_0^{2\pi} f(x)\, dx = \frac{1}{2\pi} \int_0^{2\pi} x^2\, dx = \frac{1}{2\pi}\left[\frac{x^3}{3}\right]_0^{2\pi} = \frac{1}{2\pi}\left(\frac{8\pi^3}{3}\right) = \frac{4\pi^2}{3}$$

$$a_n = \frac{1}{\pi} \int_0^{2\pi} f(x) \cos nx\, dx = \frac{1}{\pi} \int_0^{2\pi} x^2 \cos nx\, dx$$

$$= \frac{1}{\pi}\left[\frac{x^2 \sin nx}{n} + \frac{2x \cos nx}{n^2} - \frac{2 \sin nx}{n^3}\right]_0^{2\pi} \text{by parts}$$

$$= \frac{1}{\pi}\left[\left(0 + \frac{4\pi \cos 2\pi n}{n^2} - 0\right) - (0 + 0 - 0)\right]$$

$$= \frac{4}{n^2} \cos 2\pi n = \frac{4}{n^2} \text{ where } n = 1, 2, 3, \ldots$$

Hence $a_1 = \frac{4}{1^2}, a_2 = \frac{4}{2^2}, a_3 = \frac{4}{3^2}, \ldots$ and so on.

$$b_n = \frac{1}{\pi} \int_0^{2\pi} f(x) \sin nx\, dx = \frac{1}{\pi} \int_0^{2\pi} x^2 \sin nx\, dx$$

$$= \frac{1}{\pi}\left[\frac{-x^2 \cos nx}{n} + \frac{2x \sin nx}{n^2} + \frac{2 \cos nx}{n^3}\right]_0^{2\pi} \text{by parts}$$

$$= \frac{1}{\pi}\left[\left(\frac{-4\pi^2 \cos 2\pi n}{n} + 0 + \frac{2 \cos 2\pi n}{n^3}\right) - \left(0 + 0 - \frac{2 \cos 0}{n^3}\right)\right]$$

$$= \frac{1}{\pi}\left[\frac{-4\pi^2}{n} + \frac{2}{n^3} - \frac{2}{n^3}\right] = \frac{-4\pi}{n}$$

Hence $b_1 = \frac{-4\pi}{1}, b_2 = \frac{-4\pi}{2}, b_3 = \frac{-4\pi}{3}, \ldots$ and so on.

Thus $f(x) = x^2 = \dfrac{4\pi^2}{3} + \sum\limits_{n=1}^{\infty} \left(\dfrac{4}{n^2} \cos nx - \dfrac{4\pi}{n} \sin nx\right)$

i.e. $x^2 = \dfrac{4\pi^2}{3} + 4\left(\cos x + \dfrac{1}{2^2}\cos 2x + \dfrac{1}{3^2}\cos 3x + \dots\right)$

$$- 4\pi\left(\sin x + \dfrac{1}{2}\sin 2x + \dfrac{1}{3}\sin 3x + \dots\right)$$

for values of between 0 and 2π.

When $x = \pi$, $f(x) = \pi^2$.

Hence $\pi^2 = \dfrac{4\pi^2}{3} + 4\left(\cos \pi + \dfrac{1}{4}\cos 2\pi + \dfrac{1}{9}\cos 3\pi + \dfrac{1}{16}\cos 4\pi + \dots\right)$

$$- 4\pi\left(\sin \pi + \dfrac{1}{2}\sin 2\pi + \dfrac{1}{3}\sin 3\pi + \dots\right)$$

i.e. $\pi^2 - \dfrac{4\pi^2}{3} = 4\left(-1 + \dfrac{1}{4} - \dfrac{1}{9} + \dfrac{1}{16} - \dots\right) - 4\pi(0)$

$\dfrac{-\pi^2}{3} = 4\left(-1 + \dfrac{1}{4} - \dfrac{1}{9} + \dfrac{1}{16} - \dots\right)$

$\dfrac{\pi^2}{3} = 4\left(1 - \dfrac{1}{4} + \dfrac{1}{9} - \dfrac{1}{16} + \dots\right)$

Thus $\dfrac{\pi^2}{12} = 1 - \dfrac{1}{4} + \dfrac{1}{9} - \dfrac{1}{16} + \dots$

Further problems on Fourier series of non-periodic functions over a range of 2π may be found in the following Section (2) (Problems 1 to 9).

2 Further problems

1. If $f(x) = x + \pi$ when $-\pi < x < \pi$ find the Fourier series for $f(x)$ in that range. Sketch the graph of the function within and outside of the given range.
$$\left[f(x) = \pi + 2\left(\sin x - \dfrac{1}{2}\sin 2x + \dfrac{1}{3}\sin 3x - \dots\right)\right]$$

2. (a) Find the Fourier series for the function $f(x) = x^2$ when $-\pi < x < \pi$ and shetch the function within and outside of this range.
   (b) In the series obtained in (a) let $x = \pi$ and deduce the series for
   $$\sum_{n=1}^{\infty}\dfrac{1}{n^2}.$$
   (a)$\left[f(x) = \dfrac{\pi^2}{3} - 4\left(\cos x - \dfrac{1}{2^2}\cos 2x + \dfrac{1}{3^2}\cos 3x - \dots\right)\right]$
   (b)$\left[1 + \dfrac{1}{4} + \dfrac{1}{9} + \dfrac{1}{16} + \dfrac{1}{25} + \dots = \dfrac{\pi^2}{6}\right]$

3. (a) Obtain a Fourier series for the function defined by

$$f(x) = \begin{cases} x, & 0 < x < \pi \\ 0, & \pi < x < 2\pi \end{cases}$$

(b) What is the sum of the series in (a) at the point of discontinuity (i.e. at $x = \pi$)?

(c) Using the answer to (b), deduce a series for $\dfrac{\pi^2}{8}$.

(a) $\left[f(x) = \dfrac{\pi}{4} - \dfrac{2}{\pi} \left(\cos x + \dfrac{1}{3^2} \cos 3x + \dfrac{1}{5^2} \cos 5x + \ldots \right) \right.$

$$\left. + \left(\sin x - \dfrac{1}{2} \sin 2x + \dfrac{1}{3} \sin 3x - \ldots \right) \right]$$

(b) $\left[\dfrac{\pi}{2} \right]$

(c) $\left[\dfrac{\pi^2}{8} = 1 + \dfrac{1}{9} + \dfrac{1}{25} + \dfrac{1}{49} + \ldots \right]$

4. Find the Fourier series expansion to represent

$$f(x) = \begin{cases} x & \text{when } 0 < x < \pi \\ 2\pi - x & \text{when } \pi < x < 2\pi \end{cases}$$

and sketch the graph of the function within and outside of the given range, assuming the period is 2π.

$$\left[f(x) = \dfrac{\pi}{2} - \dfrac{4}{\pi} \left(\cos x + \dfrac{\cos 3x}{3^2} + \dfrac{\cos 5x}{5^2} + \ldots \right) \right]$$

5. (a) Obtain a Fourier series for the triangular wave function shown in Fig. 3.

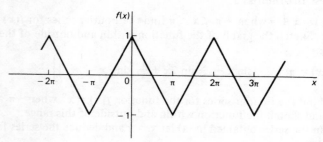

Figure 3

Note that:
$$f(x) = \begin{cases} 1 + \dfrac{2x}{\pi}, & -\pi < x < 0 \\ 1 - \dfrac{2x}{\pi}, & 0 < x < \pi \end{cases}$$

(b) Deduce from the result in (a) the series for $\dfrac{\pi^2}{8}$.

(a) $\left[f(x) = \dfrac{8}{\pi^2} \left(\cos x + \dfrac{1}{3^2} \cos 3x + \dfrac{1}{5^2} \cos 5x + \dfrac{1}{7^2} \cos 7x + \ldots \right) \right]$

(b) $\left[\dfrac{\pi^2}{8} = 1 + \dfrac{1}{3^2} + \dfrac{1}{5^2} + \dfrac{1}{7^2} + \ldots \right]$

In Problems 6 and 7 obtain the Fourier series for the given functions.

6. $f(x) = \begin{cases} 1 - x, & -\pi < x < 0 \\ 1 + x, & 0 < x < \pi. \end{cases}$

$\left[f(x) = \dfrac{\pi}{2} + 1 - \dfrac{4}{\pi} \left(\cos x + \dfrac{\cos 3x}{3^2} + \dfrac{\cos 5x}{5^2} + \ldots \right) \right]$

7. $f(x) = x, \ 0 < x < 2\pi.$

$\left[f(x) = \pi - 2 \left(\sin x + \dfrac{1}{2} \sin 2x + \dfrac{1}{3} \sin 3x + \dfrac{1}{4} \sin 4x + \ldots \right) \right]$

8. Show that the first four terms of the Fourier series for $f(x) = x^4$ in the range $-\pi$ to $+\pi$ is given by:

$f(x) = \dfrac{\pi^4}{5} + 8(6 - \pi^2) \cos x + (2\pi^2 - 3) \cos 3x$

$+ \dfrac{8}{27} (2 - 3\pi^2) \cos 3x + \ldots$

9. Show that the Fourier series expansion for $f(x) = e^x$ in the range $-\pi$ to $+\pi$ is given by:

$f(x) = \dfrac{e^\pi - e^{-\pi}}{\pi} \left[\dfrac{1}{2} + \sum_{n=1}^{\infty} \dfrac{(-1)^n}{1 + n^2} (\cos nx - n \sin nx) \right]$

What are the terms comprising the second harmonic?

$\left[\dfrac{1}{5} (\cos 2x - 2 \sin 2x) \right]$

Chapter 24

The Fourier series for even and odd functions and half range series

1 Even and odd functions

In some of the problems in the preceding two chapters either the Fourier coefficient a_n or the Fourier coefficient b_n became zero after integration. Determination of a zero coefficient is time consuming and could have been avoided, for, with a knowledge of even and odd functions, a zero coefficient may be predicted without performing the often arduous integration.

Even functions

A function $y = f(x)$ is said to be **even** if:

$f(-x) = f(x)$, for all values of x.

Examples of even functions include x^2, x^4, $\cos x$, $e^x + e^{-x}$ and so on. Four even functions are shown in Fig. 1 and it is found that a graph of such a function is always **symmetrical about the y-axis** (i.e. a mirror image).

Odd functions

A function $y = f(x)$ is said to be **odd** if:

$f(-x) = -f(x)$, for all values of x.

Examples of odd functions include x, x^3, $\sin x$, $\tan 3x$ and so on. Four odd functions are shown in Fig. 2 and it is found that a graph of such a function is always **symmetrical about the origin**.

Functions that are neither even nor odd

Many functions are neither even nor odd, two such examples being shown in Fig. 3. In these cases it cannot be predicted whether sine or cosine terms will be absent in the Fourier series.

309

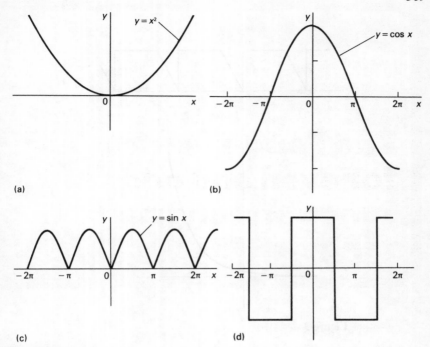

Figure 1

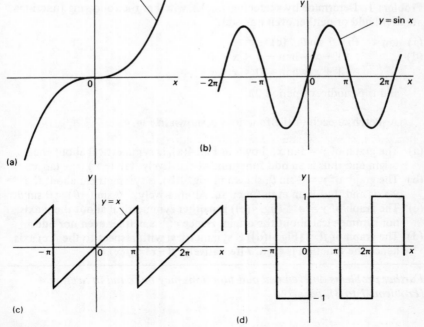

Figure 2

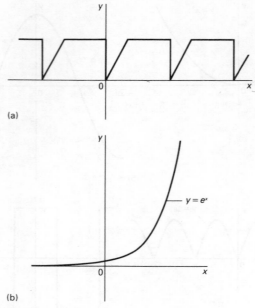

(a)

(b)

Figure 3

Worked problem on even and odd functions.

Problem 1. Determine by sketching graphs, whether the following functions are even, odd or neither even nor odd:

(a) $\tan x$ (b) $\theta \sin \theta$ (c) e^{2x}

(d) $f(x) = \begin{cases} 1 - x & \text{when } -\pi < x < 0 \\ 1 + x & \text{when } 0 < x < \pi \end{cases}$

and is periodic, of period 2π.

A sketch of each of the functions is shown in Fig. 4.

(a) The graph of $y = \tan x$, shown in Fig. 4(a), is symmetrical about the origin and thus is **an odd function**. Alternatively, $\tan(-x) = -\tan x$.

(b) The graph of $y = \theta \sin \theta$, shown in Fig. 4(b), is symmetrical about the y-axis and thus is an **even function**. Alternatively, $-\theta \sin(-\theta) = \theta \sin \theta$.

(c) The graph of $y = e^{2x}$ (fig. 4(c)) is neither symmetrical about the y-axis nor symmetrical about the origin. Hence e^{2x} is **neither even nor odd**.

(d) The graph of $f(x)$ (Fig. 4(d)) is symmetrical with respect to the $f(x)$ axis hence the function is **even**. Alternatively, $f(-x) = f(x)$.

Further problems on even and odd functions may be found in Section 4 (Problems 1 to 3), page 320.

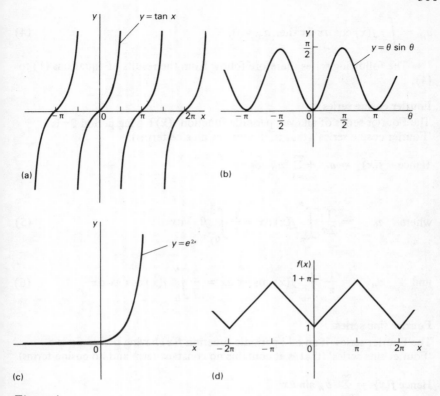

Figure 4

2 Fourier cosine and Fourier sine series

If $f(x)$ is an **even** function then $\displaystyle\int_{-\pi}^{\pi} f(x)\,\mathrm{d}x = 2\int_{0}^{\pi} f(x)\,\mathrm{d}x$ (1)

If $f(x)$ is an **odd** function then $\displaystyle\int_{-\pi}^{\pi} f(x)\,\mathrm{d}x = 0$ (2)

Equations (1) and (2) are obvious from the graphs of Figs 1 and 2.

The product $f(x)\,g(x)$ of an even function $f(x)$ and an odd function $g(x)$ is odd since $f(-x)\,g(-x) = f(x)\,[-g(x)] = -f(x)\,g(x)$.

Thus if $f(x)$ is even then $f(x)\sin nx$ is odd and since

$$b_n = \int_{-\pi}^{\pi} f(x)\sin nx\,\mathrm{d}x \text{ then } b_n = 0.$$ (3)

Similarly, if $f(x)$ is odd then $f(x)\cos nx$ is odd and since

$$a_n = \int_{-\pi}^{\pi} f(x) \cos nx \, dx \text{ then } a_n = 0. \tag{4}$$

The following series therefore follow from the results of equations (1) to (4):

Fourier cosine series

The Fourier series of an even periodic function $f(x)$ having period 2π is a 'Fourier cosine series' (that is, it contains no sine terms).

Hence $f(x) = a_0 + \sum\limits_{n=1}^{\infty} a_n \cos nx$

where
$$a_0 = \frac{1}{2\pi} \int_{-\pi}^{\pi} f(x) \, dx = \frac{1}{\pi} \int_0^{\pi} f(x) \, dx \tag{5}$$

and
$$a_n = \frac{1}{\pi} \int_{-\pi}^{\pi} f(x) \cos nx \, dx = \frac{2}{\pi} \int_0^{\pi} f(x) \cos nx \, dx \tag{6}$$

Fourier sine series

The Fourier series of an odd periodic function $f(x)$ having period 2π is a 'Fourier sine series' (that is, it contains no constant term and no cosine terms).

Hence $f(x) = \sum\limits_{n=1}^{\infty} b_n \sin nx$

where $b_n = \dfrac{1}{\pi} \int_{-\pi}^{\pi} f(x) \sin nx \, dx = \dfrac{2}{\pi} \int_0^{\pi} f(x) \sin nx \, dx$ (7)

Worked problems on Fourier cosine and Fourier sine series

Problem 1. Obtain the Fourier series for each of the square wave functions shown in Fig. 5. From the series of the waveform shown in Fig. 5(b) deduce a series for $\dfrac{\pi}{4}$.

(a) The square wave function shown in Fig. 5(a) is an **even** function since it is symmetrical about the y-axis.

Hence the Fourier series is given by $f(x) = a_0 + \sum\limits_{n=1}^{\infty} a_n \cos nx$

(i.e. the Fourier series will contain no sine terms).

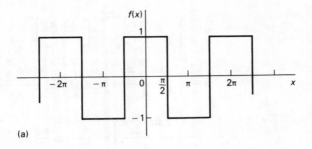

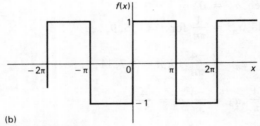

(b)

Figure 5

The function $f(x)$ is defined by:

$$f(x) = \begin{cases} -1, & -\pi < x < -\dfrac{\pi}{2} \\[2mm] 1, & -\dfrac{\pi}{2} < x < \dfrac{\pi}{2} \\[2mm] -1, & \dfrac{\pi}{2} < x < \pi \end{cases}$$

in the range $-\pi$ to $+\pi$.

From equation (5), $a_0 = \dfrac{1}{\pi} \displaystyle\int_0^{\pi} f(x)\, \mathrm{d}x = \dfrac{1}{\pi}\left\{ \int_0^{\pi/2} 1\, \mathrm{d}x + \int_{\pi/2}^{\pi} -1\, \mathrm{d}x \right\}$

$$= \dfrac{1}{\pi}\left\{ \Big[x\Big]_0^{\pi/2} + \Big[-x\Big]_{\pi/2}^{\pi} \right\}$$

$$= \dfrac{1}{\pi}\left\{ \dfrac{\pi}{2} + [(-\pi) - (-\dfrac{\pi}{2})] \right\} = 0$$

From equation (6), $a_n = \dfrac{2}{\pi} \displaystyle\int_0^{\pi} f(x) \cos nx\, \mathrm{d}x$

$$= \dfrac{2}{\pi}\left\{ \int_0^{\pi/2} 1 \cos nx\, \mathrm{d}x + \int_{\pi/2}^{\pi} -1 \cos nx\, \mathrm{d}x \right\}$$

$$= \frac{2}{\pi} \left\{ \left[\frac{\sin nx}{n} \right]_0^{\pi/2} + \left[\frac{-\sin nx}{n} \right]_{\pi/2}^{\pi} \right\}$$

$$= \frac{2}{\pi} \left\{ \left(\frac{\sin \frac{\pi}{2} n}{n} - 0 \right) + \left(0 - \frac{-\sin \frac{\pi}{2} n}{n} \right) \right\}$$

$$= \frac{2}{\pi} \left\{ \frac{2 \sin \frac{\pi}{2} n}{n} \right\} = \frac{4}{\pi n} \left(\sin \frac{n\pi}{2} \right)$$

When n is even, $a_n = 0$

When n is odd, $a_n = \frac{4}{\pi n}$ for $n = 1, 5, 9, \ldots$

and $\quad a_n = \frac{-4}{\pi n}$ for $n = 3, 7, 11, \ldots$

Hence $a_1 = \frac{4}{\pi}, a_3 = \frac{-4}{3\pi}, a_5 = \frac{4}{5\pi}, \ldots$

Hence the Fourier series for the waveform shown in Fig. 5(a) is given by:

$$f(x) = \frac{4}{\pi} \left(\cos x - \frac{1}{3} \cos 3x + \frac{1}{5} \cos 5x - \frac{1}{7} \cos 7x + \ldots \right)$$

(b) The square wave function shown in Fig. 5(b) is an **odd** function since it is symmetrical about the origin.

Hence the Fourier series is given by $f(x) = \sum_{n=1}^{\infty} b_n \sin nx$

The function $f(x)$ is defined by: $\quad f(x) = \begin{cases} -1, & -\pi < x < 0 \\ +1, & 0 < x < \pi \end{cases}$

From equation (7), $b_n = \frac{2}{\pi} \int_0^{\pi} f(x) \sin nx \, dx = \frac{2}{\pi} \int_0^{\pi} 1 \sin nx \, dx$

$$= \frac{2}{\pi} \left[\frac{-\cos nx}{n} \right]_0^{\pi}$$

$$= \frac{2}{\pi} \left[\frac{-\cos n\pi}{n} - -\frac{1}{n} \right]$$

$$= \frac{2}{\pi n} (1 - \cos n\pi)$$

When n is even, $b_n = 0$

When n is odd, $b_n = \frac{2}{\pi n} (1 - -1) = \frac{4}{\pi n}$

Hence $b_1 = \dfrac{4}{\pi}$, $b_3 = \dfrac{4}{3\pi}$, $b_5 = \dfrac{4}{5\pi}$, $\ldots$

Hence the Fourier series for the waveform shown in Fig. 5(b) is given by:

$$f(x) = \frac{4}{\pi}\left(\sin x + \frac{1}{3}\sin 3x + \frac{1}{5}\sin 5x + \frac{1}{7}\sin 7x + \ldots\right)$$

When $x = \dfrac{\pi}{2}$, $f(x) = 1$

Thus $\quad 1 = \dfrac{4}{\pi}\left(\sin\dfrac{\pi}{2} + \dfrac{1}{3}\sin\dfrac{3\pi}{2} + \dfrac{1}{5}\sin\dfrac{5\pi}{2} + \dfrac{1}{7}\sin\dfrac{7\pi}{2} + \ldots\right)$

$\qquad 1 = \dfrac{4}{\pi}\left(1 - \dfrac{1}{3} + \dfrac{1}{5} - \dfrac{1}{7} + \ldots\right)$

i.e. $\quad \dfrac{\pi}{4} = 1 - \dfrac{1}{3} + \dfrac{1}{5} - \dfrac{1}{7} + \ldots$

Problem 2. Expand the function $f(x) = x^2$ as a Fourier series in the range $-\pi < x < \pi$ given that $f(x + 2\pi) = f(x)$. Sketch the function within and outside of the given range. In the series obtained let $x = \pi$ and show that

$$\sum_{n=1}^{\infty} \frac{1}{n^2} = \frac{\pi^2}{6}.$$

A graph of $f(x) = x^2$ in the range $x = -\pi$ to $x = +\pi$ with period 2π is shown in Fig. 6 and it is seen that the function is an **even** one since it is

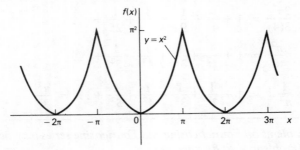

Figure 6

symmetrical about the y-axis. Thus a Fourier cosine series results and is of the form:

$$f(x) = a_0 + \sum_{n=1}^{\infty} a_n \cos nx \quad \text{(i.e. } b_n = 0)$$

From equation (5), $a_0 = \dfrac{1}{\pi}\displaystyle\int_0^{\pi} f(x)\,dx = \dfrac{1}{\pi}\displaystyle\int_0^{\pi} x^2\,dx = \dfrac{1}{\pi}\left[\dfrac{x^3}{3}\right]_0^{\pi} = \dfrac{\pi^2}{3}$

From equation (6), $a_n = \dfrac{2}{\pi} \displaystyle\int_0^\pi f(x) \cos nx \; \mathrm{d}x = \dfrac{2}{\pi} \int_0^\pi x^2 \cos nx \; \mathrm{d}x$

$$= \frac{2}{\pi} \left[\frac{x^2 \sin nx}{n} + \frac{2x \cos nx}{n^2} - \frac{2 \sin nx}{n^3} \right]_0^\pi \text{ by parts}$$

$$= \frac{2}{\pi} \left[\left(0 + \frac{2\pi \cos n\pi}{n^2} - 0 \right) - (0 + 0 - 0) \right]$$

$$= \frac{4}{n^2} \cos n\pi.$$

When n is odd, $a_n = - \dfrac{4}{n^2}$. Hence $a_1 = \dfrac{-4}{1^2}, a_3 = \dfrac{-4}{3^2}, a_5 = \dfrac{-4}{5^2}, \ldots$

When n is even, $a_n = \dfrac{4}{n^2}$. Hence $a_2 = \dfrac{4}{2^2}, a_4 = \dfrac{4}{4^2}, a_6 = \dfrac{4}{6^2}, \ldots$

Hence $f(x) = x^2$

$$= \frac{\pi^2}{3} - 4 \left(\cos x - \frac{1}{4} \cos 2x + \frac{1}{9} \cos 3x - \frac{1}{16} \cos 4x + \frac{1}{25} \cos 5x - \ldots \right)$$

When $x = \pi, f(x) = \pi^2$.

Hence $\pi^2 = \dfrac{\pi^2}{3} - 4 \left(-1 - \dfrac{1}{4} - \dfrac{1}{9} - \dfrac{1}{16} - \dfrac{1}{25} - \ldots \right)$

i.e. $\pi^2 - \dfrac{\pi^2}{3} = 4 \left(1 + \dfrac{1}{4} + \dfrac{1}{9} + \dfrac{1}{16} + \dfrac{1}{25} + \ldots \right)$

$$\frac{2\pi^2}{3(4)} = 1 + \frac{1}{4} + \frac{1}{9} + \frac{1}{16} + \frac{1}{25} + \ldots$$

$$\frac{\pi^2}{6} = 1 + \frac{1}{2^2} + \frac{1}{3^2} + \frac{1}{4^2} + \frac{1}{5^2} + \ldots$$

Hence $\displaystyle\sum_{n=1}^\infty \dfrac{1}{n^2} = 1 + \dfrac{1}{2^2} + \dfrac{1}{3^2} + \dfrac{1}{4^2} + \dfrac{1}{5^2} + \ldots = \dfrac{\pi^2}{6}$

Further problems on Fourier cosine and Fourier sine series may be found in Section 4 (Problems 4 to 8), page 321.

3 Half range Fourier series

If a function is defined over the range say 0 to π, instead of from 0 to 2π, or from $-\pi$ to $+\pi$, it may be expanded in a series of sine terms only or of cosine terms only. This is particularly useful in the solution of some partial differential equations where the boundary conditions may restrict us to a series which

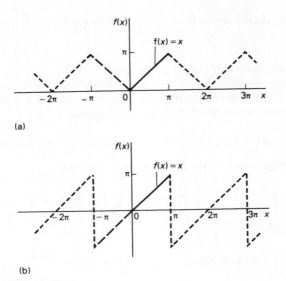

(a)

(b)

Figure 7

contains sine terms only or cosine terms only. The series produced is then
called a **half range Fourier series**. For example, if a half range cosine series is
required for the function $f(x) = x$ in the range 0 to π, then an even periodic
function is required. In Fig. 7(a), $f(x) = x$ is sketched from $x = 0$ to $x = \pi$.
An even function means that it must be symmetrical about the $f(x)$ axis and
this is shown by the broken line between $x = -\pi$ and $x = 0$. It is then assumed
that the 'triangular waveform' produced is periodic of period 2π outside of
this range as shown by the dotted lines.

When a **'half range' cosine series** is required then the same formula as
used for the Fourier cosine series (see Section 2) is applied, that is:

$$f(x) = a_0 + \sum_{n=1}^{\infty} a_n \cos nx, \text{ where } a_0 = \frac{1}{\pi} \int_0^{\pi} f(x) \, dx$$

$$\text{and } a_n = \frac{2}{\pi} \int_0^{\pi} f(x) \cos nx \, dx$$

If a **'half range' sine series** is required for the function $f(x) = x$ in the
range 0 to π, then an odd periodic function is required. In the sketch shown
in Fig. 7(b), $f(x) = x$ is sketched from $x = 0$ to $x = \pi$ as before. An odd function means that it must be symmetrical about the origin and this is shown by
the broken line between $x = -\pi$ and $x = 0$. It is then assumed that the 'saw-
tooth waveform' produced is periodic, of period 2π outside of this range as
shown by the dotted lines.

When a half range sine series is required then the same formula as used
for the Fourier sine series is applied, that is:

$$f(x) = \sum_{n=1}^{\infty} b_n \sin nx, \text{ where } b_n = \frac{2}{\pi} \int_0^{\pi} f(x) \sin nx \, dx.$$

Worked problems on half range Fourier series

Problem 1. Expand the function $f(x) = x$ in the range $0 < x < \pi$ into (a) a half range cosine series, and (b) a half range sine series.

(a) The function $f(x) = x$ is shown in Fig. 7(a) as an even function, and, for a half range cosine series, $f(x) = a_0 + \sum_{n=1}^{\infty} a_n \cos nx$.

$$a_0 = \frac{1}{\pi} \int_0^{\pi} f(x) \, dx = \frac{1}{\pi} \int_0^{\pi} x \, dx = \frac{1}{\pi} \left[\frac{x^2}{2} \right]_0^{\pi} = \frac{\pi}{2}$$

$$a_n = \frac{2}{\pi} \int_0^{\pi} f(x) \cos nx \, dx = \frac{2}{\pi} \int_0^{\pi} x \cos nx \, dx$$

$$= \frac{2}{\pi} \left[\frac{x \sin nx}{n} + \frac{\cos nx}{n^2} \right]_0^{\pi} \text{ by parts}$$

$$= \frac{2}{\pi} \left[\left(\frac{\pi \sin n\pi}{n} + \frac{\cos n\pi}{n^2} \right) - \left(0 + \frac{\cos 0}{n^2} \right) \right]$$

$$= \frac{2}{\pi} \left(0 + \frac{\cos n\pi}{n^2} - \frac{\cos 0}{n^2} \right) = \frac{2}{\pi n^2} (\cos n\pi - 1)$$

When n is odd, $a_n = \frac{2}{\pi n^2} (-1 - 1) = \frac{-4}{\pi n^2}$. Hence $a_1 = \frac{-4}{\pi}, a_3 = \frac{-4}{\pi 3^2}, \ldots$

When n is even, $a_n = \frac{2}{\pi} (1 - 1) = 0$.

Hence the Fourier half range cosine series is

$$f(x) = x = \frac{\pi}{2} - \frac{4}{\pi} \left(\cos x + \frac{1}{3^2} \cos 3x + \frac{1}{5^2} \cos 5x + \ldots \right)$$

(b) The function $f(x) = x$ is shown in Fig. 7(b) as an odd function, and, for a half range sine series, $f(x) = \sum_{n=1}^{\infty} b_n \sin nx$.

$$b_n = \frac{2}{\pi} \int_0^{\pi} f(x) \sin nx \, dx = \frac{2}{\pi} \int_0^{\pi} x \sin nx \, dx = \frac{2}{\pi} \left[\frac{-x \cos nx}{n} + \frac{\sin nx}{n^2} \right]_0^{\pi}$$

by parts

$$= \frac{2}{\pi} \left[\left(\frac{-\pi \cos n\pi}{n} + \frac{\sin n\pi}{n^2} \right) - (0 + 0) \right]$$

$$= -\frac{2}{n} \cos n\pi$$

When n is odd, $b_n = \frac{2}{n}$. Hence $b_1 = 2, b_3 = \frac{2}{3}$, $b_5 = \frac{2}{5}$, . . .

When n is even, $b_n = -\frac{2}{n}$. Hence $b_2 = -\frac{2}{2}, b_4 = -\frac{2}{4}, b_6 = -\frac{2}{6}, . . .$

Hence the Fourier half range sine series is

$$f(x) = x = 2 \left(\sin x - \frac{1}{2} \sin 2x + \frac{1}{3} \sin 3x - \frac{1}{4} \sin 4x + . . . \right)$$

Problem 2. Expand $f(x) = \sin^2 x$ as a half range Fourier sine series in the range $0 < x < \pi$ and sketch the function within and outside the given range.

When a half range sine series is required than an odd function is implied, that is a function symmetrical about the origin. The graph of $f(x) = \sin^2 x$ is shown in Fig. 8 in the range $x = 0$ to $x = \pi$. For $\sin^2 x$ to be symmetrical about the origin the function is as shown by the broken line in Fig. 8 outside of the given range.

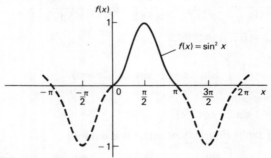

Figure 8

For a half range Fourier sine series, $f(x) = \sum\limits_{n=1}^{\infty} b_n \sin nx \ dx$

$$b_n = \frac{2}{\pi} \int_0^{\pi} f(x) \sin nx \ dx = \frac{2}{\pi} \int_0^{\pi} \sin^2 x \sin nx \ dx$$

Since $\cos 2x = 1 - 2 \sin^2 x$ then $\sin^2 x = \frac{1 - \cos 2x}{2}$

Hence $b_n = \dfrac{2}{\pi} \displaystyle\int_0^\pi \left(\dfrac{1-\cos 2x}{2}\right) \sin nx \, dx = \dfrac{1}{\pi} \int_0^\pi (\sin nx - \sin nx \cos 2x) \, dx$

$\qquad = \dfrac{1}{\pi} \displaystyle\int_0^\pi \sin nx - \dfrac{1}{2}\left(\sin (n+2)x + \sin (n-2)x \right) dx$

$\qquad = \dfrac{1}{\pi} \left[\dfrac{-\cos nx}{n} + \dfrac{\cos (n+2)x}{2(n+2)} + \dfrac{\cos (n-2)x}{2(n-2)} \right]_0^\pi \ \text{if } n \neq 2$

$\qquad = \dfrac{1}{\pi} \left[\left(\dfrac{-\cos n\pi}{n} + \dfrac{\cos (n+2)\pi}{2(n+2)} + \dfrac{\cos (n-2)\pi}{2(n-2)} \right) \right.$

$\qquad\qquad\qquad\qquad \left. -\left(\dfrac{-\cos 0}{n} + \dfrac{\cos 0}{2(n+2)} + \dfrac{\cos 0}{2(n-2)}\right) \right]$

If n is odd, $b_n = \dfrac{1}{\pi}\left[\dfrac{2}{n} - \dfrac{2}{2(n+2)} - \dfrac{2}{2(n-2)} \right]$

$\qquad\qquad = \dfrac{2(n^2-4) - n(n-2) - n(n+2)}{\pi n \, (n+2)(n-2)}$

$\qquad\qquad = \dfrac{-8}{\pi n \, (n+2)(n-2)}$

Hence, $b_1 = \dfrac{-8}{-3\pi} = \dfrac{8}{3\pi}, \ b_3 = \dfrac{-8}{\pi(3)(5)}, \ b_5 = \dfrac{-8}{\pi(5)(7)(3)}, \ \cdots$

If n is even, $b_n = 0$ if $n \neq 2$.

$\qquad b_2 = \dfrac{1}{\pi} \displaystyle\int_0^\pi \left(\sin 2x - \dfrac{1}{2}\sin 4x\right) dx = 0$

(since $\sin 4x = 2 \sin 2x \cos 2x$)

Hence the half range Fourier sine series is given by:

$f(x) = \sin^2 x = \dfrac{8}{\pi}\left\{ \dfrac{\sin x}{(1)(3)} - \dfrac{\sin 3x}{(1)(3)(5)} - \dfrac{\sin 5x}{(3)(5)(7)} - \dfrac{\sin 7x}{(5)(7)(9)} - \cdots \right\}$

Further problems on half range Fourier series may be found in the following Section 4 (Problems 9 to 16), page 322.

4 Further problems

Even and odd functions

In Problems 1 and 2 determine whether the given functions are even, odd or neither even nor odd.

1. (a) e^x (b) $\sin nx$ (c) $\ln x$ (d) $\cos 2x$ (e) $3x^3$

   (a) [neither] (b) [odd] (c) [neither] (d) [even] (e) [odd]

2. (a) e^{x^2} (b) $2x \sin x$ (c) $\dfrac{\cos x}{x}$ (d) $\sin x^2$ (e) $\sin^2 x$

   (a) [even] (b) [even] (c) [odd] (d) [even] (e) [even]

3. Are the following functions which are assumed to be periodic, of period 2π, even, odd or neither even nor odd?

   (a) $f(x) = \begin{cases} x, & -\pi < x < 0 \\ -x, & 0 < x < \pi \end{cases}$

   (b) $f(x) = \begin{cases} e^x, & -\pi < x < 0 \\ e^{-x}, & 0 < x < \pi \end{cases}$

   (c) $f(x) = \begin{cases} x, & -\dfrac{\pi}{2} < x < \dfrac{\pi}{2} \\ 0, & \dfrac{\pi}{2} < x < \dfrac{3\pi}{2} \end{cases}$

   (d) $f(t) = \begin{cases} t-1, & -\pi < t < 0 \\ t+1, & 0 < t < \pi. \end{cases}$

   (a) [even] (b) [even] (c) [odd] (d) [odd]

Fourier cosine and Fourier sine series

4. Sketch the function defined by $f(x) = \begin{cases} x + \pi, & -\pi < x < 0 \\ x - \pi, & 0 < x < \pi, \end{cases}$ which is periodic of period 2π, and obtain the Fourier series representing the given function.

$$\left[f(x) = -2 \left(\sin x + \frac{1}{2} \sin 2x + \frac{1}{3} \sin 3x + \frac{1}{4} \sin 4x + \ldots \right) \right]$$

5. Find a Fourier sine series for the function $f(x) = \begin{cases} x - 1, & -\pi < x < 0 \\ x + 1, & 0 < x < \pi \end{cases}$

   which is periodic of period 2π.

$$\left[f(x) = \frac{2}{\pi} \left((2 + \pi) \sin x - \frac{\pi}{2} \sin 2x + \frac{1}{3} (2 + \pi) \sin 3x - \frac{\pi}{4} \sin 4x + \ldots \right) \right]$$

6. Find the Fourier cosine series for the function

$$f(x) = \begin{cases} -3, & -\pi < x < -\dfrac{\pi}{2} \\ 3, & -\dfrac{\pi}{2} < x < \dfrac{\pi}{2} \\ -3, & \dfrac{\pi}{2} < x < \pi \end{cases}$$

   which is periodic outside of this range, of period 2π.

322

$$\left[f(x) = \frac{12}{\pi} \left(\cos x - \frac{1}{3} \cos 3x + \frac{1}{5} \cos 5x - \frac{1}{7} \cos 7x + \ldots \right) \right]$$

7. A function of t, which is periodic, of period 2π is defined by:

$$f(t) = \begin{cases} 1 - t, & -\pi < t < 0 \\ 1 + t, & 0 < t < \pi \end{cases}$$

Obtain a Fourier series for $f(t)$.

$$\left[f(t) = \frac{\pi}{2} + 1 - \frac{4}{\pi} \left(\cos t + \frac{1}{9} \cos 3t + \frac{1}{25} \cos 5t + \ldots \right) \right]$$

8. Find a Fourier sine series for the function $f(x) = \frac{1}{4}$ in the range $x = 0$ to $x = \pi$. By plotting graphs of the partial sums of the series show that the result approximates more and more closely to the function it represents. Deduce a series for $\frac{\pi}{4}$ from the Fourier expansion.

$$\left[\begin{array}{l} f(x) = \frac{1}{\pi} \left(\sin x + \frac{\sin 3x}{3} + \frac{\sin 5x}{5} + \frac{\sin 7x}{7} + \ldots \right) \\[2mm] \frac{\pi}{4} = 1 - \frac{1}{3} + \frac{1}{5} - \frac{1}{7} + \frac{1}{9} - \ldots \end{array} \right]$$

Half range Fourier series

9. Find (a) the half range sine series and (b) the half range cosine series for the function defined by:

$$f(x) = \begin{cases} 0, & 0 < x < \frac{\pi}{2} \\ 1, & \frac{\pi}{2} < x < \pi \end{cases}$$

(a) $\left[f(x) = \frac{2}{\pi} \left(\sin x - \sin 2x + \frac{\sin 3x}{3} + \frac{\sin 5x}{5} - \frac{\sin 6x}{3} + \ldots \right) \right]$

(b) $\left[f(x) = \frac{1}{2} - \frac{2}{\pi} \left(\cos x - \frac{\cos 3x}{3} + \frac{\cos 5x}{5} - \ldots \right) \right]$

10. Expand $f(x) = \sin^2 x$ between $x = 0$ and $x = \pi$ into a Fourier half range cosine series. $\left[f(x) = \frac{1}{2} (1 - \cos 2x) \right]$

11. Show that the first five terms of the half range Fourier sine series for $f(x) = x^2$, when $0 < x < \pi$, is given by:

$$f(x) = \left(2\pi - \frac{8}{\pi} \right) \sin x - \pi \sin 2x + \left(\frac{2\pi}{3} - \frac{8}{3^3 \pi} \right) \sin 3x - \frac{\pi}{2} \sin 4x$$

$$+ \left(\frac{2\pi}{5} - \frac{8}{5^3 \pi} \right) \sin 5x - \ldots$$

12. Show that the half range sine series for:
$$f(x) = \begin{cases} x, & 0 < x < \frac{\pi}{2} \\ 0, & \frac{\pi}{2} < x < \pi \end{cases}$$

is given by:
$$f(x) = \frac{2}{\pi} \left\{ \sin x + \frac{\pi}{4} \sin 2x - \frac{\sin 3x}{9} - \frac{\pi}{8} \sin 4x \ldots \right\}$$

13. Determine the half range Fourier cosine series for $f(x) = \sin x$ in the range 0 to π. Illustrate graphically the function represented by the series within and outside of the defined range.
$$\left[f(x) = \frac{4}{\pi} \left\{ \frac{1}{2} - \frac{\cos 2x}{(1)(3)} - \frac{\cos 4x}{(3)(5)} - \frac{\cos 6x}{(5)(7)} \ldots \right\} \right]$$

14. Find the half range Fourier sine series for $f(x) = \cos x$ in the range 0 to π.
$$\left[f(x) = \frac{4}{\pi} \left\{ \frac{2 \sin 2x}{3} + \frac{4 \sin 4x}{(3)(5)} + \frac{6 \sin 6x}{(5)(7)} + \ldots \right\} \right]$$

15. (a) Obtain the half range Fourier cosine series in the range $x = 0$ to $x = \pi$ for:
$$f(x) = \begin{cases} x, & 0 < x < \frac{\pi}{2} \\ (\pi - x), & \frac{\pi}{2} < x < \pi \end{cases}$$

(b) From the results in (a) obtain a series for $\frac{\pi^2}{8}$ by letting $x = \frac{\pi}{2}$.

(a) $\left[f(x) = \frac{\pi}{4} - \frac{2}{\pi} \left(\cos 2x + \frac{\cos 6x}{9} + \frac{\cos 10x}{25} + \ldots \right) \right]$

(b) $\left[\frac{\pi^2}{8} = 1 + \frac{1}{3^2} + \frac{1}{5^2} + \frac{1}{7^2} + \ldots \right]$

16. Show that the Fourier sine series expansion of e^x in the range 0 to π is given by:
$$\frac{2}{\pi} \sum_{n=1}^{\infty} \frac{n}{1+n^2} \left\{ 1 + (-1)^{n-1} e^{\pi} \right\} \sin nx$$

Chapter 25

Fourier series over any range

1 Expansion of a periodic function of period l

A periodic function $f(x)$ of period l repeats itself when x increases by l, that is, $f(x + l) = f(x)$. The transition from functions having period 2π to functions having any period l is straightforward since it may be achieved by a change of scale.

To find a Fourier series for a function $f(x)$ in the range $-\dfrac{l}{2} \leqslant x \leqslant \dfrac{l}{2}$ a new variable, u, is introduced such that $f(x)$, as a function of u, has period 2π. Let $u = \dfrac{2\pi x}{l}$. Then, when $x = -\dfrac{l}{2}$, $u = -\pi$ and when $x = \dfrac{l}{2}$, $u = +\pi$. Also if $f(x) = f\left(\dfrac{lu}{2\pi}\right) = F(u)$, then the Fourier series for $F(u)$ is given by:

$$F(u) = a_0 + \sum_{n=1}^{\infty} (a_n \cos nu + b_n \sin nu),$$

where, from Section 4, Chapter 22: $\quad a_0 = \dfrac{1}{2\pi} \displaystyle\int_{-\pi}^{\pi} F(u) \, du,$

$$a_n = \frac{1}{\pi} \int_{-\pi}^{\pi} F(u) \cos nu \, du$$

$$\text{and} \quad b_n = \frac{1}{\pi} \int_{-\pi}^{\pi} F(u) \sin nu \, du$$

It is usually easier to change the above formulae to terms of x. Since $u = \dfrac{2\pi x}{l}$ then $du = \dfrac{2\pi}{l} dx$ and the limits of integration are $-\dfrac{l}{2}$ to $+\dfrac{l}{2}$ instead of $-\pi$ to $+\pi$.

Hence the Fourier series expressed in terms of x is given by:

$$f(x) = a_0 + \sum_{n=1}^{\infty} \left\{ a_n \cos \left(\frac{2\pi nx}{l} \right) + b_n \sin \left(\frac{2\pi nx}{l} \right) \right\}$$

where $a_0 = \frac{1}{2\pi} \int_{-\pi}^{\pi} F(u) \, du = \frac{1}{2\pi} \int_{-\frac{l}{2}}^{\frac{l}{2}} f \left(\frac{lu}{2\pi} \right) \left(\frac{2\pi}{l} \, dx \right)$

i.e. $a_0 = \frac{1}{l} \int_{-\frac{l}{2}}^{\frac{l}{2}} f(x) \, dx$ (1)

$a_n = \frac{1}{\pi} \int_{-\pi}^{\pi} F(u) \cos nu \, du = \frac{1}{\pi} \int_{-\frac{l}{2}}^{\frac{l}{2}} f \left(\frac{lu}{2\pi} \right) \cos \left(\frac{2\pi nx}{l} \right) \left(\frac{2\pi}{l} \, dx \right)$

i.e. $a_n = \frac{2}{l} \int_{-\frac{l}{2}}^{\frac{l}{2}} f(x) \cos \left(\frac{2\pi nx}{l} \right) \, dx$ (2)

in the range $-\frac{l}{2}$ to $+\frac{l}{2}$

and $b_n = \frac{1}{\pi} \int_{-\pi}^{\pi} F(u) \sin nu \, du = \frac{1}{\pi} \int_{-\frac{l}{2}}^{\frac{l}{2}} f \left(\frac{lu}{2\pi} \right) \sin \left(\frac{2\pi nx}{l} \right) \left(\frac{2\pi}{l} \, dx \right)$

i.e. $b_n = \frac{2}{l} \int_{-\frac{l}{2}}^{\frac{l}{2}} f(x) \sin \left(\frac{2\pi nx}{l} \right) \, dx$ (3)

in the range $-\frac{l}{2}$ to $+\frac{l}{2}$.

The limits of integration in equations (1) to (3) may be replaced by any interval of length l; for example, the interval 0 to l.

Worked problems on Fourier series over any range

Problem 1. Find the Fourier series defined by: $f(x) = \begin{cases} = 0 \text{ when } -2 < x < -1 \\ = 3 \text{ when } -1 < x < 1 \\ = 0 \text{ when } 1 < x < 2 \end{cases}$

and which is periodic outside of this range, of period 4.
The function $f(x)$ is shown in Fig. 1 where $l = 4$.

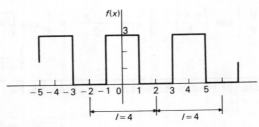

Figure 1

Since the function is symmetrical about the $f(x)$ axis it is an even function and the Fourier series contains no sine terms (i.e. $b_n = 0$).

Thus $f(x) = a_0 + \sum\limits_{n=1}^{\infty} a_n \cos \left(\dfrac{2\pi nx}{l}\right)$

From equation (1), $a_0 = \dfrac{1}{l}\displaystyle\int_{-\frac{l}{2}}^{\frac{l}{2}} f(x)\,dx = \dfrac{1}{4}\int_{-2}^{2} f(x)\,dx$

$\qquad\qquad = \dfrac{1}{4}\left[\displaystyle\int_{-1}^{1} 3\,dx + \int_{-2}^{-1} 0\,dx + \int_{1}^{2} 0\,dx\right]$

$\qquad\qquad = \dfrac{1}{4}\,[3x]_{-1}^{1} = \dfrac{1}{4}[(3) - (-3)]$

$\qquad\qquad = \dfrac{6}{4} = \dfrac{3}{2}$

From equation (2), $a_n = \dfrac{2}{l}\displaystyle\int_{-\frac{l}{2}}^{\frac{l}{2}} f(x)\cos\left(\dfrac{2\pi nx}{l}\right)\,dx$

$\qquad\qquad = \dfrac{2}{4}\displaystyle\int_{-2}^{2} f(x)\cos\left(\dfrac{2\pi nx}{4}\right)\,dx$

$= \dfrac{1}{2}\left[\displaystyle\int_{-2}^{-1} 0\cos\left(\dfrac{\pi nx}{2}\right)dx + \int_{-1}^{1} 3\cos\left(\dfrac{\pi nx}{2}\right)dx + \int_{0}^{2} 0\cos\left(\dfrac{\pi nx}{2}\right)dx\right]$

$= \dfrac{3}{2}\left[\dfrac{\sin\left(\dfrac{\pi nx}{2}\right)}{\dfrac{\pi n}{2}}\right]_{-1}^{1} = \dfrac{3}{\pi n}\left[\sin\left(\dfrac{\pi n}{2}\right) - \sin\left(\dfrac{-\pi n}{2}\right)\right]$

When n is even, $a_n = 0$

$a_1 = \dfrac{3}{\pi}(1 - -1) = \dfrac{6}{\pi}, a_3 = \dfrac{3}{3\pi}(-1 - 1) = \dfrac{-2}{\pi}, a_5 = \dfrac{3}{5\pi}(1 - -1) = \dfrac{6}{5\pi},$

and so on.

Hence $f(x) = \dfrac{3}{2} + \dfrac{6}{\pi}\cos\left(\dfrac{\pi x}{2}\right) - \dfrac{2}{\pi}\cos\left(\dfrac{3\pi x}{2}\right) + \dfrac{6}{5\pi}\cos\left(\dfrac{5\pi x}{2}\right)$

$\qquad\qquad\qquad\qquad\qquad - \dfrac{6}{7\pi}\cos\left(\dfrac{7\pi x}{2}\right) + \ldots$

$f(x) = \dfrac{3}{2} + \dfrac{6}{\pi}\left\{\cos\left(\dfrac{\pi x}{2}\right) - \dfrac{1}{3}\cos\left(\dfrac{3\pi x}{2}\right) + \dfrac{1}{5}\cos\left(\dfrac{5\pi x}{2}\right)\right.$

$\qquad\qquad\qquad\qquad\left. - \dfrac{1}{7}\cos\left(\dfrac{7\pi x}{2}\right) + \ldots\right\}$

Problem 2. Find the Fourier series expansion of the function $f(x) = x$ in the range $x = 0$ to $x = 1$.

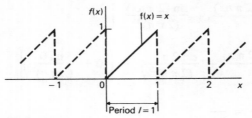

Figure 2

$f(x) = x$ is the interval 0 to 1 as shown in Fig. 2. Although the function is not periodic it may be constructed so, of period 1, outside of the given range as shown by the broken lines.

The Fourier series is given by:

$$f(x) = a_0 + \sum_{n=1}^{\infty} \left\{ a_n \cos \left(\frac{2 \pi nx}{l} \right) + b_n \sin \left(\frac{2 \pi nx}{l} \right) \right\}$$

From equation (1),

$$a_0 = \frac{1}{l} \int_{-\frac{l}{2}}^{\frac{l}{2}} f(x)\, dx = \frac{1}{l} \int_0^l f(x)\, dx = \frac{1}{1} \int_0^1 x\, dx = \left[\frac{x^2}{2} \right]_0^1 = \frac{1}{2}$$

From equation (2),

$$a_n = \frac{2}{l} \int_{-\frac{l}{2}}^{\frac{l}{2}} f(x) \cos \left(\frac{2 \pi nx}{l} \right)\, dx = \frac{2}{l} \int_0^l x \cos \left(\frac{2 \pi nx}{l} \right)\, dx$$

$$= \frac{2}{1} \int_0^1 x \cos (2 \pi nx)\, dx$$

$$= 2 \left[\frac{x \sin (2 \pi nx)}{2 \pi n} + \frac{\cos (2 \pi nx)}{(2 \pi n)^2} \right]_0^1 \quad \text{by parts,}$$

$$= 2 \left[\left(\frac{\sin (2 \pi n)}{2 \pi n} + \frac{\cos (2 \pi n)}{(2 \pi n)^2} \right) - \left(0 + \frac{\cos 0}{(2 \pi n)^2} \right) \right]$$

i.e. $a_n = 0$.

From equation (3),

$$b_n = \frac{2}{l} \int_{-\frac{l}{2}}^{\frac{l}{2}} f(x) \sin \left(\frac{2 \pi nx}{l} \right)\, dx = \frac{2}{l} \int_0^l x \sin \left(\frac{2 \pi nx}{l} \right)\, dx$$

$$= \frac{2}{1} \int_0^1 x \sin (2 \pi nx)\, dx$$

$$= 2 \left[\frac{-x \cos (2 \pi nx)}{2 \pi n} + \frac{\sin (2 \pi nx)}{(2 \pi n)^2} \right]_0^1$$

$$= 2 \left[\left(\frac{-\cos (2 \pi n)}{2 \pi n} + \frac{\sin 2 \pi n}{(2 \pi n)^2} \right) - \left(0 + \frac{\sin 0}{(2 \pi n)^2} \right) \right]$$

$$= \frac{-\cos 2 \pi n}{\pi n} = \frac{-1}{\pi n}$$

Hence, $b_1 = \frac{-1}{\pi}, b_2 = \frac{-1}{2 \pi}, b_3 = \frac{-1}{3 \pi}, \ldots$

Thus $f(x) = \frac{1}{2} - \frac{1}{\pi} (\sin 2 \pi x + \frac{1}{2} \sin 4 \pi x + \frac{1}{6} \sin 6 \pi x + \ldots)$ in the range 0 to 1.

Further problems on Fourier series over any range may be found in Section 3 (Problems 1 to 4), page 332.

2 Half-range Fourier series for functions defined over range *l*

By making the substitution $u = \frac{\pi x}{l}$ we can make the range $x = 0$ to $x = l$ correspond to the range $u = 0$ to $u = \pi$ and hence expand the function in a series of sines or cosines alone of multiples of *u*. These are called the half-range series. Thus a **half-range cosine series** in the range 0 to *l* can be expanded as:

$$f(x) = a_0 + \sum_{n=1}^{\infty} a_n \cos \left(\frac{n \pi x}{l} \right)$$

By similar substitutions as used in section (1) it may be shown that the Fourier coefficients are given by:

$$a_0 = \frac{1}{l} \int_0^l f(x) \, dx \tag{4}$$

and $a_n = \frac{2}{l} \int_0^l f(x) \cos \left(\frac{n \pi x}{l} \right) dx$ (5)

The **half-range sine series** in the range 0 to *l* can be expanded as:

$$f(x) = \sum_{n=1}^{\infty} b_n \sin \left(\frac{n \pi x}{l} \right)$$

where $b_n = \frac{2}{l} \int_0^l f(x) \sin \left(\frac{n \pi x}{l} \right) dx$ (6)

Summary

A function of x can be expanded in the range $-\frac{l}{2}$ to $\frac{l}{2}$ or from 0 to l, or any range of width l, in any of three forms. These are:

(i) as a series of sines and cosines of multiples of $\frac{2\pi x}{l}$, using equations (1)–(3) to determine the Fourier coefficients,

(ii) as a series of cosines of multiples of $\frac{\pi x}{l}$, using equations (4) and (5) to determine the Fourier coefficients, and

(iii) as a series of sines of multiples of $\frac{\pi x}{l}$, using equation (6) to determine the Fourier coefficients.

Worked problem on half-range Fourier series for functions defined over range l

Problem 1. Expand to the first 4 terms the function $f(x) = x$, when $0 < x < 4$ in a half-range Fourier (a) sine series and (b) cosine series.

(a) A half-range sine series indicates an odd function. Thus the graph of $f(x) = x$ in the range 0 to 4 is shown in Fig. 3 and is extended outside of this range so as to be symmetrical about the origin as shown by the broken lines.

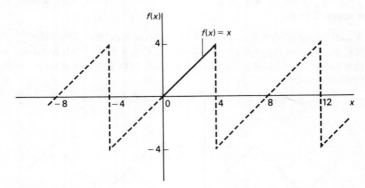

Figure 3

For a half-range sine series: $f(x) = \sum\limits_{n=1}^{\infty} b_n \sin\left(\frac{n\pi x}{l}\right)$

From equation (6),

$b_n = \frac{2}{l}\int_0^l f(x) \sin\left(\frac{n\pi x}{l}\right) \, dx = \frac{2}{4}\int_0^4 x \sin\left(\frac{n\pi x}{4}\right) \, dx$, since $l = 4$

$$= \frac{1}{2} \left[\frac{-x \cos\left(\frac{n \pi x}{4}\right)}{\frac{n \pi}{4}} - \frac{\sin\left(\frac{n \pi x}{4}\right)}{\left(\frac{n \pi}{4}\right)^2} \right]_0^4$$

$$= \frac{1}{2} \left[\left(\frac{-4 \cos n \pi}{\frac{n \pi}{4}} - \frac{\sin n \pi}{\left(\frac{n \pi}{4}\right)^2} \right) - \left(0 - \frac{\sin 0}{\left(\frac{n \pi}{4}\right)^2} \right) \right]$$

$$= \frac{1}{2} \left(\frac{-4 \cos n \pi}{\frac{n \pi}{4}} \right)$$

$$= \frac{-8}{n \pi} \cos n \pi$$

When $n = 1$, $b_1 = \frac{-8}{\pi} (-1) = \frac{8}{\pi}$

When $n = 2$, $b_2 = \frac{-8}{2\pi} (1) = \frac{-4}{\pi}$

When $n = 3$, $b_3 = \frac{-8}{3\pi} (-1) = \frac{8}{3\pi}$

When $n = 4$, $b_4 = \frac{-8}{4\pi} (1) = \frac{-2}{\pi}$

Hence $f(x) = \frac{8}{\pi} \left(\sin\left(\frac{\pi x}{4}\right) - \frac{1}{2} \sin\left(\frac{\pi x}{2}\right) + \frac{1}{3} \sin\left(\frac{3 \pi x}{4}\right) - \frac{1}{4} \sin(\pi x) + \ldots \right)$

in the range 0 to 4.

(b) A half-range cosine series indicates an even function. Thus the graph of $f(x) = x$ in the range 0 to 4 is shown in Fig. 4 and is extended outside of this range so as to be symmetrical with the $f(x)$ axis as shown by the broken lines.

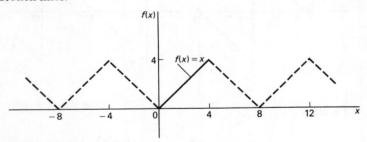

Figure 4

For a half-range cosine series: $f(x) = a_0 + \sum\limits_{n=1}^{\infty} a_n \cos\left(\frac{n \pi x}{l}\right)$

From equation (4),

$$a_0 = \frac{1}{l} \int_0^l f(x)\, dx = \frac{1}{4} \int_0^4 x\, dx = \frac{1}{4} \left[\frac{x^2}{2} \right]_0^4 = 2$$

From equation (5),

$$a_n = \frac{2}{l} \int_0^l f(x) \cos\left(\frac{n \pi x}{l} \right)\, dx = \frac{2}{4} \int_0^4 x \cos\left(\frac{n \pi x}{4} \right)\, dx$$

$$= \frac{1}{2} \left[\frac{x \sin\left(\dfrac{n \pi x}{4} \right)}{\dfrac{n \pi}{4}} - \frac{\cos\left(\dfrac{n \pi x}{4} \right)}{\left(\dfrac{n \pi}{4} \right)^2} \right]_0^4$$

$$= \frac{1}{2} \left[\left(\frac{4 \sin n \pi}{\dfrac{n \pi}{4}} - \frac{\cos n \pi}{\left(\dfrac{n \pi}{4} \right)^2} \right) - 0 \left(- \frac{\cos 0}{\left(\dfrac{n \pi}{4} \right)^2} \right) \right]$$

$$= \frac{1}{2} \left[\frac{-\cos n \pi}{\left(\dfrac{n \pi}{4} \right)^2} + \frac{1}{\left(\dfrac{n \pi}{4} \right)^2} \right]$$

$$= \frac{1}{2} \left(\frac{4}{n \pi} \right)^2 [1 - \cos n \pi]$$

$$= \frac{8}{n^2 \pi^2} (1 - \cos n \pi)$$

When n is even, $a_n = 0$.

When $n = 1$, $a_1 = \dfrac{8}{\pi^2} (2) = \dfrac{16}{\pi^2}$

When $n = 3$, $a_3 = \dfrac{8}{9\pi^2} (2) = \dfrac{16}{9\pi^2}$

When $n = 5$, $a_5 = \dfrac{8}{25\pi^2} (2) = \dfrac{16}{25\pi^2}$

When $n = 7$, $a_7 = \dfrac{8}{49\pi^2} (2) = \dfrac{16}{49\pi^2}$

Hence $f(x) = 2 + \dfrac{16}{\pi^2} \left\{ \cos\left(\dfrac{\pi x}{4} \right) + \dfrac{1}{9} \cos\left(\dfrac{3 \pi x}{4} \right) + \dfrac{1}{25} \cos\left(\dfrac{5 \pi x}{4} \right) \right.$

$$\left. + \dfrac{1}{49} \cos\left(\dfrac{7 \pi x}{4} \right) + \ldots \right\} \quad \text{in the range 0 to 4.}$$

Further problems on half-range Fourier series over range l may be found in the following Section (3) (Problems 5 to 7).

3 Further problems

Fourier series over any range

1. The voltage from a square wave generator is of the form:

$$V(t) = \begin{cases} 0, -5 < t < 0 \\ 3, 0 < t < 5 \end{cases}$$

Find the Fourier series for this periodic function having period 10.

$$\left[V(t) = \frac{3}{2} + \frac{6}{\pi} \left\{ \sin\left(\frac{\pi t}{5}\right) + \frac{1}{3} \sin\left(\frac{3\pi t}{5}\right) + \frac{1}{5} \sin\left(\frac{5\pi t}{5}\right) + \dots \right\} \right]$$

2. Figure 5(a) represents a half wave rectified sinusoidal voltage $E \sin \omega t$. Develop the Fourier series for the function.

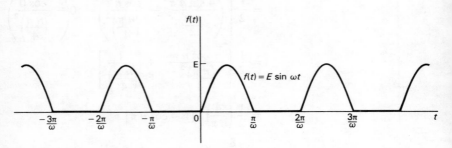

(a)

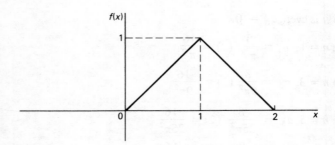

(b)

Figure 5

$$\left[f(t) = \frac{E}{\pi} + \frac{E}{2} \sin \omega t - \frac{2E}{\pi} \left(\frac{\cos 2\omega t}{(1)(3)} + \frac{\cos 4\omega t}{(3)(5)} + \dots \right) \right]$$

3. Obtain the Fourier series for $f(x)$ given that $f(x) = \begin{cases} -1, -2 < x < 0 \\ +1, 0 < x < 2 \end{cases}$.

$$\left[f(x) = \frac{4}{\pi} \left\{ \sin\left(\frac{\pi x}{2}\right) + \frac{1}{3} \sin\left(\frac{3\pi x}{2}\right) + \frac{1}{5} \sin\left(\frac{5\pi x}{2}\right) + \dots \right\} \right]$$

4. Find the Fourier series for $f(x) = x$ in the range $x = -1$ to $x = +1$.

$$\left[f(x) = \frac{2}{\pi} \left\{ \sin \pi x - \frac{1}{2} \sin 2 \pi x + \frac{1}{3} \sin 3 \pi x - \ldots \right\} \right]$$

Half-range Fourier series over range *l*

5. Find the half-range cosine series for the function $f(x) = x^2$ in the range 0 to 4. Sketch the function within and outside of the given range.

$$\left[f(x) = \frac{16}{3} - \frac{64}{\pi^2} \left\{ \cos \left(\frac{\pi x}{4} \right) - \frac{1}{4} \cos \left(\frac{\pi x}{2} \right) + \frac{1}{9} \cos \left(\frac{3 \pi x}{4} \right) - \ldots \right\} \right]$$

6. Expand $f(x) = x$ in the range $0 \leqslant x \leqslant 2$ in a half-range (a) sine series, (b) cosine series. Sketch the function in each case.

$$(a) \left[f(x) = \frac{4}{\pi} \left\{ \sin \left(\frac{\pi x}{2} \right) - \frac{1}{2} \sin (\pi x) + \frac{1}{3} \sin \left(\frac{3 \pi x}{2} \right) - \ldots \right\} \right]$$

$$(b) \left[f(x) = 1 - \frac{8}{\pi^2} \left\{ \cos \left(\frac{\pi x}{2} \right) + \frac{1}{3^2} \cos \left(\frac{3 \pi x}{2} \right) + \frac{1}{5^2} \cos \left(\frac{5 \pi x}{2} \right) + \ldots \right\} \right]$$

7. Show that the half-range sine series for the function shown in Fig. 5(b) is given by:

$$f(x) = \frac{8}{\pi^2} \left\{ \sin \left(\frac{\pi x}{2} \right) - \frac{1}{3^2} \sin \left(\frac{3 \pi x}{2} \right) + \frac{1}{5^2} \sin \left(\frac{5 \pi x}{2} \right) - \ldots \right\}$$

Chapter 26

A numerical method of harmonic analysis

1 Introduction

Many of the waveforms met in practice can be represented by simple mathematical expressions and the magnitudes of their harmonic components can be conveniently found by Fourier series. Numerical methods are used to analyse waveforms for which simple mathematical expressions cannot be obtained. **Harmonic analysis** is the process of resolving a periodic, non-sinusoidal quantity into a series of sinusoidal components of ascending order of frequency. This chapter deals with a straightforward numerical method of analysing irregular waveforms.

2 Harmonic analysis on data given in tabular or graphical form

The general Fourier series is given by:

$$f(x) = a_0 + a_1 \cos x + a_2 \cos 2x + a_3 \cos 3x + \ldots$$
$$+ b_1 \sin x + b_2 \sin 2x + b_3 \sin 3x + \ldots$$

i.e. $f(x) = a_0 + \sum\limits_{n=1}^{\infty} (a_n \cos nx + b_n \sin nx)$ (1)

and the Fourier coefficients are given by:

$$a_0 = \frac{1}{2\pi} \int_{-\pi}^{\pi} f(x) \, dx = \frac{1}{2\pi} \int_{0}^{2\pi} f(x) \, dx = \text{the mean value of } f(x) \text{ in}$$

the range $-\pi$ to π or 0 to 2π.

$$a_n = \frac{1}{\pi} \int_{-\pi}^{\pi} f(x) \cos nx \; dx = \frac{1}{\pi} \int_{0}^{2\pi} f(x) \cos nx \; dx = \text{twice the}$$
mean value of $f(x) \cos nx$ in the range 0 to 2π.

and $b_n = \frac{1}{\pi} \int_{-\pi}^{\pi} f(x) \sin nx \; dx = \frac{1}{\pi} \int_{0}^{2\pi} f(x) \sin nx \; dx = \text{twice the mean}$
value of $f(x) \sin nx$ in the range 0 to 2π.

However if an irregular graph is given we cannot evaluate the above integrals by calculus since $f(x)$ is not given by a mathematical expression. In these cases approximate methods, such as the **trapezoidal rule**, are used to evaluate the Fourier coefficients a_0, a_n and b_n. Most of the graphs to be analysed in practice are periodic. Let us assume the period of a waveform is 2π. (If it is not of period 2π then, by a simple change of variable the period may be reduced to 2π.) Let the range 0 to 2π be divided into p equal parts, the points of division being $x_0, x_1, x_2, \ldots x_p$, as shown in Fig. 1. The width of each interval is then $\frac{2\pi}{p}$. Let the ordinates at these points be labelled y_0, $y_1, y_2, \ldots y_p$. (Note that $y_0 = y_p$ since the waveform is assumed periodic.)

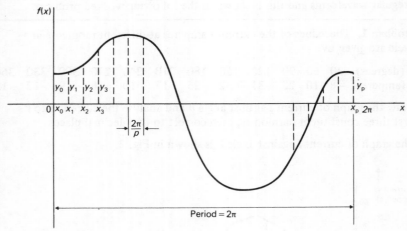

Figure 1

The traperzoidal rule states:

Area $\simeq$ (width of interval) $[\frac{1}{2}$ (first + last ordinate) + sum of remaining ordinates]

$$\simeq \left(\frac{2\pi}{p}\right) [\frac{1}{2}(y_0 + y_p) + y_1 + y_2 + y_3 + \ldots]$$

Since $y_0 = y_p$ then $\frac{1}{2}(y_0 + y_p) = y_0 = y_p$.

Hence Area $\simeq \frac{2\pi}{p} \sum_{K=1}^{p} y_K \left(\text{or Area} \simeq \frac{2\pi}{p} \sum_{K=0}^{p-1} y_K\right)$

Mean value = $\dfrac{\text{area}}{\text{length of base}} \simeq \dfrac{1}{2\pi}\left(\dfrac{2\pi}{p}\displaystyle\sum_{K=1}^{p} y_K\right)$

$$\simeq \dfrac{1}{p}\sum_{K=1}^{p} y_K$$

However, a_0 = the mean value of $f(x)$ in the range 0 to 2π.

Hence $a_0 \simeq \dfrac{1}{p}\displaystyle\sum_{K=1}^{p} y_K$ (2)

Similarly, since a_n = twice the mean value of $f(x)\cos nx$ in the range 0 to 2π then

$$a_n \simeq \dfrac{2}{p}\sum_{K=1}^{p} y_K \cos nx_K \tag{3}$$

and since b_n = twice the mean value of $f(x)\sin nx$ in the range 0 to 2π then

$$b_n \simeq \dfrac{2}{p}\sum_{K=1}^{p} y_K \sin nx_K \tag{4}$$

Equations (2) to (4) are used to determine the Fourier coefficients for irregular waveforms and this is shown in the following worked problem.

Problem 1. The values of the current i amperes at different moments in a cycle are given by:

θ (degrees)	30	60	90	120	150	180	210	240	270	300	330	360
i (amperes)	−8	10	25	31	32	25	17	10	−8	−19	−17	−13

Draw the graph of current i against angle θ and analyse the current into its first three constituent harmonics, each correct to two decimal places.

The graph of current i against angle θ is shown in Fig. 2.

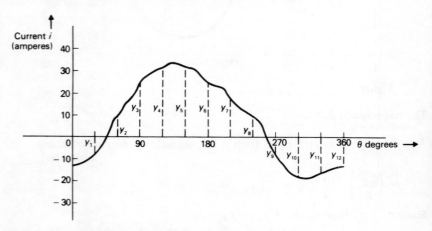

Figure 2   Graph of current against angle θ (problem 1)

The range 0 to 2π is divided into 12 equal parts (i.e. an interval width of $\frac{2\pi}{12}$, i.e. $\frac{\pi}{6}$ or $30°$) and the values of the ordinates $y_1, y_2, y_3, \ldots$ are $-8, 10, 25 \ldots$ from the given table of values. (Note that either y_0 or y_{12} is taken, but not both.)

It is usually easiest to tabulate the data as shown in Table 1.

From equation (2), $a_0 \simeq \frac{1}{p} \sum_{K=1}^{p} y_K = \frac{1}{12}(85) = 7.08$, since $p = 12$.

From equation (3), $a_n \simeq \frac{2}{p} \sum_{K=1}^{p} y_K \cos nx_K$

$\qquad$ Hence $\quad a_1 \simeq \frac{2}{12}(-127.08) = -21.18$

$\qquad\qquad\qquad a_2 \simeq \frac{2}{12}(-9) = -1.50$

$\qquad\qquad\qquad a_3 \simeq \frac{2}{12}(12) = 2.00$

From equation (4), $b_n \simeq \frac{2}{p} \sum_{K=1}^{p} y_K \sin nx_K$

$\qquad$ Hence $\quad b_1 \simeq \frac{2}{12}(88.3) = 14.72$

$\qquad\qquad\qquad b_2 \simeq \frac{2}{12}(1.72) = 0.29$

$\qquad\qquad\qquad b_3 \simeq \frac{2}{12}(-9) = -1.50$

Substituting values of the Fourier coefficients into equation (1) gives:

$i = 7.08 - 21.18 \cos\theta - 1.50 \cos 2\theta + 2.00 \cos 3\theta + \ldots$
$\quad + 14.72 \sin\theta + 0.29 \sin 2\theta - 1.50 \sin 3\theta \ldots$

From Problem 1 it can be appreciated that the arithmetic involved is not too difficult and a proforma of the type shown in Table 1 could be readily produced to tabulate results and calculations. Although 12 ordinates are used in Problem 1, the method is, in fact, independent of the number of ordinates. The more ordinates used the greater will be the accuracy of the results. Only the first three harmonics were required in Problem 1. The method can be continued for further harmonics with the arithmetic generally becoming easier the higher the harmonic.

3 Complex waveform considerations

Examination of a complex waveform often can reveal much about the harmonic content and the following five points should be considered before analysis is attempted.

(i) $a_0 = \frac{1}{2\pi} \int_0^{2\pi} f(x)\,dx$. This is the mean value of $f(x)$ in the range 0 to 2π.

Table 1

Ordinates	$\theta°$	i	$\cos\theta$	$i\cos\theta$	$\sin\theta$	$i\sin\theta$	$\cos 2\theta$	$i\cos 2\theta$	$\sin 2\theta$	$i\sin 2\theta$	$\cos 3\theta$	$i\cos 3\theta$	$\sin 3\theta$	$i\sin 3\theta$
y_1	30	-8	0.866	-6.93	0.5	-4	0.5	-4	0.866	-6.93	0	0	1	-8
y_2	60	10	0.5	5	0.866	8.66	-0.5	-5	0.866	8.66	-1	-10	0	0
y_3	90	25	0	0	1	25	-1	-25	0	0	0	0	-1	-25
y_4	120	31	-0.5	-15.5	0.866	26.85	-0.5	-15.5	-0.866	-26.85	1	31	0	0
y_5	150	32	-0.866	-27.71	0.5	16	0.5	16	-0.866	-27.71	0	0	1	32
y_6	180	25	-1	-25	0	0	1	25	0	0	-1	-25	0	0
y_7	210	17	-0.866	-14.72	-0.5	-8.5	0.5	8.5	0.866	14.72	0	0	-1	-17
y_8	240	10	-0.5	-5	-0.886	-8.66	-0.5	-5	0.866	8.66	1	10	0	0
y_9	270	-8	0	0	-1	8	-1	8	0	0	0	0	1	-8
y_{10}	300	-19	-0.5	-9.5	-0.866	16.45	-0.5	9.5	-0.866	16.45	-1	19	0	0
y_{11}	330	-17	0.866	-14.72	-0.5	8.5	0.5	-8.5	-0.866	14.72	0	0	-1	17
y_{12}	360	-13	1	-13	0	0	1	-13	0	0	1	-13	0	0
		$\sum\limits_{K=1}^{12} y_K = 85$		$\sum\limits_{K=1}^{12} y_K\cos\theta_K = -127.08$		$\sum\limits_{K=1}^{12} y_K\sin\theta_K = 88.3$		$\sum\limits_{K=1}^{12} y_K\cos 2\theta_K = -9$		$\sum\limits_{K=1}^{12} y_K\sin 2\theta_K = 1.72$		$\sum\limits_{K=1}^{12} y_K\cos 3\theta_K = 12$		$\sum\limits_{K=1}^{12} y_K\sin 3\theta_K = -9$

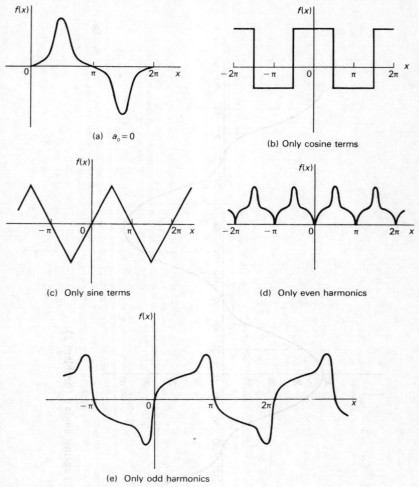

(a) $a_0 = 0$

(b) Only cosine terms

(c) Only sine terms

(d) Only even harmonics

(e) Only odd harmonics

Figure 3

Thus if the average value of $f(x)$ over one cycle is zero (i.e. the area above the x-axis is equal to the area below it) then $a_0 = 0$. Hence for a waveform of the form shown in Fig. 3(a), $a_0 = 0$.

(ii) An **even** function is defined by $f(-x) = f(x)$ and it is symmetrical about the $f(x)$-axis. **An even function contains no sine terms.** A typical even waveform is shown in Fig. 3(b).

(iii) An **odd** function is defined by $f(-x) = -f(x)$ and it is symmetrical about the origin. **An odd function contains no cosine terms.** A typical odd function is shown in Fig. 3(c).

(iv) $f(x) = f(x + \pi)$ represents a wave which repeats after half a cycle. Only even harmonics can be present in such a waveform, i.e. $a_1, a_3, a_5, \ldots b_1, b_3, b_5, \ldots$ are all zero. A typical wave of this form is shown in Fig. 3(d).

(v) $f(x) = -f(x + \pi)$ represents a wave for which the positive and negative

340

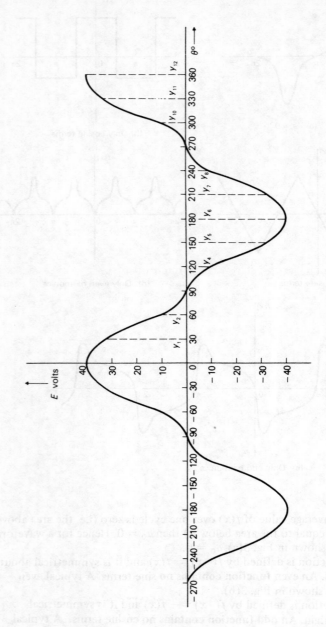

Figure 4 Graph of E volts against angle θ (problem 2)

half cycles are identical in shape. Only odd harmonics can be present in such a waveform, i.e. $a_0, a_2, a_4, \ldots b_2, b_4, \ldots$ are all zero. A typical wave of this form is shown in Fig. 3(e).

Problem 2. An alternating emf E volts is shown in Fig. 4. Analyse the waveform into its constituent harmonics as far as, and including, the fifth harmonic by taking 30° intervals.

(i) The mean value of a cycle is seen to be zero, since the area above the θ-axis is equal to the area under it. Thus the constant term, or d.c. components, $a_0 = 0$.

(ii) Since the waveform is symmetrical about the E-axis the function E is even, which means that there are no sine terms present in the Fourier series.

(iii) The waveform is of the form $f(\theta) = -f(\theta + \pi)$ which means that only odd harmonics are present.

These considerations can save a lot of time doing unnecessary calculations, for in this case we find that only odd cosine terms will appear in the Fourier series, that is, $E = a_1 \cos\theta + a_3 \cos 3\theta + a_5 \cos 5\theta + \ldots$

A proforma, similar to Table 1, but without the 'sine term' columns and without the 'even cosine term' columns is shown in Table 2 up to and including the fifth harmonic, from which the Fourier coefficients a_1, a_3 and a_5 can be determined. Twelve ordinates are chosen and labelled $y_1, y_2, \ldots y_{12}$ in Fig. 4.

Table 2

Ordinate	$\theta°$	E	$\cos\theta$	$E\cos\theta$	$\cos 3\theta$	$E\cos 3\theta$	$\cos 5\theta$	$E\cos 5\theta$
y_1	30	33	0.866	28.58	0	0	−0.866	−28.58
y_2	60	10	0.5	5	−1	−10	0.5	5
y_3	90	0	0	0	0	0	0	0
y_4	120	−10	−0.5	5	1	−10	−0.5	5
y_5	150	−33	−0.866	28.58	0	0	0.866	−28.58
y_6	180	−40	−1	40	−1	40	−1	40
y_7	210	−33	−0.866	28.58	0	0	0.866	−28.58
y_8	240	−10	−0.5	5	1	−10	−0.5	5
y_9	270	0	0	0	0	0	0	0
y_{10}	300	10	0.5	5	−1	−10	0.5	5
y_{11}	330	33	0.866	28.58	0	0	−0.866	−28.58
y_{12}	360	40	1	40	1	40	1	40
			$\sum_{K=1}^{12} E_K \cos\theta_K$ $= 214.32$		$\sum_{K=1}^{12} E_K \cos 3\theta_K$ $= 40$		$\sum_{K=1}^{12} E_K \cos 5\theta_K$ $= -14.32$	

342

From equation (3), $a_n \simeq \dfrac{2}{p} \sum\limits_{K=1}^{p} E_K \cos n\,\theta_K$, where $p = 12$

Hence
$$a_1 \simeq \frac{2}{12} \,(214.32) = 35.72$$

$$a_3 \simeq \frac{2}{12} \,(40) = 6.67$$

$$a_5 \simeq \frac{2}{12} \,(-14.32) = -2.39$$

Thus $E = 35.72 \cos\theta + 6.67 \cos 3\theta - 2.39 \cos 5\theta + \ldots$

Further problems on harmonic analysis on data given in tabular and graphical form may be found in the following Section (4) (Problems 1 to 10).

4 Further problems

1. Without performing calculations, state which harmonics will be present in the waveforms shown in Fig. 5.

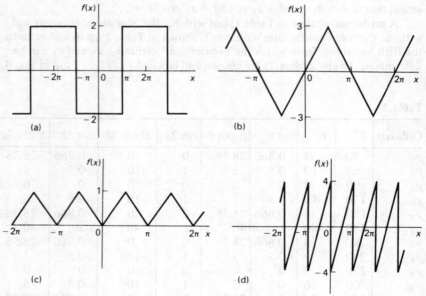

Figure 5

(a) [Only odd cosine terms present]   (b) [Only odd sine terms present]
(c) [Only even harmonics present]   (d) [Only even sine terms present]

Find the Fourier series to represent the periodic functions given by the tables of values in problems 2 to 4 up to and including the third harmonic and each coefficient correct to two decimal places. Use 12 ordinates in each case.

2.

θ	0	30	60	90	120	150	180	210	240	270	300	330
$f(\theta)$	−1.7	2.9	4.4	4.7	5.3	6.1	4.9	1.5	−2.9	−7.0	−8.5	−5.4.

$$[f(\theta) = 0.36 - 3.10 \cos \theta + 6.07 \sin \theta + 1.48 \cos 2\theta + 1.21 \sin 2\theta$$
$$- 0.02 \cos 3\theta + 0.20 \sin 3\theta]$$

3.

θ	30	60	80	120	150	180	210	240	270	300	330	360
y	14.6	17.2	21.6	28.8	42.3	51.6	62.1	49.3	31.6	21.4	9.1	7.4.

$$[y = 29.75 + 22.31 \cos \theta - 6.42 \sin \theta + 1.92 \cos 2\theta + 6.01 \sin 2\theta$$
$$- 0.78 \cos 3\theta - 0.72 \sin 3\theta]$$

4.

θ	0	30	60	90	120	150	180	210	240	270	300	330
I	11	2	−2.1	−21	−12	10	29	40	43	40	18	13.

$$[I = 12.67 - 10.89 \cos \theta - 27.15 \sin \theta + 6.58 \cos 2\theta + 5.05 \sin 2\theta$$
$$+ 2.67 \cos 3\theta + 3.33 \sin 3\theta]$$

5. Analyse the periodic waveform of y against angle θ in Fig. 6(a) into its

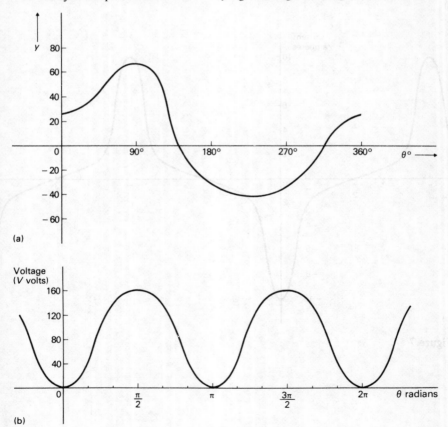

(a)

(b)

Figure 6

344

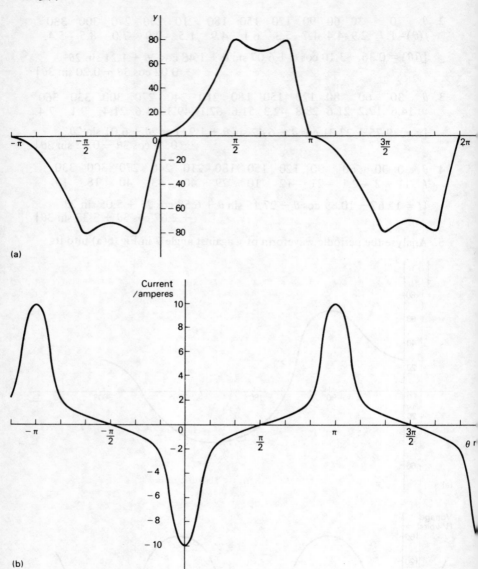

(a)

(b)

Figure 7

constituent harmonics as far as and including the third harmonic by taking 30° intervals.

$[y = 6.75 + 25.81 \cos \theta + 44.51 \sin \theta - 10.67 \cos 2\theta - 4.62 \sin 2\theta + 4.17 \cos 3\theta - 8.33 \sin 3\theta]$

6. For the waveform shown in Fig. 6(b) state why only a d.c. component and even cosine terms will be present in the Fourier series and determine the first three terms of the series using 30° intervals.
$[V = 86.67 - 86.67 \cos 2\theta - 6.67 \cos 4\theta]$

7. Find a Fourier series as far as the third harmonic to represent the periodic function y given by the waveform in Fig. 7(a). Use 30° intervals.
$[y = -25.21 \cos \theta + 73.42 \sin \theta + 10.00 \cos 3\theta + 3.33 \sin 3\theta]$

8. For the waveform shown in Fig. 7(b) state why only odd cosine harmonics appear in the Fourier series. By taking 12 intervals analyse the waveform and find the first three terms of the Fourier series.
$[I = -4.75 \cos \theta - 2.80 \cos 3\theta + 2.45 \cos 5\theta]$

9. Displacement y on a point on a pulley turned through an angle of θ degrees is given by:

θ	30	60	90	120	150	180	210	240	270
y	7.976	8.026	7.204	5.676	3.674	1.764	0.552	0.262	0.904

θ	300	330	360
y	2.492	4.736	6.824

Construct a Fourier series for the first three harmonics.
$[y = 4.17 + 2.45 \cos \theta + 0.12 \cos 2\theta + 0.08 \cos 3\theta + 3.16 \sin \theta + 0.03 \sin 2\theta + 0.01 \sin 3\theta]$

10. The anode current pulse in amperes in a class C power amplifier is given by:

$i = -1 + 2 \cos wt$, when i is positive. At all other times $i = 0$. Perform a Fourier analysis on the waveform up to and including the third harmonic.

$[i = 10^{-3} (215 + 388 \cos wt + 279 \cos 2 wt + 144 \cos 3 wt]$

Chapter 27

Introduction to Laplace transforms

1 Introduction

The solution of most electrical circuit problems reduces ultimately to the solution of differential equations. Methods of solving certain first and second order differential equations are discussed in Chapters 17 to 21. The operation, called the Laplace transformation, provides a powerful and much used alternative method of solving linear differential equations. The method has many practical applications, including aspects of control engineering and coupled electrical circuits, beam problems, various electrical systems, mechanical vibration systems and in servomechanisms, and was developed by the French mathematician Pierre Simon de Laplace (1749–1827).

Before using Laplace transforms to solve differential equations it is necessary to define the Laplace transform, to derive the transforms of elementary functions, to appreciate certain of their properties and to define and derive the inverse Laplace transform, including the use of partial fractions.

In this chapter the Laplace transform is defined and the use of a table of standard transforms to evaluate the transforms of elementary functions is discussed.

Some simple properties of Laplace transforms are discussed in Chapter 28.

In Chapter 29 the inverse Laplace transform is defined and determined for elementary functions using partial fractions and a table of standard transforms. Also included in Chapter 29 is an introduction to solving linear second order differential equations with constant coefficients.

2 Definition of the Laplace transform

Let $f(t)$ be a given function of time which is defined for all positive values of t. Let $f(t)$ be multiplied by e^{-st} (where s is a parameter introduced by us and assumed to be a real number) and then the product integrated with respect to t from zero to infinity. Then, if the resulting integral exists, it is a function of s, say $F(s)$.

That is, $F(s) = \int_0^\infty e^{-st} f(t)\, dt$

The function $F(s)$ is called the **Laplace transform** of the original function $f(t)$. There are various commonly used notations for the Laplace transform of $f(t)$ and these include:

(i) $\mathcal{L}(f)$ or $L f$

(ii) $\mathcal{L}\{f(t)\}$ or $L \{f(t)\}$

(iii) $\bar{f}(s)$ or $\tilde{f}(s)$

Also, in some texts, the letter p is used instead of s as the parameter.

The notation adopted in this text will be to use small letters for the original function (i.e. $f(t)$) and $\mathcal{L}\{f(t)\}$ for its Laplace transform.

Hence $F(s) = \mathcal{L}\{f(t)\} = \int_0^\infty e^{-st} f(t)\, dt$ (1)

where $\mathcal{L}\{\ \}$ represents symbolically the operation of taking the Laplace transform of whatever function occurs inside the bracket. The operation on the original function is called the **Laplace transformation**.

3 Linearity property of the Laplace transform

From equation (1), $\mathcal{L}\{k f(t)\} = \int_0^\infty e^{-st} k f(t)\, dt = k \int_0^\infty e^{-st} f(t)\, dt$

i.e. $\mathcal{L}\{k f(t)\} = k \mathcal{L}\{f(t)\}$ (2)

where k is any constant.

Similarly, $\mathcal{L}\{a f(t) + b g(t)\} = \int_0^\infty e^{-st} [a f(t) + b g(t)]\, dt$

$$= a \int_0^\infty e^{-st} f(t)\, dt + b \int_0^\infty e^{-st} g(t)\, dt$$

i.e. $\mathcal{L}\{a f(t) + b g(t)\} = a \mathcal{L}\{f(t)\} + b \mathcal{L}\{g(t)\}$ (3)

where a and b are any constants.

Because of the properties shown in equations (2) and (3) the Laplace transform in termed a **linear operator**.

4 Laplace transforms of elementary functions

Using the definition of the Laplace transform given in equation (1) a number
of elementary functions may be transformed.

(i) f(t) = 1

From equation (1) $\mathcal{L}\{1\} = \int_0^\infty e^{-st} (1)\, dt = \left[\dfrac{e^{-st}}{-s}\right]_0^\infty$

$$= -\frac{1}{s}\, [e^{-s(\infty)} - e^0]$$

$$= -\frac{1}{s}\, [0-1]$$

$$= \frac{1}{s}$$

(provided $s > 0$, so that $e^{-s(\infty)} = 0$.)

Hence the Laplace transform of 1 is $\dfrac{1}{s}$.

(ii) f(t) = k (where k is a constant)

From equation (1), $\mathcal{L}\{k\} = \int_0^\infty e^{-st} k\, dt$

$k\,\mathcal{L}\{1\} = k\int_0^\infty e^{-st}\, dt$, from equation (2)

$$= k\left(\frac{1}{s}\right) \text{, from (i) above.}$$

Hence the Laplace transform of k is $\dfrac{k}{s}$ (provided $s > 0$).

(iii) f(t) = e^{at} (where a is a real constant ≠ 0)

From equation (1), $\mathcal{L}\{e^{at}\} = \int_0^\infty e^{-st} (e^{at})\, dt$

$$= \int_0^\infty e^{-(s-a)t}\, dt$$

$$= \left[\frac{e^{-(s-a)t}}{-(s-a)}\right]_0^\infty$$

$$= \frac{1}{-(s-a)}\, [0-1]$$

$$= \frac{1}{s-a} \text{ (provided } s-a > 0\text{, i.e. } s > a)$$

Hence the Laplace transform of e^{at} is $\dfrac{1}{s-a}$.

(iv) f(t) = sin at (where a is a real constant)

From equation (1), $\mathcal{L}\{\sin at\} = \int_0^\infty e^{-st} \sin at \, dt$

$$= \left[\frac{e^{-st}}{s^2 + a^2} \left(-s \sin at - a \cos at \right) \right]_0^\infty$$

This is obtained by integrating by parts twice (see Chapter 16, Problem 8, page 224).

Hence $\mathcal{L}\{\sin at\} = \frac{1}{s^2 + a^2} [e^{-s(\infty)} \left(-s \sin a (\infty) - a \cos a (\infty) \right)$
$\qquad\qquad\qquad\qquad - e^0 \left(-s \sin 0 - a \cos o \right)]$

$$= \frac{1}{s^2 + a^2} [0 - 1 (-a)] = \frac{a}{s^2 + a^2} \text{ (provided } s > 0)$$

Hence the Laplace transform of sin *at* is $\dfrac{a}{s^2 + a^2}$.

(v) f(t) = cos at (where a is a real constant)

From equation (1), $\mathcal{L}\{\cos at\} = \int_0^\infty e^{-st} \cos at \, dt$

$$= \left[\frac{e^{-st}}{s^2 + a^2} \left(a \sin at - s \cos at \right) \right]_0^\infty$$

by integrating by parts twice

$$= \frac{s}{s^2 + a^2} \text{ (provided } s > 0)$$

Hence the Laplace transform of cos *at* is $\dfrac{s}{s^2 + a^2}$

(vi) f(t) = t

From equation (1), $\mathcal{L}\{t\} = \int_0^\infty e^{-st} t \, dt$

$$= \left[\frac{t e^{-st}}{-s} \right]_0^\infty - \int_0^\infty \frac{e^{-st}}{-s} \, dt, \text{ by integration by parts}$$

$$= \left[\frac{\infty e^{-s(\infty)}}{-s} - 0 \right] - \left[\frac{e^{-st}}{s^2} \right]_0^\infty$$

$$= [0 - 0] - \frac{1}{s^2} [e^{-s(\infty)} - e^0], \text{ since } \infty \times 0 = 0.$$

$$= -\frac{1}{s^2} (0 - 1) = \frac{1}{s^2} \text{ (provided } s > 0)$$

Hence the Laplace transform of *t* is $\dfrac{1}{s^2}$.

(vii) f(t) = t²

From equation (1), $\mathcal{L}\{t^2\} = \int_0^\infty e^{-st} t^2 \, dt$

$$= \left[\frac{t^2 e^{-st}}{-s} - \frac{2t e^{-st}}{s^2} - \frac{2 e^{-st}}{s^3} \right]_0^\infty$$

by integration by parts twice

$$= (0 - 0 - 0) - (0 - 0 - \frac{2}{s^3})$$

$$= \frac{2}{s^3} \text{ (provided } s > 0)$$

Hence the Laplace transform of t^2 is $\frac{2}{s^3}$.

(viii) f (t) = t³

From equation (1), $\mathcal{L}\{t^3\} = \int_0^\infty e^{-st} t^3 \, dt$

$$= \left[\frac{t^3 e^{-st}}{-s} - \frac{3t^2 e^{-st}}{s^2} - \frac{(3)(2) t e^{-st}}{s^3} \right.$$
$$\left. - \frac{(3)(2) e^{-st}}{s^4} \right]_0^\infty$$

$$= \frac{(3)(2)}{s^4} = \frac{3!}{s^4} \text{ (provided } s > 0)$$

Hence the Laplace transform of t^3 is $\frac{3!}{s^4}$.

(ix) f(t) = tⁿ (where n = 0, 1, 2, 3, . . .)
The results of (vi), (vii) and (viii) can be extended to n being any positive integer, giving:

$$\mathcal{L}\{t^n\} = \frac{n!}{s^{n+1}} \text{ (provided } s > 0)$$

Thus, for example, $\mathcal{L}\{t^4\} = \frac{4!}{s^{4+1}} = \frac{(4)(3)(2)(1)}{s^5} = \frac{24}{s^5}$

and $\mathcal{L}\{t^5\} = \frac{5!}{s^{5+1}} = \frac{(5)(4)(3)(2)(1)}{s^6} = \frac{120}{s^6}$

(x) f(t) = cosh at

$\cosh at = \frac{1}{2}(e^{at} + e^{-at})$

Hence $\mathcal{L}\{\cosh at\} = \mathcal{L}\{\frac{1}{2}(e^{at} + e^{-at})\}$

$$= \frac{1}{2}\mathcal{L}\{e^{at}\} + \frac{1}{2}\mathcal{L}\{e^{-at}\}, \text{ from equations (2) and (3)}$$

$$= \frac{1}{2}\left(\frac{1}{s-a}\right) + \frac{1}{2}\left(\frac{1}{s--a}\right) \text{, from (iii) above}$$

$$= \frac{1}{2}\left[\frac{1}{s-a} + \frac{1}{s+a}\right] = \frac{1}{2}\left[\frac{(s+a)+(s-a)}{(s-a)(s+a)}\right]$$

$$= \frac{s}{s^2 - a^2} \quad \text{(provided } s > a)$$

Hence the Laplace transform of cosh $at = \dfrac{s}{s^2 - a^2}$

(xi) f(t) = sinh at

$$\sinh at = \frac{1}{2}(e^{at} - e^{-at})$$

Hence $\mathcal{L}\{\sinh at\} = \dfrac{1}{2}\mathcal{L}\{e^{at}\} - \dfrac{1}{2}\mathcal{L}\{e^{-at}\}$, from equations (2) and (3)

$$= \frac{1}{2}\left(\frac{1}{s-a}\right) - \frac{1}{2}\left(\frac{1}{s+a}\right) = \frac{1}{2}\left(\frac{1}{s-a} - \frac{1}{s+a}\right)$$

$$= \frac{a}{s^2 - a^2} \quad \text{(provided } s > a)$$

Hence the Laplace transform of sinh $at = \dfrac{a}{s^2 - a^2}$

For ease of reference, the above Laplace transforms are summarised in Table 27.1.

With a working knowledge of the transforms listed in Table 27.1 nearly all of the transforms needed in practice can be obtained through the use of simple general properties, one of these being considered in Chapter 28.

Table 27.1 Elementary standard Laplace transforms

	Function $f(t)$	Laplace transforms $\mathcal{L}[f(t)] = \int_0^\infty e^{-st} f(t)\, dt$
(i)	1	$\dfrac{1}{s}$
(ii)	k	$\dfrac{k}{s}$
(iii)	e^{at}	$\dfrac{1}{s-a}$
(iv)	$\sin at$	$\dfrac{a}{s^2 + a^2}$
(v)	$\cos at$	$\dfrac{s}{s^2 + a^2}$
(vi)	t	$\dfrac{1}{s^2}$

	Function $f(t)$	Laplace transforms $\mathcal{L}[f(t)] = \int_0^\infty e^{-st} f(t)\, dt$
(vii)	t^2	$\dfrac{2!}{s^3}$
(viii)	t^3	$\dfrac{3!}{s^4}$
(ix)	t^n (n = positive integer)	$\dfrac{n!}{s^{n+1}}$
(x)	$\cosh at$	$\dfrac{s}{s^2 - a^2}$
(xi)	$\sinh at$	$\dfrac{a}{s^2 - a^2}$

Worked problems on standard Laplace transforms

Problem 1. Find the Laplace transforms of: (a) $1 + 3t - 2t^2 + \dfrac{t^6}{12}$, (b) $3e^{2t} - 4e^{-5t}$, (c) $2 \sin 4t - 3 \cos 2t$.

(a) $\mathcal{L}\left\{1 + 3t - 2t^2 + \dfrac{t^6}{12}\right\} = \mathcal{L}\{1\} + 3\mathcal{L}\{t\} - 2\mathcal{L}\{t^2\} + \dfrac{1}{12}\mathcal{L}\{t^6\}$

from equations (2) and (3)

$= \dfrac{1}{s} + \dfrac{3}{s^2} - \dfrac{2 \cdot 2}{s^3} + \dfrac{6 \cdot 5 \cdot 4 \cdot 3 \cdot 2 \cdot 1}{12\, s^7}$

from (1), (vi), (vii) and (ix) of Table 27.1.

$= \dfrac{1}{s} + \dfrac{3}{s^2} - \dfrac{4}{s^3} + \dfrac{60}{s^7}$

(b) $\mathcal{L}\{3e^{2t} - 4e^{-5t}\}$

$= 3\mathcal{L}\{e^{2t}\} - 4\mathcal{L}\{e^{-5t}\}$ from equations (2) and (3)

$= \dfrac{3}{(s-2)} - \dfrac{4}{(s--5)}$, from (iii) of Table 27.1

$= \dfrac{3(s+5) - 4(s-2)}{(s-2)(s+5)} = \dfrac{23-s}{(s-2)(s+5)}$

(c) $\mathcal{L}\{2 \sin 4t - 3 \cos 2t\}$ $= 2\mathcal{L}\{\sin 4t\} - 3\mathcal{L}\{\cos 2t\}$, from equations (2) and (3)

$= 2\left(\dfrac{4}{s^2 + 4^2}\right) - 3\left(\dfrac{s}{s^2 + 2^2}\right)$, from (iv) and (v) of Table 27.1

$$= \frac{8}{(s^2 + 16)} - \frac{3s}{(s^2 + 4)}$$

Problem 2. Find the Laplace transforms of:

(a) $x^5 - 2x^4$,　(b) $\cosh \theta - \sinh 2\theta$,　(c) $\cos^2 t$

(a) $\mathcal{L}\{x^5 - 2x^4\} = \mathcal{L}\{x^5\} - 2\mathcal{L}\{x^4\}$

$$= \frac{5!}{s^6} - 2\frac{4!}{s^5} \quad \text{from (ix) of Table 27.1}$$

$$= \frac{(5 \cdot 4 \cdot 3 \cdot 2 \cdot 1) - 2x(4 \cdot 3 \cdot 2 \cdot 1)}{s^6} = \frac{120 - 48x}{s^6}$$

$$= \frac{24(5 - 2x)}{s^6}$$

(b) $\mathcal{L}\{\cosh \theta - \sinh 2\theta\} = \mathcal{L}\{\cosh \theta\} - \mathcal{L}\{\sinh 2\theta\}$

$$= \frac{s}{s^2 - 1^2} - \frac{2}{s^2 - 2^2} \quad \text{from (x) and (xi) of}$$
$$\text{Table 27.1}$$

$$= \frac{s(s^2 - 4) - 2(s^2 - 1)}{(s^2 - 1)(s^2 - 4)} = \frac{s^3 - 2s^2 - 4s + 2}{(s^2 - 1)(s^2 - 4)}$$

(c) $\mathcal{L}\{\cos^2 t\} = \mathcal{L}\{\frac{1}{2}(1 + \cos 2t)\}$

$$= \frac{1}{2}\mathcal{L}\{1\} + \frac{1}{2}\mathcal{L}\{\cos 2t\}$$

$$= \frac{1}{2}\left(\frac{1}{s}\right) + \frac{1}{2}\left(\frac{s}{s^2 + 4}\right) \quad \text{from (i) and (v) of Table 27.1}$$

$$= \frac{(s^2 + 4) + (s^2)}{2s(s^2 + 4)} = \frac{2(s^2 + 2)}{2s(s^2 + 4)} = \frac{(s^2 + 2)}{s(s^2 + 4)}$$

Problem 3. Find the Laplace transforms of: (a) $\sinh^2 2t$,　(b) $2\sin(\omega t + \alpha)$, where ω and α are constants.

(a) $\mathcal{L}\{\sinh^2 2t\} = \mathcal{L}\{\frac{1}{2}(\cosh 4t - 1)\}$ since $\cosh 2t = 1 + 2\sinh^2 t$ (see Chapter 7, Section 5)

$$= \frac{1}{2}\mathcal{L}\{\cosh 4t\} - \frac{1}{2}\mathcal{L}\{1\}$$

$$= \frac{1}{2}\left(\frac{s}{s^2 - 16}\right) - \frac{1}{2}\left(\frac{1}{s}\right) \quad \text{from (x) and (i) of Table 27.1}$$

$$= \frac{(s^2) - (s^2 - 16)}{2s(s^2 - 16)} = \frac{16}{2s(s^2 - 16)} = \frac{8}{s(s^2 - 16)}$$

(b) $\mathcal{L}\{2 \sin (\omega t + \alpha)\} = \mathcal{L}\{2 (\sin \omega t \cos \alpha + \cos \omega t \sin \alpha)\}$

$$= (2 \cos \alpha) \mathcal{L}\{\sin \omega t\} + (2 \sin \alpha) \mathcal{L}\{\cos \omega t\}$$

$$= (2 \cos \alpha) \left(\frac{\omega}{s^2 + \omega^2}\right) + (2 \sin \alpha) \left(\frac{s}{s^2 + \omega^2}\right)$$

from (iv) and (v) of Table 27.1

$$= \frac{2}{s^2 + \omega^2} (\omega \cos \alpha + s \sin \alpha)$$

Further problems on standard Laplace transforms may be found in the following Section (5) (Problems 1 to 10).

5 Further problems

In Problems 1 to 9 find the Laplace transforms.

1. (a) $3t - 5$ (b) $4t^2 + 2t - 1$. (a) $\left[\dfrac{3}{s^2} - \dfrac{5}{s}\right]$ (b) $\left[\dfrac{8}{s^3} + \dfrac{2}{s^2} - \dfrac{1}{s}\right]$

2. (a) $3t^3 - 4t + 7$ (b) $t^6 - 3t^4 + t^2$.

   (a) $\left[\dfrac{18}{s^4} - \dfrac{4}{s^2} + \dfrac{7}{s}\right]$ (b) $\left[\dfrac{720}{s^7} - \dfrac{72}{s^5} + \dfrac{2}{s^3}\right]$

3. (a) $2 e^{4t}$ (b) $3 e^{-2t}$ (a) $\left[\dfrac{2}{s-4}\right]$ (b) $\left[\dfrac{3}{s+2}\right]$

4. (a) $5 \sin 2t$ (b) $4 \cos 3t$. (a) $\left[\dfrac{10}{s^2+4}\right]$ (b) $\left[\dfrac{4s}{s^2+9}\right]$

5. (a) $2 \cosh 2t$ (b) $3 \sinh t$. (a) $\left[\dfrac{2s}{s^2-4}\right]$ (b) $\left[\dfrac{3}{s^2-1}\right]$

6. (a) $\sin^2 t$ (b) $\sin t \cos t$. (a) $\left[\dfrac{2}{s(s^2+4)}\right]$ (b) $\left[\dfrac{1}{s^2+4}\right]$

7. (a) $\cosh^2 \theta$ (b) $\sinh^2 3\theta$. (a) $\left[\dfrac{s^2-2}{s(s^2-4)}\right]$ (b) $\left[\dfrac{18}{s(s^2-36)}\right]$

8. (a) $3 \sin (at + b)$, where a and b are constants,
   (b) $4 \cos (wt - \alpha)$, where w and α are constants.

   (a) $\left[\dfrac{3}{s^2+a^2} (a \cos b + s \sin b)\right]$

   (b) $\left[\dfrac{4}{s^2+w^2} (s \cos \alpha + w \sin \alpha)\right]$

9. (a) $\sin \left(t + \dfrac{\pi}{4}\right)$ (b) $2 \cos (2t + \beta)$, where β is a constant.

   (a) $\left[\dfrac{s+1}{\sqrt{2}(s^2+1)}\right]$

   (b) $\left[\dfrac{2}{s^2+4} (s \cos \beta + 2 \sin \beta)\right]$

10. Show that (a) $\mathcal{L}\left\{\dfrac{1}{w}\sin wt - \dfrac{1}{\beta}\sin\beta t\right\} = \dfrac{\beta^2 - w^2}{(s^2+w^2)(s^2+\beta^2)}$

(b) $\mathcal{L}\left\{\cos^2 2t - \sin^2 2t\right\} = \dfrac{s}{(s^2+16)}$.

Chapter 28

Properties of Laplace transforms

1 The Laplace transform of $e^{at} f(t)$

In Chapter 27 the Laplace transforms of elementary functions were considered. Further transforms can be obtained by the following important property of the Laplace transformation.

$$\text{If } F(s) = \mathcal{L}\{f(t)\} = \int_0^\infty e^{-st} f(t) \, dt \tag{1}$$

$$\text{then } \mathcal{L}\{e^{at} f(t)\} = \int_0^\infty e^{-st} (e^{at} f(t)) \, dt$$

$$= \int_0^\infty e^{-(s-a)t} f(t) \, dt$$

$$\text{i.e. } \mathcal{L}\{e^{at} f(t)\} = F(s-a) \tag{2}$$

where a is a real constant.

Hence the substitution of $s - a$ for s in the transform shown in equation (1) corresponds to the multiplication of the original function by e^{at}. This is often known as a **shift theorem**.

If a is negative it follows that:

$$\mathcal{L}\{e^{-at} f(t)\} = F(s+a) \tag{3}$$

2 Laplace transforms of functions of the type $e^{at} f(t)$

Laplace transforms of functions of the type $e^{at} f(t)$, where a is a positive real constant, may be obtained by using Table 27.1 and equations (2) and (3) above.

(i) $\mathcal{L}\{e^{at}\, t^n\}$

Since $\mathcal{L}\{t^n\}$ $= \dfrac{n!}{s^{n+1}}$, from (ix) of Table 27.1

then $\mathcal{L}\{e^{at}\, t^n\}$ $= \dfrac{n!}{(s-a)^{n+1}}$, from equation (2) (provided $s > a$)

and $\mathcal{L}\{e^{-at}\, t^n\}$ $= \dfrac{n!}{(s+a)^{n+1}}$, from equation (3) (provided $s > -a$)

(ii) $\mathcal{L}\{e^{at}\, \sin \omega t\}$

Since $\mathcal{L}\{\sin \omega t\}$ $= \dfrac{\omega}{s^2 + \omega^2}$, from (iv) of Table 27.1

then $\mathcal{L}\{e^{at} \sin \omega t\}$ $= \dfrac{\omega}{(s-a)^2 + \omega^2}$, from equation (2) (provided $s > a$)

and $\mathcal{L}\{e^{-at} \sin \omega t\}$ $= \dfrac{\omega}{(s+a)^2 + \omega^2}$, from equation (3) (provided $s > -a$)

(iii) $\mathcal{L}\{e^{at} \cos \omega t\}$

Since $\mathcal{L}\{\cos \omega t\}$ $= \dfrac{s}{s^2 + \omega^2}$, from (v) of Table 27.1

then $\mathcal{L}\{e^{at} \cos \omega t\}$ $= \dfrac{s-a}{(s-a)^2 + \omega^2}$, from equation (2) (provided $s > a$)

and $\mathcal{L}\{e^{-at} \cos \omega t\}$ $= \dfrac{s+a}{(s+a)^2 + \omega^2}$, from equation (3) (provided $s > -a$)

(iv) $\mathcal{L}\{e^{at} \cosh \omega t\}$

Since $\mathcal{L}\{\cosh \omega t\}$ $= \dfrac{s}{s^2 - \omega^2}$, from (x) of Table 27.1

then $\mathcal{L}\{e^{at} \cosh \omega t\}$ $= \dfrac{s-a}{(s-a)^2 - \omega^2}$, from equation (2) (provided $s > a$)

and $\mathcal{L}\{e^{-at} \cosh \omega t\}$ $= \dfrac{s+a}{(s+a)^2 - \omega^2}$, from equation (3) (provided $s > -a$)

(v) $\mathcal{L}\{e^{at} \sinh \omega t\}$

Since $\mathcal{L}\{\sinh \omega t\}$ $= \dfrac{\omega}{s^2 - \omega^2}$, from (xi) of Table 27.1

then $\mathcal{L}\{e^{at} \sinh \omega t\}$ $= \dfrac{\omega}{(s-a)^2 - \omega^2}$, (from equation (2) (provided $s > a$)

and $\mathcal{L}\{e^{-at} \sinh \omega t\}$ $= \dfrac{\omega}{(s+a)^2 - \omega^2}$, from equation (3) (provided $s > -a$)

For ease of reference the above Laplace transforms are summarised in Table 28.1.

Table 28.1 Laplace transforms of the form $e^{at} f(t)$

Function	Laplace transform
$e^{at} f(t)$	$\mathcal{L}\{e^{at} f(t)\}$
(a is a positive real constant)	

(i)	$e^{at} t^n$	$\dfrac{n!}{(s-a)^{n+1}}$
	$e^{-at} t^n$	$\dfrac{n!}{(s+a)^{n+1}}$
(ii)	$e^{at} \sin \omega t$	$\dfrac{\omega}{(s-a)^2 + \omega^2}$
	$e^{-at} \sin \omega t$	$\dfrac{\omega}{(s+a)^2 + \omega^2}$
(iii)	$e^{at} \cos \omega t$	$\dfrac{s-a}{(s-a)^2 + \omega^2}$
	$e^{-at} \cos \omega t$	$\dfrac{s+a}{(s+a)^2 + \omega^2}$
(iv)	$e^{at} \cosh \omega t$	$\dfrac{s-a}{(s-a)^2 - \omega^2}$
	$e^{-at} \cosh \omega t$	$\dfrac{s+a}{(s+a)^2 - \omega^2}$
(v)	$e^{at} \sinh \omega t$	$\dfrac{\omega}{(s-a)^2 - \omega^2}$
	$e^{-at} \sinh \omega t$	$\dfrac{\omega}{(s+a)^2 - \omega^2}$

Worked problems on Laplace transforms of functions of the type $e^{at} f(t)$

Problem 1. Find the Laplace transforms of: (a) $t^3 e^{2t}$, (b) $e^{3t} \sin 4t$
(c) $2 e^{-2t} \cos 2t$

(a) $\mathcal{L}\{t^3\} \quad = \dfrac{3!}{s^4} = \dfrac{6}{s^4}$, from (viii) of Table 27.1.

Hence $\mathcal{L}\{t^3 e^{2t}\} \quad = \dfrac{6}{(s-2)^4}$, from (i) of Table 28.1.

(b) $\mathcal{L}\{\sin 4t\} = \dfrac{4}{s^2 + 4^2} = \dfrac{4}{s^2 + 16}$, from (iv) of Table 27.1.

Hence $\mathcal{L}\{e^{3t} \sin 4t\}$ $= \dfrac{4}{(s-3)^2 + 16}$, from (ii) of Table 28.1.

$$= \dfrac{4}{s^2 - 6s + 25}$$

(c) $\mathcal{L}\{2 \cos 2t\}$ $= \dfrac{2s}{s^2 + 4}$, from (v) of Table 27.1

Hence $\mathcal{L}\{2e^{-2t} \cos 2t\} = \dfrac{2(s+2)}{(s+2)^2 + 4}$, from (iii) of Table 28.1

$$= \dfrac{2(s+2)}{s^2 + 4s + 8}$$

Problem 2. Find the Laplace transforms of: (a) $2e^{4t} \sin^2 t$ (b) $e^{-3t} \cosh 2t$ (c) $e^{4t}(2\cos 3t - 3\sin 3t)$

(a) $\sin^2 t = \dfrac{1}{2}(1 - \cos 2t)$ since $\cos 2t = 1 - 2\sin^2 t$

Hence $\mathcal{L}\{2e^{4t} \sin^2 t\} = \mathcal{L}\{2e^{4t} \dfrac{1}{2}(1 - \cos 2t)\}$

$$= \mathcal{L}\{e^{4t}\} - \mathcal{L}\{e^{4t} \cos 2t\}$$

$$= \dfrac{1}{s-4} - \dfrac{s-4}{(s-4)^2 + 4} \quad \text{from (iii) of Table}$$

$$\text{27.1 and (iii) of Table 28.1}$$

$$= \dfrac{1}{s-4} - \dfrac{s-4}{s^2 - 8s + 20}$$

(b) $\mathcal{L}\{e^{-3t} \cosh 2t\} = \dfrac{s+3}{(s+3)^2 - 2^2}$ from (iv) of Table 28.1

$$= \dfrac{s+3}{s^2 + 6s + 5}$$

(c) $\mathcal{L}\{e^{4t}(2\cos 3t - 3\sin 3t)\} = 2\mathcal{L}\{e^{4t} \cos 3t\} - 3\mathcal{L}\{e^{4t} \sin 3t\}$

$$= \dfrac{2(s-4)}{(s-4)^2 + 3^2} - \dfrac{3(3)}{(s-4)^2 + 3^2}$$

$$\text{from (iii) and (ii) of Table 28.1}$$

$$= \dfrac{2s - 8 - 9}{s^2 - 8s + 25} = \dfrac{2s - 17}{s^2 - 8s + 25}$$

Further problems on Laplace transforms of functions of the type $e^{at} f(t)$ may be found in Section 5 (Problems 1 to 10), page 365.

3 The Laplace transforms of derivatives

First derivative
Let the first derivative of $f(t)$ be $f'(t)$.

Since $\mathcal{L}\{f(t)\} = \int_0^\infty e^{-st} f(t)\, dt$ by definition,

then $\mathcal{L}\{f'(t)\} = \int_0^\infty e^{-st} f'(t)\, dt$

When integrating by parts, $\int u \dfrac{dv}{dt}\, dt = uv - \int v \dfrac{du}{dt}\, dt$ (see Chapter 16).

When evaluating $\int_0^\infty e^{-st} f'(t)\, dt$, let $u = e^{-st}$ and $\dfrac{dv}{dt} = f'(t)$,

from which $\dfrac{du}{dt} = -s\,e^{-st}$ and $v = \int f'(t)\, dt = f(t)$.

$$\begin{aligned}
\text{Hence } \int_0^\infty e^{-st} f'(t)\, dt &= \left[e^{-st} f(t)\right]_0^\infty - \int_0^\infty f(t)\,(-s\,e^{-st})\, dt \\
&= [0 - f(0)] + s\int_0^\infty e^{-st} f(t)\, dt \\
&= -f(0) + s\,\mathcal{L}\{f(t)\},
\end{aligned}$$

assuming that $e^{-st} f(t)$ approaches zero as t approaches zero, and $f(0)$ is the value of $f(t)$ at $t = 0$.

$$\left.\begin{aligned}
\text{Hence } \mathcal{L}\{f'(t)\} &= s\,\mathcal{L}\{f(t)\} - f(0) \\
\text{of alternatively, } \mathcal{L}\left\{\dfrac{dy}{dx}\right\} &= s\,\mathcal{L}\{y\} - y(0)
\end{aligned}\right\} \qquad (4)$$

where $y(0)$ is the value of y at $x = 0$.

Second derivative
Let the second derivative of $f(t)$ be $f''(t)$.

By definition, $\mathcal{L}\{f''(t)\} = \int_0^\infty e^{-st} f''(t)\, dt$

Integration by parts gives:

$$\begin{aligned}
\int_0^\infty e^{-st} f''(t)\, dt &= \left[e^{-st} f'(t)\right]_0^\infty + s\int_0^\infty e^{-st} f'(t)\, dt \\
&= (0 - f'(0)) + s\,\mathcal{L}\{f'(t)\}, \text{ assuming } e^{-st} f'(t) \text{ approaches} \\
&\quad\ \text{zero as } t \text{ approaches zero, and } f'(0) \text{ is the value of } f'(t) \\
&\quad\ \text{at } t = 0, \\
&= -f'(0) + s\,[s\,\mathcal{L}\{f(t)\} - f(0)], \text{ from equation 4.}
\end{aligned}$$

Hence $\mathcal{L}\{f''(t)\} = s^2 \mathcal{L}\{f(t)\} - sf(0) - f'(0)$

or alternatively, $\mathcal{L}\left\{\dfrac{d^2y}{dx^2}\right\} = s^2 \mathcal{L}\{y\} - sy(0) - y'(0)$

$$\left. \right\} \quad (5)$$

where $y'(0)$ is the value of $\dfrac{dy}{dx}$ at $x = 0$.

Higher derivatives

Laplace transforms of higher derivatives may be determined in the same way as above, and in general it is found:

$$\mathcal{L}\{f^n(t)\} = s^n \mathcal{L}\{f(t)\} - s^{n-1} f(0) - s^{n-2} f'(0) \ldots - f^{n-1}(0)$$

or $\mathcal{L}\left\{\dfrac{d^n y}{dx^n}\right\} = s^n \mathcal{L}\{y\} - s^{n-1} y(0) - s^{n-2} y'(0) \ldots y^{n-1}(0)$

$$\left. \right\} (6)$$

Equations (4), (5) and (6) are very important and are used in the solution of differential equations (see Chapter 29, Section 4).

Worked problems on the Laplace transforms of derivatives

Problem 1. Use the Laplace transform of the first derivative (i.e. equation (4)) to derive the following:

(a) $\mathcal{L}\{1\} = \dfrac{1}{s}$ (b) $\mathcal{L}\{t\} = \dfrac{1}{s^2}$ (c) $\mathcal{L}\{e^{at}\} = \dfrac{1}{s-a}$

(a) From equation (4), $\mathcal{L}\{f'(t)\} = s \mathcal{L}\{f(t)\} - f(0)$. Let $f(t) = 1$, then $f'(t) = 0$ and $f(0) = 1$.

Therefore, substituting in equation (4) gives:

$$\mathcal{L}\{0\} = s \mathcal{L}\{1\} - 1$$
$$0 = s \mathcal{L}\{1\} - 1$$
$$1 = s \mathcal{L}\{1\}$$

Hence $\mathcal{L}\{1\} = \dfrac{1}{s}$

(b) Let $f(t) = t$, then $f'(t) = 1$ and $f(0) = 0$.

Therefore, substituting in equation (4) gives:

$$\mathcal{L}\{1\} = s \mathcal{L}\{t\} - 0$$

i.e. $\dfrac{1}{s} = s \mathcal{L}\{t\}$

Hence $\mathcal{L}\{t\} = \dfrac{1}{s^2}$

(c) Let $f(t) = e^{at}$, then $f'(t) = a e^{at}$ and $f(0) = 1$.

Therefore, substituting in equation (4) gives:

$$\mathcal{L}\{a\,e^{at}\} = s\,\mathcal{L}\{e^{at}\} - 1$$
$$a\,\mathcal{L}\{e^{at}\} = s\,\mathcal{L}\{e^{at}\} - 1$$
$$1 = (s-a)\,\mathcal{L}\{e^{at}\}$$

Hence $\mathcal{L}\{e^{at}\} = \dfrac{1}{s-a}$

Problem 2. Use the Laplace transform of the second derivative (i.e. equation (5)) to derive: $\mathcal{L}\{\sin at\} = \dfrac{a}{s^2 + a^2}$

Let $f(t) = \sin at$, then $f'(t) = a\cos at$ and $f''(t) = -a^2\sin at$, $f(0) = 0$ and $f'(0) = a$.

From equation (5), $\mathcal{L}\{f''(t)\} = s^2\,\mathcal{L}\{f(t)\} - s\,f(0) - f'(0)$

$$\begin{aligned}
\text{Hence } \mathcal{L}\{-a^2\sin at\} &= s^2\,\mathcal{L}\{\sin at\} - s(0) - a \\
- a^2\,\mathcal{L}\{\sin at\} &= s^2\,\mathcal{L}\{\sin at\} - a \\
a &= (s^2 + a^2)\,\mathcal{L}\sin at
\end{aligned}$$

Hence $\mathcal{L}\{\sin at\} = \dfrac{a}{s^2 + a^2}$

Further problems on the Laplace transforms of derivatives may be found in Section 5 (Problems 11 to 13), page 366.

4 The initial and final value theorems

The initial value and final value theorems are just two of many Laplace transform theorems used to simplify and interpret the solution of certain problems.

The initial value theorem states:

$$\underset{t \to 0}{\text{limit}}\ [f(t)] = \underset{s \to \infty}{\text{limit}}\ [s\,F(s)], \text{ where } F(s) = \mathcal{L}\{f(t)\}$$

This is proved as follows:

From the previous section, equation (4):

$$\mathcal{L}\{f'(t)\} = \int_0^\infty e^{-st} f'(t)\,dt = s\,F(s) - f(0)$$

Now as s approaches infinity, e^{-st} approaches zero and $\int_0^\infty e^{-st} f'(t)\,dt$ will approach zero.

Hence $0 = \underset{s \to \infty}{\text{limit}}\ [s\,F(s) - f(0)]$

$f(0)$ is independent of s, thus

$$\underset{s \to \infty}{\text{limit}}\ [s\,F(s)] = f(0) = \underset{t \to 0}{\text{limit}}\ [f(t)]$$

For example, if $f(t) = 4e^{-2t}$

then $F(s) = \mathcal{L}\{f(t)\} = \mathcal{L}\ 4e^{-2t} = \dfrac{4}{s+2}$ from (iii) of Table 27.1

By the initial value theorem:

$$\lim_{t \to 0} [4e^{-2t}] = \lim_{s \to \infty} \left[\frac{4s}{s+2}\right]$$

i.e. $4e^0 = \dfrac{4\infty}{\infty+2}$

i.e. $\ 4 \ = \mathbf{4}$, which illustrates the theorem.

The final value theorem states:

$$\lim_{t \to \infty} [f(t)] = \lim_{s \to 0} [s\,F(s)]$$

This is proved as follows:
From the previous section, equation (4):

$$\mathcal{L}\{f'(t)\} = \int_0^\infty e^{-st} f'(t)\,dt = s\,F(s) - f(0) \qquad (7)$$

Now as s approaches zero the left-hand side of equation (7),

i.e. $\int_0^\infty e^{-st} f'(t)\,dt$, will approach $\int_0^\infty f'(t)\,dt$,

and $\int_0^\infty f'(t)\,dt = \lim_{t \to \infty} \int_0^t f'(t)\,dt$

$$= \lim_{t \to \infty} [f(t) - f(0)] = \lim_{t \to \infty} [f(t)] - f(0)$$

Now as s approaches zero the right-hand side of equation (7),

i.e. $s\,F(s) - f(0)$, approaches $\lim_{s \to 0} [s\,F(s)] - f(0)$.

Equating these results gives: $\lim_{t \to \infty} [f(t)] - f(0) = \lim_{s \to 0} [s\,F(s)] - f(0)$

i.e. $\lim_{t \to \infty} [f(t)] = \lim_{s \to 0} [s\,F(s)]$

For example, if $f(t) = 4\,e^{-2t}$

then $\lim_{t \to \infty} [4\,e^{-2t}] = \lim_{s \to 0} \left[\dfrac{4s}{s+2}\right]$

i.e. $4(0) = \dfrac{0}{2}$

i.e. $\mathbf{0} = \mathbf{0}$, which illustrates the theorem.

The initial and final value theorems are used in particular in pulse circuit applications where the response of a circuit for very small intervals of time, or the behaviour immediately after the switch is closed are of interest. The initial value theorem shows that there is a direct connection between the behaviour of a transform as s approaches infinity and the behaviour of the corresponding time function as t approaches zero. The final value theorem is particularly useful in investigating the stability of systems and is concerned with the steady state response for large values of t, i.e. after all transient effects have died away. The behaviour of $s\,F(s)$ as s approaches zero is seen to be the same as that of $f(t)$ as t approaches infinity. A particular practical application of the final value theorem is in automatic aircraft-landing systems.

Worked problems on the initial and final value theorems

Problem 1. Verify the initial value theorem for the following:

(a) $4 - 2 \cos t$,   (b) $(3t + 4)^2$

(a) $f(t) = 4 - 2 \cos t$.

$$F(s) = \mathcal{L}\{f(t)\} = \mathcal{L}\{4 - 2 \cos t\} = \frac{4}{s} - \frac{2s}{s^2 + 1} , \text{ from (ii) and (v)}$$
$$\text{of Table 27.1}$$

By the initial value theorem, $\quad \underset{t \to 0}{\text{limit}} \ [f(t)] = \underset{s \to \infty}{\text{limit}} \ [s\,F(s)]$

$$\text{i.e.} \quad \underset{t \to 0}{\text{limit}} \ [4 - 2 \cos t] = \underset{s \to \infty}{\text{limit}} \left[\frac{4s}{s} - \frac{2s^2}{s^2 + 1} \right]$$

$$= 4 - \underset{s \to \infty}{\text{limit}} \left[\frac{2s^2}{s^2 + 1} \right]$$

As $s \to \infty$, $s^2 + 1 \to s^2$

Thus $4 - 2 \cos 0 = 4 - \underset{s \to \infty}{\text{limit}} \left[\frac{2s^2}{s^2} \right]$

i.e. $4 - 2 = 4 - 2$

i.e. $\quad 2 = $ **2**, which verifies the theorem in this case.

(b) $f(t) = (3t + 4)^2 = 9t^2 + 24t + 16$

$$F(s) = \mathcal{L}\{f(t)\} = \mathcal{L}\{9t^2 + 24t + 16\} = 9\left(\frac{2}{s^3}\right) + \frac{24}{s^2} + \frac{16}{s}$$
$$\text{from (vii), (vi) and (ii) of Table 27.1.}$$

By the initial value theorem, $\underset{t \to 0}{\text{limit}} \ [(3t + 4)^2] = \underset{s \to \infty}{\text{limit}} \left[\frac{18s}{s^3} + \frac{24s}{s^2} + \frac{16s}{s} \right]$

i.e. $(0 + 4)^2 = (0 + 0 + 16)$

i.e. $\quad 16 = $ **16**, which verifies the theorem in this case.

Problem 2. Verify the final value theorem for the function $3 + e^{-t} (\sin t + \cos t)$.

$$f(t) = 3 + e^{-t} (\sin t + \cos t)$$

$$F(s) = \mathcal{L}\{f(t)\} = \mathcal{L}\{3 + e^{-t} \sin t + e^{-t} \cos t\}$$

$$= \frac{3}{s} + \frac{1}{(s+1)^2 + 1} + \frac{s+1}{(s+1)^2 + 1}$$

from (ii) of Table 27.1 and (ii) and (iii) of Table 28.1

By the final value theorem, $\underset{t \to \infty}{\text{limit}} [f(t)] = \underset{s \to 0}{\text{limit}} [s \, F(s)]$

i.e. $\underset{t \to \infty}{\text{limit}} [3 + e^{-t} (\sin t + \cos t)]$

$$= \underset{s \to 0}{\text{limit}} \left[3 + \frac{s}{(s+1)^2 + 1} + \frac{s(s+1)}{(s+1)^2 + 1}\right]$$

i.e. $(3 + 0) = (3 + 0 + 0)$

i.e. $3 = 3$, which verifies the theorem.

Further problems on the initial and final value theorems may be found in the following Section (5) (Problems 14 to 17), page 366.

5 Further problems

Laplace transforms of functions of the type $e^{at} f(t)$

In problems 1 to 8, find the Laplace transforms.

1. (a) $t \, e^t$ (b) $t^2 \, e^{3t}$.   (a) $\left[\dfrac{1}{(s-1)^2}\right]$ (b) $\left[\dfrac{2}{(s-3)^3}\right]$

2. (a) $2t^3 \, e^{-2t}$ (b) $e^t \sin t$.   (a) $\left[\dfrac{12}{(s+2)^4}\right]$ (b) $\left[\dfrac{1}{s^2 - 2s + 2}\right]$

3. (a) $3e^{2t} \cos 3t$ (b) $e^{-2t} \sin 2t$.   (a) $\left[\dfrac{3(s-2)}{s^2 - 4s + 13}\right]$ (b) $\left[\dfrac{2}{s^2 + 4s + 8}\right]$

4. (a) $4e^{-3t} \cos 2t$ (b) $2e^t \cos^2 t$.   (a) $\left[\dfrac{4(s+3)}{s^2 + 6s + 13}\right]$

   (b) $\left[\dfrac{1}{s-1} + \dfrac{s-1}{s^2 - 2s + 5}\right]$

5. (a) $e^t \sin^2 t$ (b) $2e^{2t} \cosh t$.   (a) $\left[\dfrac{1}{2}\left\{\dfrac{1}{s-1} - \dfrac{s-1}{s^2 - 2s + 5}\right\}\right]$

   (b) $\left[\dfrac{2(s-2)}{s^2 - 4s + 3}\right]$

6. (a) $e^t \sinh 2t$ (b) $3e^{-2t} \cosh 3t$.   (a) $\left[\dfrac{2}{s^2 - 2s - 3}\right]$ (b) $\left[\dfrac{3(s+2)}{s^2 + 4s - 5}\right]$

7. (a) $4e^{-3t} \sinh 4t$ (b) $e^t (2 \cos 3t - 4 \sin 3t)$.   (a) $\left[\dfrac{16}{s^2 + 6s - 7}\right]$

(b) $\left[\dfrac{2s - 14}{s^2 - 2s + 10}\right]$

8. (a) $e^{-2t} (\cosh t - \sinh t)$ (b) $3e^{3t} (\sinh t - \sin t)$.

(a) $\left[\dfrac{s+1}{s^2 + 4s + 3}\right]$

(b) $\left[\dfrac{6}{(s^2 - 6s + 10)(s^2 - 6s + 8)}\right]$

9. Show that $\mathcal{L}\{(\sin t - \cos t)^2\} = \dfrac{s^2 - 2s + 4}{s(s^2 + 4)}$.

10. Show that $\mathcal{L}\{\cosh^2 4t\} = \dfrac{s^2 - 32}{s(s^2 - 64)}$.

Laplace transforms of derivatives

11. Use the Laplace transform of the first derivative to derive the following transforms: (a) $\mathcal{L}\{t^2\} = \dfrac{2}{s^3}$ (b) $\mathcal{L}\{e^{-at}\} = \dfrac{1}{s+a}$
(c) $\mathcal{L}\{3t^3\} = \dfrac{18}{s^4}$.

12. Prove that $\mathcal{L}\left\{\dfrac{d^3 y}{dt^3}\right\} = s^3 \mathcal{L}\{y\} - s^2 y(0) - s y'(0) - y''(0)$, where $y(0), y'(0)$ and $y''(0)$ are the values of y, $\dfrac{dy}{dt}$ and $\dfrac{d^2 y}{dt^2}$ respectively when $t = 0$.

13. Use the Laplace transform of the second derivative to derive the following transforms: (a) $\mathcal{L}\{\cos at\} = \dfrac{s}{s^2 + a^2}$

(b) $\mathcal{L}\{\cosh at\} = \dfrac{s}{s^2 - a^2}$ (c) $\mathcal{L}\{\sinh at\} = \dfrac{a}{s^2 - a^2}$.

Initial and final value theorems

14. Verify the initial value theorem for the functions:
(a) $3 - 2 \sin t$ (b) $(t + 3)^2$.

15. Verify the initial value theorem for the functions:

    (a) $t + \sin 2t$ (b) $1 + 3\cos t$.

16. Verify the final value theorem for the function $t^2\, e^{-2t}$.

17. Verify the final value theorem for the function $2 + e^{-2t} \sin 2t$.

Chapter 29

Inverse Laplace transforms and the use of Laplace transforms to solve differential equations

1 Definition of the inverse Laplace transform

If the Laplace transform of a function $f(t)$ is $F(s)$, i.e. $\mathcal{L}\{f(t)\} = F(s)$, then $f(t)$ is called **the inverse Laplace transform** of $F(s)$ and is written as $f(t) = \mathcal{L}^{-1}\{F(s)\}$, where $\mathcal{L}^{-1}\{\ \}$ is called the inverse Laplace transformation operator. The inverse Laplace transform is used in the solution of differential equations (see Section 4).

For example, since $\mathcal{L}\{1\} = \dfrac{1}{s}$

$$\text{then } \mathcal{L}^{-1}\left\{\frac{1}{s}\right\} = 1$$

Similarly, since $\mathcal{L}\{e^{at}\} = \dfrac{1}{s-a}$

$$\text{then } \mathcal{L}^{-1}\left\{\frac{1}{s-a}\right\} = e^{at}$$

2 Inverse Laplace transforms of simple functions

Inverse Laplace transforms may be most conveniently found from tables of standard transforms such as Table 27.1 and Table 28.1. For ease of reference Table 29.1 lists some inverse Laplace transforms.

Table 29.1 Inverse Laplace transforms.

	$F(s) = \mathcal{L}\{f(t)\}$	$\mathcal{L}^{-1}\{F(s)\} = f(t)$
1.	$\dfrac{1}{s}$	1
2.	$\dfrac{k}{s}$	k
3.	$\dfrac{1}{s-a}$	e^{at}
4.	$\dfrac{a}{s^2 + a^2}$	$\sin at$
5.	$\dfrac{s}{s^2 + a^2}$	$\cos at$
6.	$\dfrac{1}{s^2}$	t
7.	$\dfrac{2!}{s^3}$	t^2
8.	$\dfrac{3!}{s^4}$	t^3
9.	$\dfrac{n!}{s^{n+1}}$	t^n (n = positive integer)
10.	$\dfrac{s}{s^2 - a^2}$	$\cosh at$
11.	$\dfrac{a}{s^2 - a^2}$	$\sinh at$
12.	$\dfrac{n!}{(s-a)^{n+1}}$	$e^{at} t^n$
13.	$\dfrac{\omega}{(s-a)^2 + \omega^2}$	$e^{at} \sin \omega t$
14.	$\dfrac{s-a}{(s-a)^2 + \omega^2}$	$e^{at} \cos \omega t$
15.	$\dfrac{s-a}{(s-a)^2 - \omega^2}$	$e^{at} \cosh \omega t$
16.	$\dfrac{\omega}{(s-a)^2 - \omega^2}$	$e^{at} \sinh \omega t$

Worked problems on finding inverse Laplace transforms

Problem 1. Find the following inverse Laplace transforms:

(a) $\mathcal{L}^{-1}\left\{\dfrac{1}{s^4+4}\right\}$ (b) $\mathcal{L}^{-1}\left\{\dfrac{3}{4s-2}\right\}$ (c) $\mathcal{L}^{-1}\left\{\dfrac{1}{s^3}\right\}$ (d) $\mathcal{L}^{-1}\left\{\dfrac{2}{s^4}\right\}$

(a) Since, from 4 of Table 29.1, $\mathcal{L}^{-1}\left\{\dfrac{a}{s^2+a^2}\right\} = \sin at$, $\mathcal{L}^{-1}\left\{\dfrac{1}{s^2+a^2}\right\}$

$= \dfrac{1}{a}\sin at$

Thus $\mathcal{L}^{-1}\left\{\dfrac{1}{s^2+4}\right\} = \mathcal{L}^{-1}\left\{\dfrac{1}{s^2+2^2}\right\} = \dfrac{1}{2}\sin 2t.$

(b) $\mathcal{L}^{-1}\left\{\dfrac{3}{4s-2}\right\} = \mathcal{L}^{-1}\left\{\dfrac{3}{4(s-\frac{1}{2})}\right\} = \dfrac{3}{4}\mathcal{L}^{-1}\left\{\dfrac{1}{s-\frac{1}{2}}\right\} = \dfrac{3}{4}e^{\frac{1}{2}t},$

from 3 of Table 29.1

(c) Since, from 7 of Table 29.1, $\mathcal{L}^{-1}\left\{\dfrac{2}{s^3}\right\} = t^2$ then $\mathcal{L}^{-1}\left\{\dfrac{1}{s^3}\right\} = \dfrac{1}{2}t^2$

(d) Since, from 8 of Table 29.1, $\mathcal{L}^{-1}\left\{\dfrac{3!}{s^4}\right\} = t^3$ then $\mathcal{L}^{-1}\left\{\dfrac{1}{s^4}\right\} = \dfrac{t^3}{3!}$

and $\mathcal{L}^{-1}\left\{\dfrac{2}{s^4}\right\} = \dfrac{2t^3}{3!} = \dfrac{1}{3}t^3$

Problem 2. Find the following inverse Laplace transforms:

(a) $\mathcal{L}^{-1}\left\{\dfrac{2s}{s^2+9}\right\}$ (b) $\mathcal{L}^{-1}\left\{\dfrac{3s}{s^2-25}\right\}$ (c) $\mathcal{L}^{-1}\left\{\dfrac{4}{s^2-5}\right\}$ (d) $\mathcal{L}^{-1}\left\{\dfrac{1}{(s-2)^4}\right\}$

(a) $\mathcal{L}^{-1}\left\{\dfrac{2s}{s^2+9}\right\} = 2\,\mathcal{L}^{-1}\left\{\dfrac{s}{s^2-3^2}\right\} = 2\cos 3t$, from 5 of Table 29.1

(b) $\mathcal{L}^{-1}\left\{\dfrac{3s}{s^2-25}\right\} = 3\,\mathcal{L}^{-1}\left\{\dfrac{s}{s^2-5^2}\right\} = 3\cosh 5t$, from 10 of Table 29.1

(c) Since, from 11 of Table 29.1, $\mathcal{L}^{-1}\left\{\dfrac{a}{s^2-a^2}\right\} = \sinh at$

then $\mathcal{L}^{-1}\left\{\dfrac{1}{s^2-a^2}\right\} = \dfrac{1}{a}\sinh at$

Thus $\mathcal{L}^{-1}\left\{\dfrac{4}{s^2-5}\right\} = 4\,\mathcal{L}^{-1}\left\{\dfrac{1}{s^2-(\sqrt{5})^2}\right\} = \dfrac{4}{\sqrt{5}}\sinh\sqrt{5}t$

(d) Since, from 12 of Table 29.1, $\mathcal{L}^{-1}\left\{\dfrac{n!}{(s-a)^{n+1}}\right\} = e^{at}t^n$

then $\mathcal{L}^{-1}\left\{\dfrac{1}{(s-a)^{n+1}}\right\} = \dfrac{1}{n!}e^{at}t^n$

Thus, given $\mathcal{L}^{-1}\left\{\dfrac{1}{(s-2)^4}\right\}$ means that $n = 3$ and $a = 2$.

Hence $\mathcal{L}^{-1}\left\{\dfrac{1}{(s-2)^4}\right\} = \dfrac{1}{3!} e^{2t} t^3 = \dfrac{1}{6} e^{2t} t^3$

Problem 3. Find the following inverse Laplace transforms:

(a) $\mathcal{L}^{-1}\left\{\dfrac{2}{(s^2 - 6s + 13)}\right\}$ (b) $\mathcal{L}^{-1}\left\{\dfrac{s+2}{(s^2 + 4s + 20)}\right\}$

(c) $\mathcal{L}^{-1}\left\{\dfrac{3s-1}{(s^2 - 2s - 24)}\right\}$ (d) $\mathcal{L}^{-1}\left\{\dfrac{2}{(s^2 + 4s + 3)}\right\}$

(a) $\mathcal{L}^{-1}\left\{\dfrac{2}{(s^2 - 6s + 13)}\right\} = \mathcal{L}^{-1}\left\{\dfrac{2}{(s-3)^2 + 2^2}\right\} = e^{3t} \sin 2t$, from

13 of Table 29.1

(b) $\mathcal{L}^{-1}\left\{\dfrac{s+2}{(s^2 + 4s + 20)}\right\} = \mathcal{L}^{-1}\left\{\dfrac{s+2}{(s+2)^2 + 4^2}\right\} = e^{-2t} \cos 4t$, from

14 of Table 29.1

(c) $\mathcal{L}^{-1}\left\{\dfrac{3s-1}{(s^2 - 2s - 24)}\right\} = \mathcal{L}^{-1}\left\{\dfrac{3s-1}{(s-1)^2 - 5^2}\right\} = \mathcal{L}^{-1}\left\{\dfrac{3(s-1)+2}{(s-1)^2 - 5^2}\right\}$

$= \mathcal{L}^{-1}\left\{\dfrac{3(s-1)}{(s-1)^2 - 5^2}\right\} + \mathcal{L}^{-1}\left\{\dfrac{2}{(s-1)^2 - 5^2}\right\}$

since $\mathcal{L}^{-1}\{\ \}$ is a linear operator.

$= 3 e^t \cosh 5t + \mathcal{L}^{-1}\left\{\dfrac{\frac{2}{5}(5)}{(s-1)^2 - 5^2}\right\}$,

from 15 of Table 29.1

$= 3 e^t \cosh 5t + \dfrac{2}{5} e^t \sinh 5t$, from 16 of

Table 29.1

(d) $\mathcal{L}^{-1}\left\{\dfrac{2}{(s^2 + 4s + 3)}\right\} = \mathcal{L}^{-1}\left\{\dfrac{2}{(s+2)^2 - 1^2}\right\} = 2 e^{-2t} \sinh t$,

from 16 of Table 29.1.

Further problems on finding inverse Laplace transforms may be found in Section 5 (Problems 1 to 13), page 380.

3 Inverse Laplace transforms using partial fractions

From Section 1, if the Laplace transform of a function $f(t)$ is $F(s)$, then the inverse Laplace transform of $F(s)$ is given by:

$$f(t) = \mathcal{L}^{-1}\{F(s)\}$$

When $F(s)$ is a rational function of s but is not immediately recognisable as a standard type (from Table 29.1), it may often be expressed, using partial fractions, as the sum of a number of terms which may be inverted at sight.

The process of resolving an algebraic expression into partial fractions is discussed in Chapter 2.

Summarising:

Provided that the numerator $f(s)$ is of less degree than the relevant denominator, the following identities are typical examples of the form of partial fraction used:

$$\frac{f(s)}{(s+a)(s+b)(s+c)} = \frac{A}{(s+a)} + \frac{B}{(s+b)} + \frac{C}{(s+c)}$$

$$\frac{f(s)}{(s-a)^3(s+b)} \equiv \frac{A}{(s-a)} + \frac{B}{(s-a)^2} + \frac{C}{(s-a)^3} + \frac{D}{(s+b)}$$

$$\frac{f(s)}{(as^2+bs+c)(s-d)} \equiv \frac{As+B}{(as^2+bs+c)} + \frac{D}{(s-d)}$$

The technique of finding inverse Laplace transforms using partial fractions is demonstrated in the following worked problems.

Worked problems on finding inverse Laplace transforms using partial fractions

Problem 1. Find $\mathcal{L}^{-1}\left\{\dfrac{3s+1}{s^2+2s-3}\right\}$

$$\frac{3s+1}{s^2+2s-3} \equiv \frac{3s+1}{(s-1)(s+3)} \equiv \frac{A}{(s-1)} + \frac{B}{(s+3)}$$

Hence $3s+1 \equiv A(s+3) + B(s-1)$,

from which $A = 1$ and $B = 2$ (for method, see Chapter 2).

Hence $\mathcal{L}^{-1}\left\{\dfrac{3s+1}{s^2+2s-3}\right\} \equiv \mathcal{L}^{-1}\left\{\dfrac{1}{s-1} + \dfrac{2}{s+3}\right\}$

$$= \mathcal{L}^{-1}\left\{\frac{1}{s-1}\right\} + \mathcal{L}^{-1}\left\{\frac{2}{s+3}\right\}$$

$$= e^t + 2e^{-3t}, \text{ from 3 of Table 29.1.}$$

Problem 2. Find $\mathcal{L}^{-1} \left\{ \dfrac{5 s^2 - 2s - 4}{s\,(s + 1)\,(s - 2)} \right\}$

$$\frac{5 s^2 - 2s - 4}{s\,(s + 1)\,(s - 2)} \equiv \frac{A}{s} + \frac{B}{(s + 1)} + \frac{C}{(s - 2)}$$

Hence $5 s^2 - 2s - 4 \equiv A\,(s + 1)\,(s - 2) + B\,(s)\,(s - 2) + C\,(s)\,(s + 1)$ from which $A = 2$, $B = 1$ and $C = 2$.

Hence $\mathcal{L}^{-1} \left\{ \dfrac{5 s^2 - 2s - 4}{s\,(s + 1)\,(s - 2)} \right\} \equiv \mathcal{L}^{-1} \left\{ \dfrac{2}{s} + \dfrac{1}{(s + 1)} + \dfrac{2}{(s - 2)} \right\}$

$$= 2 + e^{-t} + 2\,e^{2t}, \text{ from 2 and 3 of Table 29.1.}$$

Problem 3. Find $\mathcal{L}^{-1} \left\{ \dfrac{4 s^3 - 14s^2 + 3s + 13}{(s + 2)(s - 1)^3} \right\}$

$$\frac{4s^3 - 14s^2 + 3s + 13}{(s + 2)(s - 1)^3} \equiv \frac{A}{(s + 2)} + \frac{B}{(s - 1)} + \frac{C}{(s - 1)^2} + \frac{D}{(s - 1)^3}$$

Hence $4s^3 - 14s^2 + 3s + 13 \equiv A\,(s - 1)^3 + B\,(s + 2)\,(s - 1)^2 + C\,(s + 2)(s - 1)$
$\qquad + D\,(s + 2)$ from which $A = 3$, $B = 1$, $C = -5$ and $D = 2$.

Hence $\mathcal{L}^{-1} \left\{ \dfrac{4s^3 - 14s^2 + 3s + 13}{(s + 2)(s - 1)^3} \right\} \equiv \mathcal{L}^{-1} \left\{ \dfrac{3}{(s + 2)} + \dfrac{1}{(s - 1)} - \dfrac{5}{(s - 1)^2} \right.$

$$\left. + \frac{2}{(s - 1)^3} \right\}$$

$$= 3\,e^{-2t} + e^t - 5\,e^t\,t + e^t\,t^2$$

from 3 and 12 of Table 29.1.

Problem 4. Find $\mathcal{L}^{-1} \left\{ \dfrac{5 s^2 + 7s - 1}{(s + 4)\,(s^2 + 1)} \right\}$

$$\frac{5 s^2 + 7s - 1}{(s + 4)\,(s^2 + 1)} \equiv \frac{A}{(s + 4)} + \frac{B s + C}{(s^2 + 1)}$$

Hence $5 s^2 + 7s - 1 \equiv A\,(s^2 + 1) + (B s + C)\,(s + 4)$
from which $A = 3$, $B = 2$ and $C = -1$.

Hence $\mathcal{L}^{-1} \left\{ \dfrac{5 s^2 + 7s - 1}{(s + 4)\,(s^2 + 1)} \right\} \equiv \mathcal{L}^{-1} \left\{ \dfrac{3}{s + 4} + \dfrac{2s - 1}{(s^2 + 1)} \right\}$

$$= \mathcal{L}^{-1} \left\{ \frac{3}{s + 4} \right\} + \mathcal{L}^{-1} \left\{ \frac{2s}{s^2 + 1} \right\} - \mathcal{L}^{-1} \left\{ \frac{1}{s^2 + 1} \right\}$$

$$= 3\,e^{-4t} + 2 \cos t - \sin t, \text{ from 3, 5 and 4}$$
of Table 29.1.

Problem 5. Find $\mathcal{L}^{-1}\left\{\dfrac{3s+10}{s\,(s^2+2s+10)}\right\}$

$$\frac{3s+10}{s\,(s^2+2s+10)} \equiv \frac{A}{s} + \frac{Bs+C}{(s^2+2s+10)}$$

Hence $3s+10 \equiv A\,(s^2+2s+10)+(B\,s+C)\,(s)$
from which $A=1, B=-1$ and $C=1$.

Hence $\mathcal{L}^{-1}\left\{\dfrac{3s+10}{s\,(s^2+2s+10)}\right\} \equiv \mathcal{L}^{-1}\left\{\dfrac{1}{s} + \dfrac{1-s}{(s^2+2s+10)}\right\}$

$$= \mathcal{L}^{-1}\left\{\frac{1}{s}\right\}+\mathcal{L}^{-1}\left\{\frac{1-s}{(s+1)^2+3^2}\right\}$$

$$= \mathcal{L}^{-1}\left\{\frac{1}{s}\right\}+\mathcal{L}^{-1}\left\{\frac{-(s+1)+2}{(s+1)^2+3^2}\right\}$$

$$= \mathcal{L}^{-1}\left\{\frac{1}{s}\right\} -\mathcal{L}^{-1}\left\{\frac{(s+1)}{(s+1)^2+3^2}\right\}$$

$$+ \mathcal{L}^{-1}\left\{\frac{\frac{2}{3}\,(3)}{(s+1)^2+3^2}\right\}$$

$$= 1-e^{-t}\cos 3t + \frac{2}{3}\,e^{-t}\sin 3t,$$

from 1, 14 and 13 of Table 29.1.

Further problems on finding inverse Laplace transforms using partial fractions may be found in Section 5 (Problems 14 to 23), page 381.

4 Use of Laplace transforms to solve second order differential equations with constant coefficients

Consider the second order differential equation

$$\frac{d^2 y}{dx^2} - 4\,\frac{dy}{dx} + 3y = 0,$$

with given boundary conditions that when $x=0, y=2$ and $\dfrac{dy}{dx}=1$.

Taking the Laplace transform of both sides gives:

$$\mathcal{L}\left\{\frac{d^2y}{dx^2}\right\} -4\,\mathcal{L}\left\{\frac{dy}{dx}\right\} +3\,\mathcal{L}\,\{y\} = \mathcal{L}\,\{0\}$$

From Chapter 28, Section 3, equations (4) and (5) we have:

$$\mathcal{L}\left\{\frac{dy}{dx}\right\} = s\,\mathcal{L}\{y\}\ -y(0), \text{ where } y(0) \text{ is the value of } y \text{ at } x = 0 \text{ and}$$

$$\mathcal{L}\left\{\frac{d^2y}{dx^2}\right\} = s^2\,\mathcal{L}\{y\} - s\,y(0) - y'(0), \text{ where } y'(0) \text{ is the value of } \frac{dy}{dx}$$

at $x = 0$.

Hence, substituting these equations gives:

$$[s^2\,\mathcal{L}\{y\} - s\,y(0) - y'(0)] - 4\,[s\,\mathcal{L}\{y\}\ - y(0)] + 3\,[\mathcal{L}\{y\}] = 0$$

However, from the given boundary conditions, $y(0) = 2$ and $y'(0) = 1$.

Hence $s^2\,\mathcal{L}\{y\}\ - 2s - 1 - 4s\,\mathcal{L}\{y\}\ + 8 + 3\,\mathcal{L}\{y\}\ = 0$.

The given differential equation has now been converted into an algebraic equation in $\mathcal{L}\{y\}$.

Rearranging gives: $(s^2 - 4s + 3)\,\mathcal{L}\{y\} = 2s - 7$

i.e. $\mathcal{L}\{y\} = \dfrac{2s - 7}{(s^2 - 4s + 3)}$

Thus $y = \mathcal{L}^{-1}\left\{\dfrac{2s - 7}{s^2 - 4s + 3}\right\}$

$\dfrac{2s - 7}{s^2 - 4s + 3}$ may be resolved into partial fractions.

Thus $\dfrac{2s - 7}{s^2 - 4s + 3} \equiv \dfrac{2s - 7}{(s - 3)(s - 1)} \equiv \dfrac{A}{(s - 3)} + \dfrac{B}{(s - 1)}$

Hence $2s - 7 \equiv A(s - 1) + B(s - 3)$,

from which $A = -\dfrac{1}{2}$ and $B = \dfrac{5}{2}$.

Hence $\mathcal{L}^{-1}\left\{\dfrac{2s - 7}{s^2 - 4s + 3}\right\} \equiv \mathcal{L}^{-1}\left\{\dfrac{-\frac{1}{2}}{s - 3}\right\} + \mathcal{L}^{-1}\left\{\dfrac{\frac{5}{2}}{s - 1}\right\}$

Therefore $y = -\dfrac{1}{2}e^{3x} + \dfrac{5}{2}e^x$, which is the solution of the differential equation $\dfrac{d^2y}{dx^2} - 4\dfrac{dy}{dx} + 3y = 0$.

This demonstrates how Laplace transforms are used to solve differential equations, thus providing an alternative method of solution to those discussed in Chapters 20 and 21. One of the main advantages of Laplace transforms in the solution of differential equations is that the initial conditions are automatically included, thus making the use of the general solution with arbitrary constants unnecessary.

Procedure to solve differential equations by using Laplace transforms

From the above example, the following procedure for solving differential equations of any order, with constant coefficients and given initial conditions may be used:

1. Take the Laplace transform of both sides of the differential equation by

    (a) applying the formula for the Laplace transforms of derivatives (i.e. by equations (4) and (5) of Section 3, Chapter 28), and
    (b) using a list of standard transforms (such as Tables 27.1 and 28.1).

2. Insert the given initial conditions, i.e. $y(0), y'(0)$, etc. (At this stage the given differential equation has been converted into an algebraic equation in $\mathcal{L}\{y\}$ and is often called the **subsidiary equation** of the given differential equation.)

3. Rearrange the equation to make $\mathcal{L}\{y\}$ the subject.

4. Find y by

    (a) using partial fractions to rearrange the expression to give standard forms (as in Tables 27.1 and 28.1) and then
    (b) taking the inverse of each term by using Table 29.1.

This procedure is demonstrated in the following worked problems.

Worked problems on solving differential equations using Laplace transforms

Problem 1. Use Laplace transforms to solve $4\dfrac{d^2y}{dt^2} - 12\dfrac{dy}{dt} + 9y = 0$,

given that when $t = 0, y = 2$ and $\dfrac{dy}{dt} = 4$.

This is the same problem as problem 2 of Chapter 20, page 261, and a comparison of methods can be made.

Following the above procedure:

1. $4\mathcal{L}\left\{\dfrac{d^2y}{dt^2}\right\} - 12\mathcal{L}\left\{\dfrac{dy}{dt}\right\} + 9\mathcal{L}\{y\} = 0.$

   $4[s^2\mathcal{L}\{y\} - sy(0) - y'(0)] - 12[s\mathcal{L}\{y\} - y(0)] + 9\mathcal{L}\{y\} = 0.$

2. $y(0) = 2$ and $y'(0) = 4$.

   Thus $[4s^2\mathcal{L}\{y\} - 8s - 16] - [12s\mathcal{L}\{y\} - 24] + 9\mathcal{L}\{y\} = 0.$

3. $(4s^2 - 12s + 9)\mathcal{L}\{y\} = 8s + 16 - 24$

   i.e. $\mathcal{L}\{y\} = \dfrac{8s - 8}{4s^2 - 12s + 9}$

4. $\qquad y = \mathcal{L}^{-1}\left\{\dfrac{8s - 8}{4s^2 - 12s + 9}\right\}$

$\dfrac{8s - 8}{4s^2 - 12s + 9} \equiv \dfrac{8s - 8}{(2s - 3)^2} \equiv \dfrac{A}{(2s - 3)} + \dfrac{B}{(2s - 3)^2}$

Hence $8s - 8 \equiv (A(2s - 3) + B$

from which $A = B = 4.$

Hence $y = \mathcal{L}^{-1}\left\{\dfrac{8s - 8}{4s^2 - 12s + 9}\right\} = \mathcal{L}^{-1}\left\{\dfrac{4}{(2s - 3)} + \dfrac{4}{(2s - 3)^2}\right\}$

$\qquad = \mathcal{L}^{-1}\left\{\dfrac{4}{2\left(s - \dfrac{3}{2}\right)}\right\} + \mathcal{L}^{-1}\left\{\dfrac{4}{\left[2\left(s - \dfrac{3}{2}\right)\right]^2}\right\}$

$\qquad = \mathcal{L}^{-1}\left\{\dfrac{2}{\left(s - \dfrac{3}{2}\right)}\right\} + \mathcal{L}^{-1}\left\{\dfrac{1}{\left(s - \dfrac{3}{2}\right)^2}\right\}$

i.e. $\quad y = 2 e^{\frac{3}{2}t} + e^{\frac{3}{2}t}\, t$, from 3 and 12 of Table 29.1

i.e. $\quad y = (t + 2)\, e^{\frac{3}{2}t}$.

Problem 2. Use Laplace transforms to solve $\dfrac{d^2y}{dx^2} + 4\dfrac{dy}{dx} = 6$, given that

when $x = 0$, $y = 0$ and $\dfrac{dy}{dx} = 0$.

This is the same problem as problem 2 of Chapter 21, page 270

1. $\mathcal{L}\left\{\dfrac{d^2y}{dx^2}\right\} + 4\,\mathcal{L}\left\{\dfrac{dy}{dx}\right\} = \mathcal{L}\{6\}$

$[s^2\,\mathcal{L}\{y\} - s\,y(0) - y'(0)] + 4\,[s\,\mathcal{L}\{y\} - y(0)] = \dfrac{6}{s}$

2. $y(0) = 0$ and $y'(0) = 0$.

Hence $s^2\,\mathcal{L}\{y\} + 4s\,\mathcal{L}\{y\} = \dfrac{6}{s}$.

3. $(s^2 + 4s)\,\mathcal{L}\{y\} = \dfrac{6}{s}$

i.e. $\mathcal{L}\{y\} = \dfrac{6}{s(s^2 + 4s)} = \dfrac{6}{s^2(s + 4)}$

4. $y = \mathcal{L}^{-1}\left\{\dfrac{6}{s^2(s + 4)}\right\}$

$\dfrac{6}{s^2(s + 4)} \equiv \dfrac{A}{s} + \dfrac{B}{s^2} + \dfrac{C}{s + 4}$

Hence $6 \equiv A\,s(s + 4) + B(s + 4) + C s^2$

from which $A = -\dfrac{3}{8}$, $B = \dfrac{3}{2}$ and $C = \dfrac{3}{8}$.

Hence $y = \mathcal{L}^{-1}\left\{\dfrac{6}{s^2(s + 4)}\right\} = \mathcal{L}^{-1}\left\{\dfrac{-\dfrac{3}{8}}{s} + \dfrac{\dfrac{3}{2}}{s^2} + \dfrac{\dfrac{3}{8}}{(s + 4)}\right\}$

i.e. $\quad y = -\dfrac{3}{8} + \dfrac{3}{2}x + \dfrac{3}{8}e^{-4x}$, from 2, 6 and 3 of Table 29.1

i.e. $\quad y = \dfrac{3}{8}(e^{-4x} - 1) + \dfrac{3}{2}x$

Problem 3. Solve, using Laplace transforms, $\dfrac{d^2\theta}{dt^2} - 6\dfrac{d\theta}{dt} + 10\,\theta = 20 - e^{2t}$

given that when $t = 0$, $\theta = 4$ and $\dfrac{d\theta}{dt} = \dfrac{25}{2}$.

1. $\quad \mathcal{L}\left\{\dfrac{d^2\theta}{dt^2}\right\} - 6\mathcal{L}\left\{\dfrac{d\theta}{dt}\right\} + 10\,\mathcal{L}\{\theta\} = \mathcal{L}\{20\} - \mathcal{L}\{e^{2t}\}$

$[s^2\,\mathcal{L}\{\theta\} - s\,\theta(0) - \theta'(0)] - 6\,[s\,\mathcal{L}\{\theta\} - \theta(0)] + 10\,\mathcal{L}\{\theta\} =$

$$\dfrac{20}{s} - \dfrac{1}{(s-2)}$$

2. $\quad [s^2\,\mathcal{L}\{\theta\} - 4s - \dfrac{25}{2}] - [6s\,\mathcal{L}\{\theta\} - 24] + 10\,\mathcal{L}\{\theta\} = \dfrac{20}{s} - \dfrac{1}{(s-2)}$

since $\theta(0) = 4$ and $\theta'(0) = \dfrac{25}{2}$.

3. $\quad (s^2 - 6s + 10)\,\mathcal{L}\{\theta\} = \dfrac{20}{s} - \dfrac{1}{(s-2)} + 4s + \dfrac{25}{2} - 24$

Hence $\mathcal{L}\{\theta\} = \left\{ \dfrac{\dfrac{20}{s} - \dfrac{1}{(s-2)} + 4s - \dfrac{23}{2}}{(s^2 - 6s + 10)} \right\}$

$$= \dfrac{20\,(s-2) - s + 4s\,(s)\,(s-2) - \dfrac{23}{2}\,(s)\,(s-2)}{s\,(s-2)\,(s^2 - 6s + 10)}$$

4. $\quad \theta = \mathcal{L}^{-1}\left\{ \dfrac{4s^3 - \dfrac{39}{2}s^2 + 42s - 40}{s\,(s-2)\,(s^2 - 6s + 10)} \right\}$

$$\dfrac{4s^3 - \dfrac{39}{2}s^2 + 42s - 40}{s\,(s-2)\,(s^2 - 6s + 10)} \equiv \dfrac{A}{s} + \dfrac{B}{(s-2)} + \dfrac{Cs + D}{(s^2 - 6s + 10)}$$

Hence $4s^3 - \dfrac{39}{2}s^2 + 42s - 40 \equiv A\,(s-2)\,(s^2 - 6s + 10)$

$$+ B(s)\,(s^2 - 6s + 10) + (Cs + D)(s)(s-2)$$

from which $A = 2$, $B = -\dfrac{1}{2}$, $C = \dfrac{5}{2}$ and $D = -\dfrac{3}{2}$.

Hence $\theta = \mathcal{L}^{-1}\left\{ \dfrac{2}{s} - \dfrac{\dfrac{1}{2}}{(s-2)} + \dfrac{\dfrac{5}{2}s - \dfrac{3}{2}}{(s^2 - 6s + 10)} \right\}$

$$= \mathcal{L}^{-1} \left\{ \frac{2}{s} - \frac{\frac{1}{2}}{(s-2)} + \frac{\frac{5}{2}(s-3)+6}{(s-3)^2+1^2} \right\}$$

$$= \mathcal{L}^{-1} \left\{ \frac{2}{s} - \frac{\frac{1}{2}}{(s-2)} + \frac{\frac{5}{2}(s-3)}{(s-3)^2+1^2} + \frac{6}{(s-3)^2+1^2} \right\}$$

i.e. $\theta = 2 - \frac{1}{2} e^{2t} + \frac{5}{2} e^{3t} \cos t + 6 e^{3t} \sin t$,

from 2, 3, 14 and 13 of Table 29.1.

Problem 4. The current i flowing in the circuit shown in Fig. 1 is given by the differential equation $L \dfrac{di}{dt} + Ri = E$, where E is the constant applied voltage, L the inductance and R the resistance. Using Laplace transforms solve the equation for i given that when $t = 0$, $i = 0$.

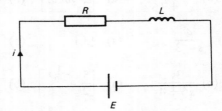

Figure 1.

This is the same problem as problem 3 of Chapter 17, page 235, and a comparison of methods can be made.

1. $\mathcal{L} \left\{ L \dfrac{di}{dt} \right\} + \mathcal{L} \{ Ri \} = \mathcal{L} \{ E \}$

   $L [s \mathcal{L} \{ i \} - i(0)] + R \mathcal{L} \{ i \} = \dfrac{E}{s}$

2. $i(0) = 0$

   Hence $L s \mathcal{L}\{ i \} + R \mathcal{L}\{ i \} = \dfrac{E}{s}$

3. $\mathcal{L}\{ i \} [Ls + R] = \dfrac{E}{s}$

   i.e. $\mathcal{L}\{ i \} = \dfrac{E}{s(Ls+R)}$

4. $i = \mathcal{L}^{-1} \left\{ \dfrac{E}{s(Ls+R)} \right\}$

   $= \mathcal{L}^{-1} \left\{ \dfrac{E}{Rs} - \dfrac{LE}{R(Ls+R)} \right\}$, by partial fractions

$$= \mathcal{L}^{-1}\left\{\frac{E}{R}\left(\frac{1}{s}\right) - \frac{E}{R}\left(\frac{1}{s+\frac{R}{L}}\right)\right\}$$

i.e. $i = \dfrac{E}{R} - \dfrac{E}{R}\, e^{-\frac{R}{L}t}$ from 1 and 3 of Table 29.1

Hence $i = \dfrac{E}{R}\left(1 - e^{-\frac{R}{L}t}\right)$

Further problems on solving differential equations using Laplace transforms may be found in the following Section (5) (Problems 24 to 33).

5 Further problems

Inverse Laplace transforms

Find the inverse Laplace transforms of the functions in Problems 1 to 13.

1. (a) $\dfrac{1}{s^2 + 9}$ (b) $\dfrac{2}{s^2 + 16}$. (a) $\left[\dfrac{1}{3}\sin 3t\right]$ (b) $\left[\dfrac{1}{2}\sin 4t\right]$

2. (a) $\dfrac{3}{s}$ (b) $\dfrac{4}{s-3}$. (a) $[3\,]$ (b) $[4\,e^{3t}]$

3. (a) $\dfrac{5}{2s+1}$ (b) $\dfrac{3s}{s^2 + 4}$. (a) $\left[\dfrac{5}{2}\,e^{-\frac{1}{2}t}\right]$ (b) $[3\cos 2t]$

4. (a) $\dfrac{2s}{2s^2 + 8}$ (b) $\dfrac{3}{s^2}$. (a) $[\cos 2t]$ (b) $[3t]$

5. (a) $\dfrac{4}{s^3}$ (b) $\dfrac{5}{s^4}$. (a) $[2\,t^2]$ (b) $\left[\dfrac{5}{6}\,t^3\right]$

6. (a) $\dfrac{6}{s^5}$ (b) $\dfrac{3s}{s^2 - 16}$. (a) $\left[\dfrac{1}{4}\,t^4\right]$ (b) $[3\cosh 4t]$

7. (a) $\dfrac{4s}{3s^2 - 27}$ (b) $\dfrac{2}{s^2 - 4}$. (a) $\left[\dfrac{4}{3}\cosh 3t\right]$ (b) $[\sinh 2t]$

8. (a) $\dfrac{3}{2s^2 - 6}$ (b) $\dfrac{2}{(s-2)^3}$. (a) $\left[\dfrac{\sqrt{3}}{2}\sinh\sqrt{3}t\right]$ (b) $[e^{2t}\,t^2]$

9. (a) $\dfrac{1}{(s-3)^4}$ (b) $\dfrac{3}{(s+2)^5}$. (a) $\left[\dfrac{1}{6}\,e^{3t}\,t^3\right]$ (b) $\left[\dfrac{1}{8}\,e^{-2t}\,t^4\right]$

10. (a) $\dfrac{4}{s^2 - 4s + 12}$ (b) $\dfrac{3}{2s^2 - 8s + 10}$.

(a) $[\sqrt{2}\,e^{2t}\sin 2\sqrt{2}t]$ $\left[(b)\ \dfrac{3}{2}\,e^{2t}\sin t\right]$